Lecture Notes of the Institute for Computer Sciences, Social Informatics and Telecommunications Engineering

293

More information about this series at http://www.springer.com/series/8197

Trung Quang Duong · Nguyen-Son Vo ·
Loi K. Nguyen · Quoc-Tuan Vien ·
Van-Dinh Nguyen (Eds.)

Industrial Networks and Intelligent Systems

5th EAI International Conference, INISCOM 2019
Ho Chi Minh City, Vietnam, August 19, 2019
Proceedings

Springer

Editors
Trung Quang Duong
Queen's University Belfast
Belfast, UK

Loi K. Nguyen
Nong Lam University Ho Chi Minh City
Ho Chi Minh City, Vietnam

Van-Dinh Nguyen
School of Electronic Engineering
Soongsil University
Seoul, Korea (Republic of)

Nguyen-Son Vo
Faculty of Electrical and Electronics
Engineering
Duy Tan University
Da Nang, Vietnam

Quoc-Tuan Vien
School of Science and Technology
Middlesex University
London, UK

ISSN 1867-8211   ISSN 1867-822X (electronic)
Lecture Notes of the Institute for Computer Sciences, Social Informatics
and Telecommunications Engineering
ISBN 978-3-030-30148-4   ISBN 978-3-030-30149-1 (eBook)
https://doi.org/10.1007/978-3-030-30149-1

This Springer imprint is published by the registered company Springer Nature Switzerland AG
The registered company address is: Gewerbestrasse 11, 6330 Cham, Switzerland

Preface

We are delighted to introduce the proceedings of the 2019 European Alliance for Innovation (EAI) International Conference on Industrial Networks and Intelligent Systems (INISCOM). This conference has brought researchers, developers, and practitioners from around the world who are leveraging and developing industrial networks and intelligent systems. The theme of INISCOM 2019 was "Vertical IoT Solutions and Its Applications for Better Citizens' Lives."

The technical program of INISCOM 2019 consisted of 24 full papers in oral presentation sessions at the main conference tracks. The conference tracks were: Track 1, Telecommunications Systems and Networks; Track 2, Industrial Networks and Applications; Track 3, Hardware and Software Design, Information Processing and Data Analysis; Track 4, Security and Privacy. Aside from the high-quality technical paper presentations, the technical program also featured three keynote speeches. The three keynote speakers were Norman Apsley and Mahmoda Ali from the United Kingdom and Prof. Tolga M. Duman from Turkey.

Coordination with the steering chairs, Prof. Imrich Chlamtac, Prof. Dr. Nguyen Hay, Prof. Carlo Cecati, and Prof. Song Guo, was essential for the success of the conference. We sincerely appreciate their constant support and guidance. It was also a great pleasure to work with such an excellent Organizing Committee who worked hard in organizing and supporting the conference. In particular, we thank the Technical Program Committee, led by our TPC co-chairs, Dr. Quoc-Tuan Vien and Dr. Van-Dinh Nguyen, who have completed the peer-review process of technical papers and complied a high-quality technical program. We are also grateful to conference manager, Kitti Szilagyiova, and all the authors who submitted their papers to INIS-COM 2019.

We strongly believe that INISCOM provides a good forum for all researcher, developers, and practitioners to discuss all science and technology aspects that are relevant to industrial networks and intelligent systems. We also expect that the future INISCOM events will be as successful and stimulating, as indicated by the contributions presented in this volume.

Trung Q. Duong
Nguyen-Son Vo
Loi K. Nguyen
Quoc-Tuan Vien
Van-Dinh Nguyen

Conference Organization

Steering Committee

Imrich Chlamtac	University of Trento, Italy
Nguyen Hay	Nong Lam University, Vietnam
Carlo Cecati	University of L'Aquila, Italy
Song Guo	The University of Aizu, Japan

Organizing Committee

General Chair

Trung Q. Duong   Queen's University Belfast, UK

General Co-chair

Nguyen Kim Loi   Nong Lam University, Vietnam

TPC Chair and Co-chair

Quoc Tuan Vien	Middlesex University, UK
Van-Dinh Nguyen	Soongsil University, South Korea

Web Chair

Nguyen-Son Vo   Duy Tan University, Vietnam

Publicity and Social Media Chairs

Tomohiko Taniguchi	Fujitsu Laboratories, Japan
Al-Sakib Khan Pathan	Islamic University in Madinah, Saudi Arabia

Workshop Chairs

Wei Liu	University of Sheffield, UK
Chinmoy Kundu	University of Texas at Dallas, USA

Sponsorship and Exhibits Chair

Hoa Le-Minh   Northumbria University, UK

Publications Chairs

Nguyen-Son Vo	Duy Tan University, Vietnam
Che Minh Tung	Nong Lam University, Vietnam

Panels Chairs

Nguyen Gia Nhu   Duy Tan University, Vietnam
Berk Canberk   Istanbul Technical University, Turkey

Tutorial Chairs

Antonino Masaracchia   Queen's University Belfast, UK
Tuan Le   Middlesex University, UK

Demos Chairs

Nguyen Quang Sang   Duy Tan University, Vietnam
Zoran Hadzi-Velkov   Ss. Cyril and Methodius University, Macedonia

Posters and PhD Track Chairs

Le-Nam Tran   University College Dublin, Ireland
Daniel Benevides da Costa Federal University of Cearas, Brazil

Local Chair

Nguyen Ngoc Thuy   Nong Lam University, Vietnam

Conference Manager

Kitti Szilagyiova   European Alliance for Innovation (EAI), Belgium

Technical Program Committee

Abbas Kouzani   Deakin University, Australia
Abdel-Hamid Ayman   Arab Academy for Science, Technology, and Maritime
                         Transport, Egypt
Adriana Giret   Universidad Politécnica de Valencia, Spain
Aguiar Rui   University of Aveiro, Portugal
Alexandropoulos George   Athens Information Technology, Greece
Ali Hilal Al-Bayati   De Montfort University, UK
Al-Qahtani Fawaz   Texas A&M University at Qatar, Qatar
Al-Sakib Khan Pathan   Southeast University, Bangladesh
Anpalagan Alagan   Ryerson University, Canada
Antonino Masaracchia   Queens University Belfast, UK
Atif Yacine   UAE University, UAE
Ayse Kortun   Queens University Belfast, UK
Ban Nguyen   Posts and Telecommunications Institute of Technology,
                         Vietnam
Bang Giang Truong Vu   VNU, University of Engineering and Technology,
                         Vietnam
Bejaoui Tarek   University of Paris-Sud 11, France
Ben Horan   Deakin University, Australia
Ben-Othman Jalel   University of Paris 13, France

Binh Duong Nguyen	International University, Vietnam
Boggia Gennaro	Politecnico di Bari, Italy
Bonnie Law	Hong Kong Polytechnic University, SAR China
Bouras Christos	University of Patras and RACTI, Greece
Bouzguenda Lotfi	Université de Sfax, Tunisia
Bowen Wang	China University of Mining and Technology, China
Bui Loc	Tan Tao University, Vietnam
Chau Yuen	Singapore University of Technology and Design, Singapore
Canberk Berk	Istanbul Technical University, Turkey
Chatzimisios Periklis Alexander	TEI of Thessaloniki, Greece
Che Anh Ho Chi Minh City	University of Technology, France
Chee Peng Lim	Deakin University, Australia
Chen Jiming	Zhejiang University, China
Chan Sammy	City University of Hong Kong, SAR China
Chen Thomas	City University London, UK
Chen Yuanfang	Institut Mines-Telecom, Telecom SudParis, France
Cheng Long	Singapore University of Technology and Design, Singapore
Cheng Yin	Queens University Belfast, UK
Chunsheng Zhu	University of British Columbia, Canada
Cho Sungrae	Chung-Ang University, South Korea
Cong-Kha Pham	The University of Electro-Communications, Japan
Constandinos Mavromoustakis	University of Nicosia, Cyprus
Da Costa Daniel Benevides	Federal University of Ceara (UFC), Brazil
Dang Ngoc	Posts and Telecommunications Institute of Technology, Vietnam
Dao Chien	Hanoi University of Science and Technology, Vietnam
De-Thu Huynh	HCMC University of Transport, Vietnam
Dezhong Peng	Sichuan University, China
Dinh-Duc Anh-Vu	University of Information Technology, Vietnam
Duong Trung Q.	Queen's University Belfast, UK
Duy Tran Trung	Posts and Telecommunications Institute of Technology, Vietnam
Dzung Nguyen	Hanoi University of Science and Technology, Vietnam
Edmundo Monteiro	University of Coimbra, Portugal
Edmund Lai	Massey University, New Zealand
El-Hajj Wassim	American University of Beirut, Lebanon
Elkashlan Maged	Queen Mary, University of London, UK
Fei Richard Yu	Carleton University, Canada
Feng-Tsun Chien	National Chiao Tung University, Taiwan
Fiedler Markus	Blekinge Institute of Technology, Sweden
George Grispos	Lero – The Irish Software Research Centre, Ireland
George Pallis	University of Cyprus, Cyprus

Li Xinrong	University of North Texas, USA
Linh-Trung Nguyen	Vietnam National University, Vietnam
Liu Yang	Guangzhou University, China
Lloret Jaime	Universidad Politécnica de Valencia, Spain
Logothetis Michael	University of Patras, Greece
Lucia Lo Bello	University of Catania, Italy
Lu Chih-Wen	National Tsing Hua University, Taiwan
Lu Liu	University of Derby, UK
Md. Apel Mahmud	Deakin University, Australia
Maham Behrouz	University of Tehran, Iran
Manoj Panda	Swinburne University of Technology, Australia
Mai Linh	International University of Viet Nam, Vietnam
Malone David	NUI Maynooth, Ireland
Matthaiou Michail	Queen's University Belfast, UK
Michalis Mavrovouniotis	De Montfort University, UK
Michele Minichino	ENEA, Italy
Milos Manic	University of Idaho, USA
Memon Qurban	United Arab Emirates University, UAE
Mojtaba Ahmadieh Khanesar	Semnan University, Iran
Morteza Biglari-Abhari	University of Auckland, New Zealand
Moussa Ouedraogo	Luxembourg Institute of Science and Technology, Luxembourg
Muhammad Azhar Iqbal	Capital University of Science & Technology, Pakistan
Naccache David	ENS, France
Nallanathan Arumugam	King's College London, UK
Nam Pham	HUST, Vietnam
Nasser Nidal	Alfaisal University, Saudi Arabia
Ng Derrick Wing Kwan	University Erlangen-Nürnberg, Germany
Ngo Duc	Hanoi University of Science and Technology, Vietnam
Ngo Hien	Linkoping University, Sweden
Nguyen Dinh-Thong	Sydney University of Technology, Australia
Nguyen Ha	University of Saskatchewan, Canada
Nguyen Huan	Middlesex University, UK
Nguyen Hung	University of Southampton, Vietnam
Nguyen Minh	International University, Vietnam
Nguyen Minh Son	International University at Ho Chi Minh, Vietnam
Nguyen Mui	Kyung Hee University, South Korea
Nguyen Nam Tran	University of Saskatchewan, Canada
Nguyen Tuan-Duc	International University, HCMC VNU, Vietnam
Nguyen-Le Hung	The University of Danang, Vietnam
Nguyen-Son Vo	Duy Tan University, Vietnam
Nguyen-Thanh Nhan	Telecom Paris Tech, France
Noël Crespi	Telecom SudParis, France
Panlong Yang	Institute of Communication Engineering, China
Panayotis Kikiras	AGT International, Switzerland

Tran Nam	Le Quy Don University (LQDTU), Vietnam
Tran Quang	National Institute of Informatics, Japan
Tran Thien Thanh	Ho Chi Minh City University of Transport, Vietnam
Tran Xuan-Tu	Vietnam National University, Vietnam
Trong Tu Bui	Ho Chi Minh University of Science, Vietnam
Truong Kien	Posts and Telecommunications Institute of Technology, Vietnam
Tsiftsis Theodoros	Technological Educational Institute of Lamia, Greece
Tuan Pham	Da Nang University of Technology, Vietnam
Van-Ca Phan	HCMC University of Technology and Education, Vietnam
Vien Quoc-Tuan	Middlesex University, UK
Vinel Alexey	Tampere University of Technology, Finland
Vinh Phan Cong	NTT University in Vietnam, Vietnam
Vollero Luca	Università Campus Bio-Medico (Roma), Italy
Vo Nguyen Quoc Bao	Posts and Telecommunications Institute of Technology, Vietnam
Vu Van Yem	Hanoi University of Science and Technology, Vietnam
Vu Xuan-Thang	LSS-SUPELEC, France
Wang Kun Nanjing	University of Posts and Telecommunications, China
Wang Wenwu	University of Surrey, UK
Wong Kai Kit	University College London, UK
Wu Xiaoling	Chinese Academy of Sciences, China
Wymeersch Henk	Chalmers University of Technology, Sweden
Xia Minghua	Institut National de la Recherche Scientifique (INRS), Canada
Xuan-Kien Dang	HCMC University of Transport, Vietnam
Yang Nan	The University of New South Wales, Australia
Ye Lu	Broadcom Corporation, USA
Yu Rong	Guangdong University of Technology, China
Yuan Jinhong	University of New South Wales, Australia
Yuen Chau	Singapore University of Technology and Design, Singapore
Zeadally Sherali	University of Kentucky, USA
Zhang Yan	Simula Research Laboratory and University of Oslo, Norway
Zhichao Sheng	Queen's University Belfast, UK
Zhong Caijun	Zhejiang University, China
Zhou Zhangbing	Institute Telecom, France
Zhu Chunsheng	The University of British Columbia, Canada
Zhu Zuqing	University of Science and Technology of China, China

Contents

Industrial Networks and Applications

Hardware and Software Design, Information Processing and Data Analysis

Telecommunications Systems and Networks

Outage Analysis of MIMO-NOMA Relay System with User Clustering and Beamforming Under Imperfect CSI in Nakagami-m Fading Channels

Tran Manh Hoang[1,2], Xuan Nam Tran[2], Ba Cao Nguyen[2], and Le The Dung[3,4(✉)]

[1] Telecommunication University, Ho Chi Minh City, Vietnam
tranmanhhoang@tcu.edu.vn
[2] Faculty of Radio Electronics, Le Quy Don Technical University, Hanoi, Vietnam
namtx@mta.edu.vn, bacao.sqtt@gmail.com
[3] Divison of Computational Physics, Institute for Computational Science, Ton Duc Thang University, Ho Chi Minh City, Vietnam
lethedung@tdtu.edu.vn
[4] Faculty of Electrical and Electronics Engineering, Ton Duc Thang University, Ho Chi Minh City, Vietnam

Abstract. In this paper, we propose and analyze a downlink multiple-user MIMO-NOMA relay system where all users are grouped into several clusters. To mitigate the inter-cluster interference, the superposition signals at the base station are beamed at the relay nodes. Then, the relays communicate with the users through superposition coding in power domain. We derive the exact closed-form expressions of the outage probability of each user in every cluster for the cases of perfect successive-interference cancellation (SIC) and imperfect SIC. The outage performance is analyzed the over Nakagami-m fading channels, taking into account the imperfection of channel state information (CSI). All analysis results are compared with Monte-Carlo simulation result to verify the correctness of the derived mathematical expressions. The results show that the channel fading severity, channel estimation error, and the quality of SIC structure have strong influences on the outage performance of the proposed MIMO-NOMA relay system.

Keywords: Multiple-input multiple-output ·
Non-orthogonal multiple-access · User clustering · Beamforming ·
Successive interference cancellation · Outage probability

1 Introduction

The fifth generation (5G) networks will be deployed in 2020 to provide the Internet of Things (IoT) service. IoT basically connects people, processes, data, and

T. Q. Duong et al. (Eds.): INISCOM 2019, LNICST 293, pp. 3–17, 2019.
https://doi.org/10.1007/978-3-030-30149-1_1

every possible things together. The key challenge in IoT is to maintain reliable communication in the condition of the limited spectrum and low cost [1]. Meanwhile, the non-orthogonal multiple-access (NOMA) is considered as a promising multiple-access technique for 5G and beyond mobile networks due to its superior spectral efficiency [2–4].

The main feature of NOMA is to use the power domain and code domain for multiple access, which adopts the superposition code at the transmitter and successive interference cancellation (SIC) method at the receiver to detect the desired signals [5]. NOMA can be applied for both uplink and downlink. In the NOMA systems, less transmission power is allocated to users which have better channel conditions while more transmission power is allocated to users which have worse channel conditions [6–8]. The purpose of this strategy is to achieve a balance between the system throughput and the fairness among users [9]. On the other hand, the power allocation in NOMA systems can be based on the priority of users, i.e. users with higher priority are allocated more power while users with lower priority are allocated less power [10,11]. The disadvantage of [10,11] is that the users with lower priority will have higher interference in the same the channel conditions.

In the literatures, there have been many research works on analyzing the performance of NOMA in various scenarios. The authors of [12] investigated a NOMA downlink system where all users locate randomly. They derived the closed-form expressions of the outage probability as well as the ergodic capacity. In [13], both downlink and uplink of NOMA systems were studied. The dynamic power allocation with undertaking QoS for different users was proposed to provide more performance fairness. Since the NOMA enhances the bandwidth efficiency and throughput, it is suitable for the multiple-user systems. The authors of [14] proposed and analyzed a multi-beam multiple-input single-output (MISO)-NOMA system. Unfortunately, the authors assumed that each cluster had only two users, namely near user and far user. A NOMA system under the condition of user quality fairness was studied in [15]. The authors concluded that it is important to allocate power according to the channel gains. In [16], the impact of user pairing on the performance of two NOMA systems, i.e. NOMA with fixed power allocation (F-NOMA) and cognitive-radio-inspired NOMA (CR-NOMA), was characterized. The power allocation coefficient was chosen to satisfy the predefined QoS requirements of users. In aforementioned works, the authors assumed that the transmitter was able to communicate with all users. However, in many cases it may not possible due to power limitation. Thus, the deployment of relay nodes is necessary. The authors of [17–19] proposed the multiple-user multiple-input-multiple-output (MIMO)-NOMA systems. The results showed that the performance is improved when multiple users are gathered into a cluster. However, the authors in these works studied the scenario where the base station (BS) directs beams towards users without having the CSI. In fact, the NOMA systems always require the accurate CSI to allocate power for users, but the variation in wireless environment may cause the imperfect CSI [20,21]. Moreover, when designing the transmission beamforming vectors, the BS needs to know the accurate CSI to precode the signals. Therefore, we are interested in examining the influence of the imperfect CSI on the performance of MIMO-NOMA downlink relay system.

Motivated by the above issues, in this paper we combine NOMA and beamforming into a downlink multiple-user MIMO relay system. This model is extremely useful in urban environment where several users may be blocked by high buildings or mountains. The contributions of our paper can be summarized as follows:

- We propose a downlink MIMO-NOMA relay system to improve the spectrum utilization efficiency and performance. We divide all users into subgroups to not only mitigate the extra-user interference but also reduce the complexity of the system. On the other hand, we use relays to forward signals to destination which helps to overcome the effect of channel fading and to significantly reduce the training sequences to estimate uplink CSI.
- We derive the exact closed-form expressions of the outage probability of each user in every clusters. For the practical purpose, we investigate the proposed MIMO-NOMA relay system over Nakagami fading channels with various fading levels.
- We analyze the system in the case of imperfect CSI which is caused by the channel estimation error. The results show that the outage performance is reduced remarkably under the influence of imperfect CSI. All analysis results are compared with simulation results to confirm the correctness of the derived mathematical expressions.

The rest of the paper is organized as follows. Section 2 presents the system model of the proposed MIMO-NOMA relay system with beamforming. Mathematical analysis of the outage probability of the proposed system is given in detail in Sect. 3. Numerical results are presented in Sect. 4. Finally, Sect. 5 concludes the paper.

2   System Model

We consider a downlink multiple-user relay system as shown in Fig. 1. In this system model, N users (D_n, $n = 1, 2, 3, ..., N$) are grouped into L clusters according to their spatial positions which are obtained by using the spatial direction methods such as Global Positioning System (GPS) technique or user location tracking algorithms. We denote $D_{l,n}$ as the nth user in the lth cluster. There are L relays (R) which are deployed according to the geographical location of the clusters to assist the BS in forwarding signals to users. The direct link from BS to each user in the groups is assumed not available because the distance from BS to users is larger than of of the covering area of BS or because of deep shadow fading. Moreover, the system operates in half-duplex mode, i.e. the communication from BS to users takes two time slots. To improve the spectral usage efficiency, the NOMA method is used for users in each cluster. Particularly, the signals from all users in a cluster is superposed at the BS, then they are decoded by SIC algorithms at the relays and users to reduce the interference from other signals to the desired signal. Moreover, the users in each cluster have different propagation distances, thus the channel gains between the relay and users are different. The channel state information (CSI) is imperfect. Based on these properties, it is possible to do the signal superposition in power domain.

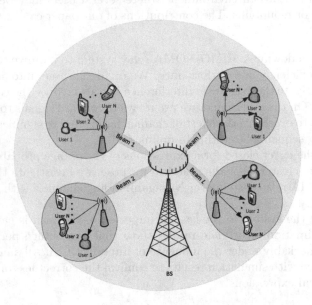

Fig. 1. System model of the proposed multiple-user MIMO-NOMA relay system with beamforming.

Regarding to the antenna configuration, the BS and each relay are equipped with N_t and N_r antennas, respectively. Meanwhile, the users only employ single antennas. This antenna configuration is absolutely reasonable because the BS and relay are big enough and have fixed locations to ensure the uncorrelated electromagnetic coupling between antennas. In this paper, we set the number of antennas of BS less than or equal to the number of antennas of relay because when BS has more antennas than relay, different approaches need to be used to implement MIMO-NOMA[1].

2.1 CSI Requests and Channel Model

To do beamforming and power allocation, the BS needs to know the statistical CSI. It is assumed that CSI is available at the BS because at the beginning of each time slot the relays send symbol pilots to the BS to estimate the uplink

---

[1] For example, one possible way is to allocate different beamforming vector to each user, then the precoding matrices at the base station can be optimized by taking user fairness into consideration.

channel. Since the BS and relay are imperfectly synchronized, the pilot signals of all relay transmissions to the BS are non-orthogonal and imperfect. Thus, perfect CSI at the BS is very difficult to be obtained due to two reasons, i.e. channel estimation error and feedback delay. It is assumed that the minimum mean squared error (MMSE) estimation is used [22]. Then, the relation between the estimated channel matrix $\hat{\mathbf{h}}_l$ and the actual channel matrix $\mathbf{h}_l$ can be given by

$$\mathbf{h}_l = \hat{\mathbf{h}}_l + \mathbf{e}_l, \tag{1}$$

where $\mathbf{e}_l \sim \mathcal{CN}(0, \sigma_{\mathrm{SR}_l}^2 \mathbf{I}_l)$ is the channel estimation error vector with i.i.d. zero mean and unit variance complex Gaussian distributed elements.

To allocate power for users, the relays also estimate the CSIs of the links from relays to users. The relation between the estimated channel coefficient $\hat{g}_n$ and the actual channel coefficient g_n can be expressed as

$$g_n = \hat{g}_n + e_n, \tag{2}$$

where e_n denotes the error of the channel between R_l and $\mathbf{D}_n$.

Moreover, $\hat{\Omega}_{\mathcal{A}} = \Omega_{\mathcal{A}} - \sigma_{e_{\mathcal{A}}}^2$ with $\mathcal{A} \in \{\mathrm{SR}_m, \mathrm{RD}_n\}$ are the normalized channel gains of $\hat{\mathbf{h}}_l$ and g_n. We should note that the elements of $\hat{\mathbf{h}}_l$, $\hat{g}_i$ are statistically independent of $\mathbf{e}_l$ and e_n. Denote $\rho_{\mathcal{A}}$, $0 \le \rho_{\mathcal{A}} \le 1$, as the estimation error coefficient which indicates the difference between the channel with estimation error and the perfect channel. Then, the normalized variance of the estimation error is $\sigma_{e_{\mathcal{A}}}^2 = \rho_{\mathcal{A}} \Omega_{\mathcal{A}}$ and the of estimation channel error is $\hat{\Omega}_{\mathcal{A}} = (1 - \rho_{\mathcal{A}}) \Omega_{\mathcal{A}}$.

To balance the implementation complexity and the system performance, we use zero-force beamforming (ZFBF) at the BS. We design a weight $\mathbf{w}_l \in \mathbb{C}^{N_t \times L}$ for the lth cluster to mitigate the interference from other clusters. $\mathbf{w}_l$ can be represented as the projection of $\mathbf{h}_l$ in the null space of the lth cluster interference channels so that the channel gain is maximized and the inter-cluster interference is canceled. Mathematically, $\mathbf{w}_l$ is computed as

$$\mathbf{w}_l = \frac{\Pi_l \mathbf{h}_l}{\|\Pi_l \mathbf{h}_l\|}, \tag{3}$$

where $\Pi_l = \mathbf{I}_N - \mathbf{H}_l (\mathbf{H}_l^H \mathbf{H}_l)^{-1} \mathbf{H}_l^H$ and $\mathbf{H}_l$ is

$$\mathbf{H}_l = [\mathbf{h}_1, \mathbf{h}_2, \cdots, \mathbf{h}_{l-1}, \mathbf{h}_{l+1}, \cdots, \mathbf{h}_L]^T, \tag{4}$$

$\mathbf{H}_l$ is an extended channel matrix which excludes only $\mathbf{h}_l^H$. $[\cdots]^T$ and $(\cdot)^H$ represent the transpose matrix and the conjugate transpose, respectively. The signals can be transmitted to lth relay if the null space of $\mathbf{H}_l$ has one greater-than-zero dimension. Thus, we have

$$\mathbf{h}_l^H \mathbf{w}_j = 0, \forall l \ne j. \tag{5}$$

In other words, the relay must completely cancel the interference of the first hop. Moreover, it is required that the size of the matrix for beamforming design is not too large and the computational complexity is bearable[2].

---

[2] This system can be extended to massive MIMO-NOMA.

The channel matrix between the BS and lth relay is denoted as $\mathbf{h}_l$, where $\mathbf{h}_l = [h_{l,1}, h_{l,2}, \cdots, h_{l,k}, \cdots, h_{l,N_r}] \in \mathbb{C}^{N_r \times 1}$ with $l \in \{1, \cdots, L\}$ and $k \in \{1, \cdots, N_r\}$, $h_{l,k} \sim \mathcal{G}(m_1, m_1/\Omega_{l,k})$ refers to the element complex channel coefficient of the channel from the BS to lth relay, and $\mathbb{E}\{|h_{l,k}|^2\} = \Omega_{l,k}$ is the variance of channel gain. All channel coefficients follow i.n.i.d Nakagami-m distribution where m denotes the fading parameter, which models the scale of line-of-sight (LoS) and multipath scatters. Moreover, it is well-known that the square of Nakagami distribution random variable is Gamma distribution [23], where $\frac{1}{2} \leq m = \frac{\mathbb{E}^2\{Z\}}{var\{Z\}} \leq \infty$ is the inverse of the normalized variance of Z with $Z \in \{|h_{l,k}|^2, |g_n|^2\}$.

2.2   Signal Model

A set of signals $\mathbf{x}_S = [\mathbf{x}_{S,1}, \cdots, \mathbf{x}_{S,L}]^T \in \mathbb{C}^{L \times 1}$ is constructed at the BS and broadcasted to all relays. Then, it is multiplied by beamforming vector at the output antenna according to the principle of zero-forcing (ZF) method.

The superposition modulation in NOMA multiple antennas system consists of two steps, i.e. power allocation and beamforming. Every antenna of the BS transmits the superposition codebook, which includes N signals of the lth cluster and can be described as $\mathbf{w}_l \mathbf{x}_{S,l} = \mathbf{w}_l[x_{l,1}, \cdots, x_{l,N}]^T$. This superposition code book is taken from $\mathbf{x}_S$, with $x_{l,n}$ is the signal of $D_{l,n}$, $\mathbb{E}\{|\mathbf{x}_{S,l}|^2\} = P_S$.

Consequently, the signal which is transmitted to the lth cluster can be expressed as

$$\mathbf{x}_{S,l} = \mathbf{w}_l \sum_{n=1}^{N} \sqrt{a_n P_S} x_n, \tag{6}$$

where a_n and x_n denote the power allocation coefficient and signal of n-th user with $\sum_{n=1}^{N} a_n = 1$ and the normalization of $\mathbf{w}_l$, $\|\mathbf{w}_l\|^2 = 1$.

Then, the received signal at the relay R_l in the first time slot can be given by

$$y_{R_l} = \mathbf{h}_l^H \mathbf{w}_l \sum_{n=1}^{N} \sqrt{a_n P_S} x_l + w_{l,n}$$

$$= \underbrace{\mathbf{h}_l^H \mathbf{w}_l \sqrt{a_n P_S} x_l}_{\text{desired signal of } (l,n)\text{-th user}} + \underbrace{\mathbf{h}_l^H \mathbf{w}_l \sum_{i=n+1}^{N} \sqrt{a_{l,i} P_S} x_{l,i}}_{\text{interference of other users}}$$

$$+ \underbrace{\mathbf{h}_l^H \mathbf{w}_l \sum_{k=1}^{n-1} \sqrt{\xi_1 a_{l,k} P_S} x_{l,k}}_{\text{interference of imperfect SIC}} + w_{l,n}, \tag{7}$$

where $w_{l,n} \sim \mathcal{CN}(0, \sigma_l^2)$ is an i.i.d. additive white Gaussian noise (AWGN) at the lth relay. It should be noted that the term $\mathbf{h}_l^H \mathbf{w}_l \sum_{k=1}^{n-1} \sqrt{a_k P_S} x_k$ in (7) equals to zero in the case of perfect SIC.

Denote the large-scale fading coefficient of n-th user by $\sqrt{d_n^{-\beta}}$, where d_n is the distance between relay and nth user and β is the path-loss factor. Moreover,

$\tilde{g}_n$ represents the small-scale fading coefficient of nth user, i.e. $g_n = \tilde{g}_n \sqrt{d_n^{-\beta}}$, then $g_n \sim \mathcal{G}(m_t, m_t/\Omega_{\mathrm{RD}_n})$ where $t \in \{2, 3, 4\}$, $\Omega_{\mathrm{RD}_n} = \mathbb{E}\{|g_n|^2\}$. The additive white Gaussian noise (AWGN) at D_n is $w_{D_n} \sim \mathcal{CN}(0, N_0)$ with N_0 is the noise variance. All channel gains follow i.n.i.d. Rayleigh distribution. Without loss of generality, it is assumed that $d_1 > d_2, \cdots, > d_N$. Therefore, the channel gains are sorted according to an ascending order $|g_1|^2 <, \cdots, < |g_N|^2$.

During the second time slot, the relays re-encode and forward the messages to the users. Hence, the output signal at the SIC architecture of $D_{l,n}$ is

$$y_{D_{l,n}} = \underbrace{g_n \sqrt{b_n P_{\mathrm{R}}} x_l}_{\text{desired signal of } n\text{-th user}} + \underbrace{g_n \sum_{i=n+1}^{N} \sqrt{b_i P_{\mathrm{R}}} x_i}_{\text{interference of other users}}$$

$$+ \underbrace{\sum_{k=1}^{n-1} \sqrt{\xi_2 b_k P_{\mathrm{R}}} g_n x_k}_{\text{interference of imperfect SIC}} + w_{D_n}. \tag{8}$$

At the relays, SIC is used to remove the interference from the signals of $(i = n + 1)$th users, which have higher power. In the case of perfect SIC, the signal-to-interference-plus-noise ratio (SINR) of $D_{l,n}$ at the relay is denoted by γ_{R_n}. From (7), the SINR is calculated as

$$\gamma_{R_n} = \frac{P_{\mathrm{S}} a_n |\mathbf{h}_l^H \mathbf{w}_l|^2}{\sum_{i=n+1}^{N} P_{\mathrm{S}} a_i |\mathbf{h}_l^H \mathbf{w}_l|^2 + \sigma_{R_n}^2}. \tag{9}$$

In the case of imperfect SIC, the term $\mathbf{h}_l^H \mathbf{w}_l \sum_{k=1}^{n-1} \sqrt{\xi_1 a_{l,k} P_{\mathrm{S}}} x_{l,k}$ in (7) is not equal to zero. It depends on the quality of SIC structure. As a result of this feature, the instantaneous SINR is

$$\gamma_{R_n}^{\mathrm{ipSIC}} = \frac{P_{\mathrm{S}} a_n |\mathbf{h}_l^H \mathbf{w}_l|^2}{\sum_{i=n+1}^{N} P_{\mathrm{S}} a_i |\mathbf{h}_l^H \mathbf{w}_l|^2 + |\mathbf{h}_l^H \mathbf{w}_l|^2 \xi_1 \sum_{k=1}^{n-1} a_{l,k} P_{\mathrm{S}} + \sigma_R^2}, \tag{10}$$

where ξ_1 represents the impact level of the residual interference at the relay.

From (8), the SINR of $D_{l,n}$ when $b_i < b_n$ and under perfect SIC is given by

$$\gamma_{D_n} = \frac{P_{\mathrm{R}} b_n |g_n|^2}{\sum_{i=n+1}^{N} b_i P_{\mathrm{R}} |g_n|^2 + \sigma_{D_n}^2}. \tag{11}$$

When the signal power of $D_{l,i}$ is larger than $D_{l,n}$, i.e. $b_i > b_n$, the SINR for $D_{l,n}$ to detect the signals of $D_{l,i}$ can be expressed as

$$\gamma_{D_{l,i}} = \begin{cases} \dfrac{P_{\mathrm{R}} b_i |g_n|^2}{\sum_{k=i+1}^{N} b_k P_{\mathrm{R}} |g_n|^2 + \sigma_{D_n}^2}, & \text{if } i < N, \tag{12a} \\[4mm] \dfrac{P_{\mathrm{R}} b_i |g_n|^2}{\sigma_{D_n}^2} & \text{if } i = N. \tag{12b} \end{cases}$$

After x_i is decoded successfully, i.e. $\gamma_{D_{l,i}} \geq \gamma_{\mathrm{th}i}$, it will be canceled out at the SIC structure of $D_{l,n}$, where $\gamma_{\mathrm{th}i}$ is the targeted SINR of $D_{l,i}$. SIC is carried out continuously until all signals of $D_{l,n}$ are decoded successfully.

In the case of imperfect SIC, the SINR is

$$\gamma_{D_n}^{\text{ipSIC}} = \frac{P_R b_n |g_n|^2}{\sum_{i=n+1}^{N} b_i P_R |g_n|^2 + \sum_{k=1}^{n-1} \xi_2 b_k P_R |g_n|^2 + \sigma_{D_n}^2}, \tag{13}$$

where ξ_2 represents the impact level of the residual interference at the D_n.

We should note that for the DF protocol, the end-to-end SINR of the system is the minimum of the SINRs of BS – R and R – D_n links, i.e.,

$$\gamma_{e2e} = \min(\gamma_{R_n}, \gamma_{D_n}). \tag{14}$$

3 Performance Analysis

3.1 Outage Probability

In this section, the outage probability (OP) of $D_{l,n}$ in the proposed system is derived in two cases, perfect SIC and imperfect SIC. The OP is defined as the probability that the transmission rate of the system falls below the minimum required data rate. Let r_1, r_n (bit/s/Hz) be the minimum required data rate from BS to R and from R to D_n, respectively. For simply, we set $r_1 = r_n = r$, then the OP of the system is calculated as

$$\begin{aligned} \text{OP}_{D_n} &= \Pr\left[\min(\gamma_{R_n}, \gamma_{D_n}) < \gamma_{\text{th}}\right] \\ &= F_{\gamma_{R_n}}(\gamma) + F_{\gamma_{D_n}}(\gamma) - F_{\gamma_{R_n}}(\gamma) F_{\gamma_{D_n}}(\gamma), \end{aligned} \tag{15}$$

where the threshold $\gamma_{\text{th}} = 2^{2r} - 1$ is used as the protected value of the SINR to ensure the quality of service of the network and satisfy the target data rate r of R_l or $D_{l,n}$.

From (9) and (11), we have the $F_{\gamma_{R_n}}(\gamma)$ and $F_{\gamma_{D_n}}(\gamma)$ in the case of perfect SIC as

$$F_{\gamma_{D_n}}^{\text{ipSIC}}(\gamma) = \Pr\left(\frac{P_S a_n |\mathbf{h}_l^H \mathbf{w}_l|^2}{\sum_{i=n+1}^{N} P_S a_i |\mathbf{h}_l^H \mathbf{w}_l|^2 + \sigma_R^2} < \gamma_{\text{th}}\right), \tag{16}$$

$$F_{\gamma_{D_n}}^{\text{ipSIC}}(\gamma) = \Pr\left(\frac{P_R b_n |g_n|^2}{\sum_{i=n+1}^{N} b_i P_R |g_n|^2 + \sigma_{D_n}^2} < \gamma_{\text{th}}\right). \tag{17}$$

Then, based on (10) and (13), we can rewrite (16) and (17) as

$$F_{\gamma_{R_n}}^{\text{ipSIC}}(\gamma) = \Pr\left(\frac{P_S a_n |\mathbf{h}_l^H \mathbf{w}_l|^2}{\sum_{i=n+1}^{N} P_S a_i |\mathbf{h}_l^H \mathbf{w}_l|^2 + |\mathbf{h}_l^H \mathbf{w}_l|^2 \xi_1 \sum_{k=1}^{n-1} a_k P_S + \sigma_R^2} < \gamma_{\text{th}}\right). \tag{18}$$

$$F_{\gamma_{D_n}}^{\text{ipSIC}}(\gamma) = \Pr\left(\frac{P_R b_n |g_n|^2}{\sum_{i=n+1}^{N} b_i P_R |g_n|^2 + \sum_{k=1}^{n-1} \xi_2 b_k P_R |g_n|^2 + \sigma_{D_n}^2} < \gamma_{\text{th}}\right). \tag{19}$$

Let us denote $X = |\mathbf{h}_l^H \mathbf{w}_l|^2$ and $Y = |g_n|^2$ for the notation convenience. Since the normalized $\mathbf{w}_l$ is independent of $\mathbf{h}_l^H$, $|\mathbf{h}_{l,k}^H \mathbf{w}_l|^2 = \|\mathbf{h}_{l,k}\|^2$ is Chi-square distributed[3] with the degree of freedom is $2K$, where $k \in \{1, \cdots, K\}$ and $K = N_r(N_t - (L-1))$ [25], $|g_n|^2$ is Chi-square distributed with the degree of freedom is 2. From (16), the CDF of γ_{R_n} can be calculated as

$$F_{\gamma_{R_n}}(\gamma) = \Pr\left(X < \frac{\gamma_{th}\sigma_R^2}{P_S(a_n - \gamma_{th}\tilde{a})}\right)$$
$$= \frac{1}{\Gamma(m_1 K)}\gamma\left(m_1 K, \frac{m_1 \gamma_{th}\sigma_R^2}{P_S(a_n - \gamma_{th}\tilde{a})\hat{\Omega}_{SR}}\right). \tag{20}$$

where $\tilde{a} = \sum_{i=n+1}^{N} a_i$.

From (17), the CDF of the SINR at $D_{l,n}$ is given by

$$F_{\gamma_{D_n}}(\gamma) = \Pr\left(\max_{n,\cdots,N} Y_n < \frac{\gamma_{th}\sigma_{D,n}^2}{P_R(b_n - \gamma_{th}\tilde{b})}\right)$$
$$= 1 - \prod_{n}^{N} \Pr\left(Y_n \geq \frac{\gamma_{th}\sigma_{D,n}^2}{P_R(b_n - \gamma_{th}\tilde{b})}\right), \tag{21}$$

where $\tilde{b} = \sum_{i=n+1}^{N} b_i$.

It is required that $a_n > \gamma_{th}\tilde{a}$ and $b_n > \gamma_{th}\tilde{b}$. Otherwise, the outage always occurs because we only consider $X, Y \in (0, \infty]$. Hence, the allocated power for D_n should be more than the total power of other users. Finally, the outage probability in the case of perfect SIC is.

$$\begin{aligned}
OP_{D_n} &= \frac{1}{\Gamma(m_1 K)}\gamma\left(m_1 K, \frac{m_1 \gamma_{th}\sigma_R^2}{P_S(a_n - \gamma_{th}\tilde{a})\hat{\Omega}_{SR}}\right) \\
&+ \sum_{j=0}^{N}\binom{N}{j}(-1)^j e^{-\frac{jm_t\gamma_{th}\sigma_{D,n}^2}{P_R(b_n-\gamma_{th}\tilde{b})\hat{\Omega}_{RD_j}}} \sum_{\ell=0}^{j(m_t-1)} c_\ell^j \left(\frac{m_t\gamma_{th}\sigma_{D,n}^2}{P_R(b_n - \gamma_{th}\tilde{b})\hat{\Omega}_{RD_j}}\right)^\ell \\
&- \frac{1}{\Gamma(m_1 K)}\gamma\left(m_1 K, \frac{m_1 \gamma_{th}\sigma_R^2}{P_S(a_n - \gamma_{th}\tilde{a})\hat{\Omega}_{SR}}\right) \\
&\times \sum_{j=0}^{N}\binom{N}{j}(-1)^j e^{-\frac{jm_t\gamma_{th}\sigma_{D,n}^2}{P_R(b_n-\gamma_{th}\tilde{b})\hat{\Omega}_{RD_j}}} \sum_{\ell=0}^{j(m_t-1)} c_\ell^j \left(\frac{m_t\gamma_{th}\sigma_{D,n}^2}{P_R(b_n - \gamma_{th}\tilde{b})\hat{\Omega}_{RD_j}}\right)^\ell.
\end{aligned} \tag{22}$$

---

[3] If a random variable has Chi-square distribution with the degree of freedom is K, it can be presented as the summation of K Rayleigh distribution random variables [24, p. 16].

Next, we consider the case of imperfect SIC. From (18) and (19), we can rewrite the CDF of γ_{R_n} and γ_{D_n} as

$$
\begin{aligned}
F_{\gamma_{R_n}}^{\text{ipSIC}}(\gamma) &= \Pr\left(X < \frac{\gamma_{\text{th}}\sigma_R^2}{P_S[a_n - \gamma_{\text{th}}(\beta_1 + \beta_2)]} \right) \\
&= \frac{1}{\Gamma(m_1 K)}\gamma\left(m_1 K, \frac{m_1\gamma_{\text{th}}\sigma_R^2}{P_S[a_n - \gamma_{\text{th}}(\beta_1 + \beta_2)]\hat{\Omega}_{\text{SR}}} \right),
\end{aligned}
\tag{23}
$$

$$
\begin{aligned}
F_{\gamma_{D_n}}^{\text{ipSIC}}(\gamma) &= \Pr\left(\max_{n,\cdots,N} Y_n < \frac{\gamma_{\text{th}}\sigma_{D,n}^2}{P_R[b_n - \gamma_{\text{th}}(\psi_1 + \psi_2)]} \right) \\
&= 1 - \prod_n^N \Pr\left(Y_n \geq \frac{\gamma_{\text{th}}\sigma_{D,n}^2}{P_R[b_n - \gamma_{\text{th}}(\psi_1 + \psi_2)]} \right),
\end{aligned}
\tag{24}
$$

where $\gamma(,\cdot,)$ is the lower incomplete Gamma function [26], $\beta_1 = \sum_{i=n+1}^{N} a_i$, $\beta_2 = \xi_1 \sum_{k=1}^{n-1} a_k$, and $\psi_1 = \sum_{i=n+1}^{N} b_i$, $\psi_2 = \xi_2 \sum_{k=1}^{n-1} b_k$. It is also required that $a_n > \gamma_{\text{th}}(\beta_1 + \beta_2)$ and $b_n > \gamma_{\text{th}}(\psi_1 + \psi_2)$ in (23) and (24), respectively. Otherwise, the outage always occurs.

Similar for the case of perfect SIC, we obtain the outage probability in the case of imperfect SIC as

$$
\begin{aligned}
\text{OP}_{D_n}^{\text{ipSIC}} &= \frac{1}{\Gamma(m_1 K)}\gamma\left(m_1 K, \frac{m_1\gamma_{\text{th}}\sigma_R^2}{P_S \Delta(n)\hat{\Omega}_{\text{SR}}} \right) \\
&+ \sum_{j=0}^{N}\binom{N}{j}(-1)^j \exp\left(-\frac{jm_t\gamma_{\text{th}}\sigma_{D,n}^2}{P_R\Delta_1(n)\hat{\Omega}_{\text{RD}_j}} \right) \sum_{\ell=0}^{j(m_t-1)} c_\ell^j \left(-\frac{m_t\gamma_{\text{th}}\sigma_{D,n}^2}{P_R\Delta_1(n)\hat{\Omega}_{\text{RD}_j}} \right)^\ell \\
&- \frac{1}{\Gamma(m_1 K)}\gamma\left(m_1 K, \frac{m_1\gamma_{\text{th}}\sigma_R^2}{P_S \Delta(n)\hat{\Omega}_{\text{SR}}} \right) \\
&\times \sum_{j=0}^{N}\binom{N}{j}(-1)^j \exp\left(-\frac{jm_t\gamma_{\text{th}}\sigma_{D,n}^2}{P_R\Delta_1(n)\hat{\Omega}_{\text{RD}_j}} \right) \sum_{\ell=0}^{j(m_t-1)} c_\ell^j \left(-\frac{m_t\gamma_{\text{th}}\sigma_{D,n}^2}{P_R\Delta_1(n)\hat{\Omega}_{\text{RD}_j}} \right)^\ell,
\end{aligned}
\tag{25}
$$

where $\Delta(n) = [a_n - \gamma_{\text{th}}(\beta_1 + \beta_2)]$ and $\Delta_1(n) = [b_n - \gamma_{\text{th}}(\psi_1 + \psi_2)]$.

4   Numerical Results

We provide some typical numerical results to evaluate the performance in terms of the outage probability and ergodic rate of the proposed MIMO-NOMA relay system. The system parameters are set as follows. The number of relay nodes, transmission antennas, and reception antennas are $L = 3$, $N_t = N_r = 3$. There are three users in each cluster with the threshold data rates $r_1 = r_2 = r_3 = 1$ b/s/Hz, respectively. The number of users in each cluster is set to three because

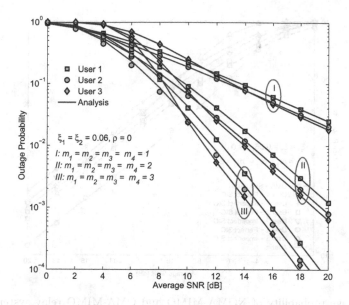

Fig. 2. Outage probability versus SNRs for different fading levels. $a_1 = 0.6, a_2 = 0.3, a_3 = 0.1, \xi_1 = \xi_2 = 0.06$.

the complexity and performance degradation of each user is proportional to the number of users [27]. All channels between the BS and each user are assumed to be Nakagami-m distributions. More specifically, the fading parameter m of each channel is denoted as follows: m_1 for BS – R_l link, m_t for R_l – $D_{l,1}$ link, and m_3, m_4 for R_l – $D_{l,2}$ and R_l – $D_{l,3}$ links, respectively. Without loss of generality, we assume the first user is farthest from the BS while the third user is nearest to the BS. Consequently, we can choose the average channel gains as $\Omega_{\text{BS}-R_l} = \Omega_{l,1} = 2$, $\Omega_{l,2} = 3$, and $\Omega_{l,3} = 6$. Then, the power allocation coefficients are in an increasing order $a_1 > a_2 > a_3$. In general, the power allocation coefficient is given by $a_n = (N - n + 1)/\mu$, where μ is chosen such that $\sum_{n=1}^{N} \sqrt{a_n} = 1$. To simplify the system design and settings, we use similar power allocation coefficients for the BS and the relay, i.e. $a_n = b_n$.

Figure 2, presents the outage probability (OP) of each user versus the SNR in dB for different fading levels. We use three cases of fading, i.e. case I: $m_1 = m_2 = m_3 = m_4 = 1$, case II: $m_1 = m_2 = m_3 = m_4 = 2$, case III: $m_1 = m_2 = m_3 = m_4 = 3$, and fixed power allocation. From Fig. 2, we can see that the outage performance of the user 3 is always the best among three users although the allocated power for user 3 is the lowest. The reason is user 3 is the closest to the relay, then the channel gain from the relay to $D_{l,3}$ is the largest[4]. On the other hand, we can also see that the diversity gain of case III is the largest while case I is the lowest. Particularly, the diversity gain of case I is one while

---

[4] The decay of the magnitude power signal is proportional to the squared distance of the multipath fading rays.

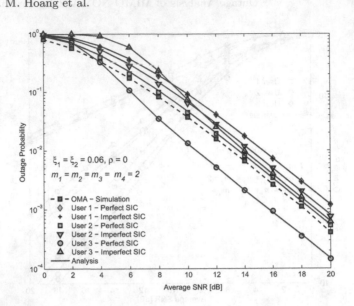

Fig. 3. Outage probability of NOMA-MIMO and OMA-MIMO relay systems versus the SNR for perfect/imperfect SIC. $a_1 = 0.6, a_2 = 0.3, a_3 = 0.1, \xi_1 = \xi_2 = 0.06$.

the diversity gains of case II and the case III are two and three, respectively. Therefore, we can conclude that the OP is improved as m increases because m represents the strength of the LoS.

Figure 3 shows the outage probability of each user versus the average SNR in dB for both perfect SIC and imperfect SIC. The fading parameters are set as $m_1 = m_t = m_3 = 2$. For simplicity, we assume that the coefficients of imperfect SIC at the relay and users are similar. We also compare the OPs of the proposed NOMA-MIMO relay systems and OMA-MIMO relay systems. To ensure the fairness between the NOMA and OMA relay systems, we use the same threshold γ_{th} for these two systems. As observed from Fig. 3, the OP of user 1 does not change for both cases of perfect SIC and imperfect SIC because user 1 does not use SIC. Additionally, it is the worst outage performance compared with the OPs of other users in high average SINR regime. We should notice that the OP of user 3 in the case of imperfect SIC is the worst when the average SNR is less than 9 dB. On the other hand, for user 2 and user 3, the OP increases significantly when SIC is imperfect. This is because the residual intra-cluster interference which appears after SIC will be increased if the number of users in a cluster is large. Due to the existence of SIC error (or can be called as error propagation [28, p. 242]) in the case of imperfect SIC, the remaining power of signal pre-decoding process impacts the demodulation of next signals. Thus, when designing the NOMA systems high quality SIC structure is needed. Moreover, the OP of user 3 in NOMA relay system is lower than the OMA relay system, but the OPs of user 1 and user 2 are higher.

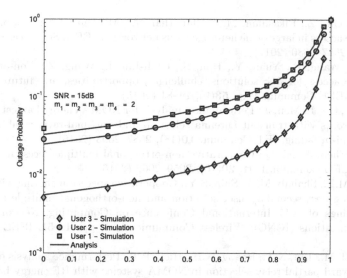

Fig. 4. Outage probability of each users versus estimation error coefficient. $a_1 = 0.6, a_2 = 0.3, a_3 = 0.1, \xi_1 = \xi_2 = 0.05$.

Figure 4 plots the outage probability of each user versus the estimation error coefficient in the case of imperfect SIC. We aim to investigate how the imperfect CSI impacts the outage performance of the proposed MIMO-NOMA relay system. As shown in Fig. 4, the outage probability increases when the estimation error coefficient is higher. In the worst case $\rho = 1$, the outage probability always occur. When $\rho < 0.6$, the OP reduces slowly but when $\rho > 0.6$ it drops rapidly. Another feature is that the OP of user 3 is the lowest compared with the OPs of user 1 and user 2.

5   Conclusions

In this paper, we proposed and investigated a downlink multiple-user MIMO-NOMA relay system under perfect/imperfect CSI over Nakagami-m fading channels. We provided the exact closed-form expression of the outage probability of each user in every cluster of the proposed system. All analysis results closely match with the Monte-Carlo simulation results, confirming the accuracy of the derived mathematical expressions. It is indicated that the channel fading severity, channel estimation error, and the quality of SIC structure greatly impact the outage performance of the proposed MIMO-NOMA relay system.

References

1. Ge, X., Cheng, H., Guizani, M., Han, T.: 5G wireless backhaul networks: challenges and research advances. IEEE Netw. **28**(6), 6–11 (2014)
2. Liu, Y., Qin, Z., Elkashlan, M., Ding, Z., Nallanathan, A., Hanzo, L.: Non orthogonal multiple access for 5G and beyond. IEEE J. Sel. Top. Sig. Process. **105**(12), 2347–2381 (2017)

3. Liu, Y., Qin, Z., Elkashlan, M., Nallanathan, A., McCann, J.A.: Non-orthogonal multiple access in large-scale heterogeneous networks. IEEE J. Sel. Areas Commun. **35**(12), 2667–2680 (2017)
4. Dai, L., Wang, B., Yuan, Y., Han, S., Chih-Lin, I., Wang, Z.: Non-orthogonal multiple access for 5G: solutions, challenges, opportunities, and future research trends. IEEE Commun. Mag. **53**(9), 74–81 (2015)
5. Han, W., Ge, J., Men, J.: Performance analysis for NOMA energy harvesting relaying networks with transmit antenna selection and maximal-ratio combining over Nakagami-m fading. IET Commun. **10**(18), 2687–2693 (2016)
6. Lv, L., Chen, J., Ni, Q.: Cooperative non-orthogonal multiple access in cognitive radio. IEEE Commun. Lett. **20**(10), 2059–2062 (2016)
7. Kader, M.F., Shahab, M.B., Shin, S.Y.: Cooperative spectrum sharing with energy harvesting best secondary user selection and non-orthogonal multiple access. In: Proceedings of 2017 International Conference on Computing, Networking and Communications (ICNC): Wireless Communications, pp. 46–51. IEEE, January 2017
8. Hoang, T.M., Tan, N.T., Hoang, N.H., Hiep, P.T.: Performance analysis of decode-and-forward partial relay selection in NOMA systems with RF energy harvesting. In: Hoang, T.M., Tan, N.T., Hoang, N.H., Hiep, P.T. (eds.) Wireless Networks, pp. 1–11 (2018). https://doi.org/10.1007/s11276-018-1746-8
9. Emam, S., Çelebi, M.: Non-orthogonal multiple access protocol for overlay cognitive radio networks using spatial modulation and antenna selection. AEU-Int. J. Electr. Commun. **86**, 171–176 (2018)
10. Deng, P., Wang, B., Wu, W., Guo, T.: Transmitter design in MISO-NOMA system with wireless-power supply. IEEE Commun. Lett. **22**, 844–847 (2018)
11. Shin, W., Vaezi, M., Lee, B., Love, D.J., Lee, J., Poor, H.V.: Non-orthogonal multiple access in multi-cell networks: theory, performance, and practical challenges. IEEE Commun. Mag. **55**(10), 176–183 (2017)
12. Ding, Z., Yang, Z., Fan, P., Poor, H.V.: On the performance of non-orthogonal multiple access in 5G systems with randomly deployed users. IEEE Sig. Process. Lett. **21**(12), 1501–1505 (2014)
13. Yang, Z., Ding, Z., Fan, P., Al-Dhahir, N.: A general power allocation scheme to guarantee quality of service in downlink and uplink NOMA systems. IEEE Trans. Commun. **15**(11), 7244–7257 (2016)
14. Choi, J.: Minimum power multicast beamforming with superposition coding for multiresolution broadcast and application to NOMA systems. IEEE Trans. Commun. **63**(3), 791–800 (2015)
15. Timotheou, S., Krikidis, I.: Fairness for non-orthogonal multiple access in 5G systems. IEEE Sign. Process. Lett. **22**(10), 1647–1651 (2015)
16. Ding, Z., Fan, P., Poor, H.V.: Impact of user pairing on 5G nonorthogonal multiple-access downlink transmissions. IEEE Trans. Veh. Technol. **65**(8), 6010–6023 (2016)
17. Ding, Z., Adachi, F., Poor, H.V.: The application of MIMO to non-orthogonal multiple access. IEEE Trans. Commun. **15**(1), 537–552 (2016)
18. Tam, H.H.M., Tuan, H.D., Nasir, A.A., Duong, T.Q., Poor, H.V.: Mimo energy harvesting in full-duplex multi-user networks. IEEE Trans. Wirel. Commun. **16**(5), 3282–3297 (2017)
19. Tuan, H., Nasir, A.A., Nguyen, H.H., Duong, T.Q., Poor, H.V.: Non-orthogonal multiple access with improper Gaussian signaling. IEEE J. Sel. Areas Commun. **13**, 496–507 (2019)

20. Hoang, T.M., Van Son, V., Dinh, N.C., Hiep, P.T.: Optimizing duration of energy harvesting for downlink NOMA full-duplex over nakagami-m fading channel. AEU-Int. J. Electron. Commun. **95**, 199–206 (2018)
21. Hoang, T.M., Tran, X.N., Thanh, N., Dung, L.T.: Performance Analysis of MIMO SWIPT Relay Network with Imperfect CSI. Mob. Netw. Appl. **24**(2), 630–642 (2019)
22. Cheng, H.V., Björnson, E., Larsson, E.G.: Performance analysis of NOMA in training-based multiuser MIMO systems. IEEE Trans. Commun. **17**(1), 372–385 (2018)
23. Nguyen, B.C., Hoang, T.M., Tran, P.T.: Performance analysis of full-duplex decode-and-forward relay system with energy harvesting over Nakagami-m fading channels. AEU-Int. J. Electron. Commun. **98**, 114–122 (2019)
24. Shankar, P.M.: Fading and Shadowing in Wireless Systems. Springer, Heidelberg (2017)
25. Mukkavilli, K.K., Sabharwal, A., Erkip, E., Aazhang, B.: On beamforming with finite rate feedback in multiple-antenna systems. IEEE Trans. Inf. Theor. **49**(10), 2562–2579 (2003)
26. Zwillinger, D.: Table of Integrals, Series, and Products. Elsevier, Amsterdam (2014)
27. Bariah, L., Muhaidat, S., Al-Dweik, A.: Error probability analysis of non-orthogonal multiple access over nakagami-m fading channels. IEEE Trans. Commun. **64**(1), 76–88 (2018)
28. Tse, D., Viswanath, P.: Fundamentals of Wireless Communication. Cambridge University Press, Cambridge (2005)

User-Pairing Scheme in NOMA Systems: A PSO-Based Approach

Antonino Masaracchia[1(✉)], Long D. Nguyen[2], Trung Q. Duong[1],
Daniel B. da Costa[3], and Thuong Le-Tien[4]

[1] School of Electronics, Electrical Engineering and Computer Science,
Queen's University Belfast, Belfast BT3 9DT, UK
A.Masaracchia@qub.ac.uk
[2] Technology Development, Dong Nai University, Dong Nai 810000, Vietnam
[3] Department of Computer Engineering, Federal University of Ceará,
Sobral, CE 62010-560, Brazil
[4] Faculty of Electrical and Electronics Engineering,
Ho Chi Minh City University of Technology, Ho Chi Minh City 700000, Vietnam

Abstract. Non-orthogonal multiple access (NOMA) is considered a promising technology for improving the spectral efficiency in fifth generation communication systems. In contrast to orthogonal multiple access (OMA), NOMA allows to allocate one frequency channel to multiple users at the same time within the same cell. Basically, this is possible through power-domain superposition coding (SC) multiplexing at transmitter and successive interference cancellation (SIC) at receiver. For this reason, either an optimal power allocation scheme and an optimal user-aggregation policy result to have a key role on NOMA systems, especially in power constrained scenarios like disaster communications. In this paper, a particle swarm optimization (PSO)-based approach for user aggregation in NOMA systems is presented. The efficiency of this approach in finding the optimal aggregation scheme which require the minimum transmission power, maintaining the quality of service (QoS) constraint of each user, is evaluated through simulations, providing comments and remarks about the obtained results.

Keywords: NOMA · PSO · Sub-channel mapping · User-pairing

1 Introduction

During the last decade, the diffusion of powerful multimedia devices, such as smartphones and tablets, has grown exponentially, creating the need for a new cellular technology referred to as 5G [7,11,18].

This work was supported in part by the Newton Prize 2017 and the Newton Fund Institutional Link through the Fly-by-Flood Monitoring Project under Grant ID 428328486, which is delivered by the British Council.

© ICST Institute for Computer Sciences, Social Informatics and Telecommunications Engineering 2019
Published by Springer Nature Switzerland AG 2019. All Rights Reserved
T. Q. Duong et al. (Eds.): INISCOM 2019, LNICST 293, pp. 18–25, 2019.
https://doi.org/10.1007/978-3-030-30149-1_2

An important aspect, used to improve the system capacity in cellular mobile communications, is the design of the multiple radio access technology (M-RAT). Nowadays, such multiple access technologies can be categorized into two different classes: *(i)* orthogonal multiple access (OMA) and *(ii)* non-orthogonal multiple access (NOMA).

Frequency division multiple access (FDMA), time division multiple access (TDMA), code division multiple access (CDMA), and orthogonal frequency-division multiple access (OFDMA) are examples of OMA schemes. In contrast to OMA, NOMA allows to allocate one frequency channel to multiple users at the same time within the same cell, offering a number of advantages which permit to label NOMA as a promising multiple access scheme for future radio access networks [2,9,14–16,19,20,23].

Since the basic principle of NOMA is to serve multiple users by power-domain superposition coding (SC) multiplexing at transmitter and successive interference cancellation (SIC) at receiver, one of the main challenges of this multiple access technique is represented by the power allocation scheme adopted by the transmitter. The problem of optimal power allocation for NOMA systems, with respect to different network performances maximization like energy efficiency maximization and maximum throughput, has been widely investigated in literature [5,6,21,22,24–26]. However, another aspect which represents a key factor for NOMA system performance, is the user-aggregation policy adopted for multiplexing users along different sub-channels [3].

To the best of our knowledge, at date, most of the works on NOMA face this aspect pairing at most two users per sub-channel [4,8,17]. One of the most extensive study can be recognized in [27], where a general scheme for aggregate more than two users into a single sub-channel is provided. Generally, the optimization process for user-pairing and sub-channel mapping in NOMA systems is represented by a mixed integer-linear problem (MILP) which, even if small, may be hard to solve. Under this perspective, this paper proposes and evaluates the performance of a particle swarm optimization (PSO) approach for user-aggregation which require the minimum transmitting power.

2   Introduction to NOMA Systems

In this section, some NOMA basics are presented. It is assumed that a base station (BS) serves N users located within its coverage area. Without loss of generality, it is also supposed that *(i)* both transmitter and receivers are equipped with a single antenna, and *(ii)* users' channel coefficient are ordered in a ascending manner, i.e., $0 < |h_1|^2 \leq |h_2|^2 \cdots \leq |h_N|^2$. In downlink the BS serves the N users employing power-domain SC multiplexing. Then, the signal received by user i can be expressed as:

$$y_i = h_i \cdot x + w_i; \quad \forall i = 1 \cdots N; \tag{1}$$

where $x = \sum_{i=1}^{N} \sqrt{P\beta_i} S_i$ is the superimposed signal containing all S_i messages, h_i denotes the channel coefficient, and w_i represents the noise term with spectral density σ^2. In particular, since $\sum_{i=1}^{N} \beta_i = 1$, the transmitter employ a total amount of transmitting power equal to P. Each user implement the SIC iteratively, decoding signals transmitted to users with weaker channel condition firstly and subtracting them from superimposed received signal. Then, the signal obtained from this subtracting process is used to decode its own related message. Taking that into account and supposing that $\|S_i\|^2 = 1$, the achievable rate in downlink for user i can be expressed as:

$$R_{i,DL} = \log_2 \left(1 + \frac{\beta_i P |h_i|^2}{P |h_i|^2 \sum_{k=i+1}^{N} \beta_k + \sigma^2} \right), \tag{2}$$

As one can note, only the noise spectral density is present in Eq. (2) when $i = N$, since the messages of users $i < N$ have been deleted through SIC.

3   User-Aggregation Problem Formulation

Considering Eq. (2), in order to guarantee a minimum quality of service (QoS) to user i, i.e., $R_{i,DL} \geq R_i^{min}$, the minimum amount of power P_i^{min} which should be allocated to that user is formulated as:

$$P_i^{min} \geq A_i \times \left(\sum_{k=i+1}^{N} P_k + \frac{\sigma^2}{|h_i^2|} \right), \tag{3}$$

in which $P_i = P\beta_i$ and $A_i = \left(2^{R_i^{min}} - 1 \right)$. Supposing that all the users have the same QoS requirements, i.e., $A_N = A_{N-1} = \cdots = A_1 = A$, Eq. (3) can be written as:

$$P_i^{min} \geq \begin{cases} A \times \frac{\sigma^2}{|h_N|^2} = P_N^{min}, & i = N; \\ A \times \left(\sum_{k=i+1}^{N} P_k + \frac{\sigma^2}{|h_k^2|} \right), & i < N; \end{cases} \tag{4}$$

In particular, after some mathematical manipulations, the second case can be expressed as follow:

$$P_i^{min} \geq A \times P_N + A \times \sum_{k=i+1}^{N-1} P_k + P_N^{min} \frac{|h_N|^2}{|h_k|^2}. \tag{5}$$

Then, the total amount of power required in order to guarantee the QoS of all users is:

$$P_{tot} = \sum_{i=1}^{N} P_i^{min} \geq \sum_{i=1}^{N-1} A \times P_N^{min} + A \times \sum_{i=1}^{N-1} \sum_{k=i+1}^{N-1} P_k + P_N^{min} \times \sum_{i=1}^{N-1} \frac{|h_N|^2}{|h_i|^2} + P_N^{min}. \tag{6}$$

Grouping by common factors and observing that the first term is independent of index i, the following expression is obtained:

$$P_{tot} \geq P_N^{min} \times \left((N-1) \times A + 1 + \sum_{i=1}^{N-1} \frac{|h_N|^2}{|h_i|^2} \right) + A \times \sum_{i=1}^{N-1} \sum_{k=i+1}^{N-1} P_k. \quad (7)$$

Then, this represents the minimum amount of power which is necessary to use in order to guarantee the QoS of all users multiplexed within the same sub-channel. This amount of energy strongly depends from user aggregation and sub-channel mapping process. Supposing that N users should be multiplexed along M independent sub-channels, and indicating with $\mathbf{U} \in \{0;1\}^{N \times M}$ the sparse matrix in which the element $u_{i,j}$ is equal to 1 if user i is allocated to sub-carrier j and 0 otherwise, the optimization problem is formulated as:

$$\min_{\mathbf{U}} \quad P_{tot} ; \quad (8a)$$

$$\text{s.t.} \quad R_{i,DL} \geq R_i^{min}, \quad \forall\, i = 1 \cdots N; \quad (8b)$$

$$\sum_{j=1}^{M} u_{i,j} = 1, \quad \forall\, i = 1 \cdots N; \quad (8c)$$

The constraint (8b) represents the minimum QoS requirement of each user. The constraint (8c) makes sure that each user will be multiplexed only into one sub-channel. Since this type of problem represents a MILP problem, in order to find an optimal solution, a PSO-based approach, which respect to other heuristic approaches has shown a more promising behaviour [10], is proposed.

4 A Particle Swarm Optimization (PSO) Approach for Optimal User-Pairing

PSO is one of metaheuristic optimization technique inspired by natural life behaviour like bird flocking and fish schooling [1,12]. It consists in a set of a predefined number, say N_p, of particles with a position X_i and a velocity V_i in a dimensional space of dimension D. Iteratively, each particle, which represents a solution of the optimization problem, is evaluated through a fitting function, obtaining the personal best of the particle, i.e., $Pbest_i$. This $Pbest_i$ is compared with the global best value, i.e., $Gbest$. After this comparison each particle adjusts its own position and velocity along each dimension according with the following equations:

$$V_{i,d}(t) = w \cdot V_{i,d}(t-1) + c_1 \cdot r_1 \cdot (Xpbset_{i,d} - X_{i,d}(t-1)) \\ + c_2 \cdot r_2 \cdot (Xgbest_{i,d} - X_{i,d}(t-1)) , \quad (9)$$

and

$$X_{i,d}(t) = X_{i,d}(t-1) + V_{i,d}(t) , \quad (10)$$

where (9) and (10) represent velocity and position along dimension d, respectively, w is the inertial weight, c_1 and c_2 are two non-negative constants and r_1 and r_2 are two different uniformly random distributed numbers in the range $[0, 1]$.

As in [13], in this paper the initial set of particle has been created in a random fashion. The fitting function for each particle is the total required power expressed by Eq. (7). Moreover, no consistent changes to the solution happened after 500 iterations. Then, in order to ensure a consistent result, the number of 700 iterations has been set as PSO stop criterion. The most important parameters for (9) have been chosen as the same in [13] and are provided in Table 1.

5   Simulation Results

As simulation scenario, it is considered a scenario in which an available bandwidth B is divided equally into M independent sub-channels used to multiplex N users. These users are distributed into a circular area of radius R according with a poisson point process (PPP). The transmitter is supposed at the center of this area. It is assumed that the channel statistics of all users along the whole bandwidth are known. The channel gain of Eq. (1) has been supposed as $h_i = d_i^{-\alpha/2} \times g_i$ where g_i follows a Rayleigh distribution, d_i represents the distance between transmitter and receiver and α is the path-loss exponent. The noise power along the whole bandwidth is $N_0 = 290 \cdot k \cdot B \cdot NF$, where k and NF are Boltzmann constants and noise figure at 9 dB, respectively. Then, the noise power in each sub-channel is N_0/M. The most relevant simulation parameters are summarized in Table 1 and all results represent the average of 10 different simulation runs. The policy efficiency (PE) has been used as index for

Table 1. Simulation parameters

Parameter	Value	Parameter	value
Cell radius (m)	200	M (average number of nodes)	100
Bandwidth (MHz)	40	Pathloss exponent α	4
N_p	50	$N_{iterations}$	200
C_1	1.4962	C_2	1.4962
w	0.7968	N number of sub-channels	[25:50]
V_{max}	0.5	V_{min}	−0.5
QoS threshold [bps/Hz]	[1:5]	σ Rayleigh	1

performance evaluation. In particular, indicating with $P_{av,R}$ the average power required through random policy, and with $O_{av,i}$ the power required through PSO policy, the PE is defined as follow:

$$PE_i = \frac{P_{av,R} - O_{av,i}}{P_{av,R}} \tag{11}$$

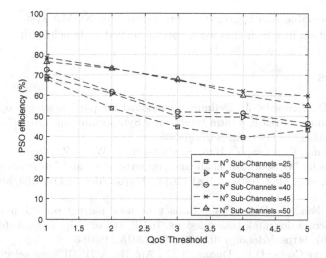

Fig. 1. Policy efficiency gain over different QoS thresholds.

In summary, this represents the reduction in power requirements by using the configuration from PSO output instead of random policy assignment.

Figure 1 shows the variation of the PE gain, expressed in percentage, by varying the QoS thresholds and the number of sub-channels. From these graphics one can note how the PE ranges from a minimum of 45% to a maximum of 80%. In addition, the PE increases by increasing the number of available sub-channels and decreases by increasing the QoS constraint. These results are in line with Eq. (7). Indeed, an increase of the QoS constraint results in an exponentially increase of the minimum required power for all the users. Moreover, reducing the number of sub-channels more users will be multiplexed in each sub-band and then, according with Eq. (5), the minimum required power for each of them increase as well. As a consequence, the total required power increase. These results confirm the efficiency of the PSO in finding the optimal configuration which require the minimum power, satisfying the QoS requirements of each user.

6   Conclusions and Future Works

Due to its advantages which can contribute to reach some requirements of next generation 5G networks, during the last few years NOMA technology has attracted the attention of the research community. In line with NOMA principle, i.e., power-domain SC multiplexing at transmitter, this paper presents a performance analysis of a PSO-based approach for user-aggregation along different sub-channels. In particular, through simulations, one can note how this user-pairing scheme is able to find the optimal configuration that permits to require the minimum transmission power, satisfying the QoS requirements of each user. However, depending on the considered scenario, the PSO-based algorithm can result in a high computational cost procedure. Thus, the design of explicit and scalable user aggregation procedures for NOMA systems represents a future direction in which this work can be served as benchmark.

References

1. Bratton, D., Kennedy, J.: Defining a standard for particle swarm optimization. In: Proceedings of IEEE Swarm Intelligence Symposium, pp. 120–127, April 2007. https://doi.org/10.1109/SIS.2007.368035
2. Dai, L., Wang, B., Yuan, Y., Han, S., Chih-Lin, I., Wang, Z.: Non-orthogonal multiple access for 5G: solutions, challenges, opportunities, and future research trends. IEEE Commun. Mag. **53**(9), 74–81 (2015). https://doi.org/10.1109/MCOM.2015.7263349
3. Ding, Z., Fan, P., Poor, H.V.: Impact of user pairing on 5G nonorthogonal multiple-access downlink transmissions. IEEE Trans. Veh. Technol. **65**(8), 6010–6023 (2016). https://doi.org/10.1109/TVT.2015.2480766
4. Do, N.T., da Costa, D.B., Duong, T.Q., An, B.: A BNBF user selection scheme for NOMA-based cooperative relaying systems with SWIPT. IEEE Commun. Lett. **21**(3), 664–667 (2017). https://doi.org/10.1109/LCOMM.2016.2631606
5. Do, T.N., da Costa, D.B., Duong, T.Q., An, B.: Improving the performance of cell-edge users in MISO-NOMA systems using TAS and SWIPT-based cooperative transmissions. IEEE Trans. Green Commun. Netw. **2**(1), 49–62 (2018). https://doi.org/10.1109/TGCN.2017.2777510
6. Do, T.N., da Costa, D.B., Duong, T.Q., An, B.: Improving the performance of cell-edge users in NOMA systems using cooperative relaying. IEEE Trans. Commun. **66**(5), 1883–1901 (2018). https://doi.org/10.1109/TCOMM.2018.2796611
7. Fettweis, G., Alamouti, S.: 5G: personal mobile internet beyond what cellular did to telephony. IEEE Commun. Mag. **52**(2), 140–145 (2014). https://doi.org/10.1109/MCOM.2014.6736754
8. Hojeij, M., Farah, J., Nour, C.A., Douillard, C.: Resource allocation in downlink non-orthogonal multiple access (NOMA) for future radio access. In: Proceedings of IEEE 81st Vehicular Technology Conference (VTC Spring), pp. 1–6, May 2015. https://doi.org/10.1109/VTCSpring.2015.7146056
9. Islam, S.M.R., Avazov, N., Dobre, O.A., Kwak, K.: Power-domain non-orthogonal multiple access (NOMA) in 5G systems: potentials and challenges. IEEE Commun. Surv. Tutorials **19**(2), 721–742 (2017). https://doi.org/10.1109/COMST.2016.2621116
10. Jordehi, A.R., Jasni, J.: Particle swarm optimisation for discrete optimisation problems: a review. Artif. Intell. Rev. **43**(2), 243–258 (2015)
11. Jovović, I., Husnjak, S., Forenbacher, I., Maček, S.: Innovative application of 5G and blockchain technology in industry 4.0. EAI Endorsed Trans. Ind. Netw. Intell. Syst. **6**(18) (2019). https://doi.org/10.4108/eai.28-3-2019.157122
12. Kennedy, J., Eberhart, R.: Particle swarm optimization. In: Proceedings of ICNN 1995 - International Conference on Neural Networks, vol. 4, pp. 1942–1948, November 1995. https://doi.org/10.1109/ICNN.1995.488968
13. Kuila, P., Jana, P.K.: Energy efficient clustering and routing algorithms for wireless sensor networks: particle swarm optimization approach. Eng. Appl. Artif. Intell. **33**, 127 – 140 (2014). https://doi.org/10.1016/j.engappai.2014.04.009. http://www.sciencedirect.com/science/article/pii/S0952197614000852

14. Lee, S.: Cooperative non-orthogonal multiple access for future wireless communications. EAI Endorsed Trans. Ind. Netw. Intell. Syst. **5**(17) (2018). https://doi.org/10.4108/eai.19-12-2018.156078
15. Lee, S., da Costa, D.B., Vien, Q., Duong, T.Q., de Sousa, Jr., R.T., : Non-orthogonal multiple access schemes with partial relay selection. IET Commun. **11**(6), 846–854 (2017). https://doi.org/10.1049/iet-com.2016.0836
16. Lee, S., Duong, T.Q., da Costa, D.B., Ha, D., Nguyen, S.Q.: Underlay cognitive radio networks with cooperative non-orthogonal multiple access. IET Commun. **12**(3), 359–366 (2018). https://doi.org/10.1049/iet-com.2017.0559
17. Liang, W., Ding, Z., Li, Y., Song, L.: User pairing for downlink non-orthogonal multiple access networks using matching algorithm. IEEE Trans. Commun. **65**(12), 5319–5332 (2017). https://doi.org/10.1109/TCOMM.2017.2744640
18. Mavromatis, I., Tassi, A., Rigazzi, G., Piechocki, R.J., Nix, A.: Multi-radio 5G architecture for connected and autonomous vehicles: application and design insights. EAI Endorsed Trans. Ind. Netw. Intell. Syst. **4**(13) (2018). https://doi.org/10.4108/eai.20-3-2018.154368
19. Nasir, A.A., Tuan, H.D., Duong, T.Q., Poor, H.V.: UAV-enabled communication using NOMA. IEEE Trans. Commun. 1 (2019). https://doi.org/10.1109/TCOMM.2019.2906622
20. Nasir, A.A., Tuan, H.D., Duong, T.Q., Debbah, M.: NOMA throughput and energy efficiency in energy harvesting enabled networks. IEEE Transactions. Commun. 1, ISSN 0090-6778 (2019). https://doi.org/10.1109/TCOMM.2019.2919558
21. Nguyen, L.D.: Resource allocation for energy efficiency in 5G wireless networks. EAI Endorsed Trans. Ind. Netw. Intell. Syst. **5**(14) (2018). https://doi.org/10.4108/eai.27-6-2018.154832
22. Nguyen, V., Tuan, H.D., Duong, T.Q., Poor, H.V., Shin, O.: Precoder design for signal superposition in MIMO-NOMA multicell networks. IEEE J. Sel. Areas Commun. **35**(12), 2681–2695 (2017). https://doi.org/10.1109/JSAC.2017.2726007
23. Saito, Y., Kishiyama, Y., Benjebbour, A., Nakamura, T., Li, A., Higuchi, K.: Non-orthogonal multiple access (NOMA) for cellular future radio access. In: Proceedings of IEEE 77th Vehicular Technology Conference (VTC Spring), pp. 1–5, June 2013. https://doi.org/10.1109/VTCSpring.2013.6692652
24. Tuan, H.D., Nasir, A.A., Nguyen, H.H., Duong, T.Q., Poor, H.V.: Non-orthogonal multiple access with improper Gaussian signaling. J. Sel. Top. Sign. Process. **13**(3), 496–507 (2019). https://doi.org/10.1109/JSTSP.2019.2901993
25. Zhang, Y., Wang, H., Zheng, T., Yang, Q.: Energy-efficient transmission design in non-orthogonal multiple access. IEEE Trans. Veh. Technol. **66**(3), 2852–2857 (2017). https://doi.org/10.1109/TVT.2016.2578949
26. Zhu, J., Wang, J., Huang, Y., He, S., You, X., Yang, L.: On optimal power allocation for downlink non-orthogonal multiple access systems. IEEE J. Sel. Areas Commun. **35**(12), 2744–2757 (2017). https://doi.org/10.1109/JSAC.2017.2725618
27. Zhu, L., Zhang, J., Xiao, Z., Cao, X., Wu, D.O.: Optimal user pairing for downlink non-orthogonal multiple access (NOMA). IEEE Wirel. Commun. Lett. 1 (2018). https://doi.org/10.1109/LWC.2018.2853741

Resource Sharing and Segment Allocation Optimized Video Streaming over Multi-hop Multi-path in Dense D2D 5G Networks

Quang-Nhat Tran[1,2], Nguyen-Son Vo[1(✉)], Minh-Phung Bui[3], Van-Ca Phan[4], Zeeshan Kaleem[5], and Trung Q. Duong[6]

[1] Duy Tan University, Da Nang, Vietnam
vonguyenson@duytan.edu.vn
[2] Ho Chi Minh City University of Transport, Ho Chi Minh City, Vietnam
nhat.tran@ut.edu.vn
[3] Van Lang University, Ho Chi Minh City, Vietnam
buiminhphung@vanlanguni.edu.vn
[4] HCMC University of Technology and Education, Ho Chi Minh City, Vietnam
capv@hcmute.edu.vn
[5] COMSATS University Islamabad, Wah Campus, Islamabad, Pakistan
zeeshankaleem@gmail.com
[6] Queen's University Belfast, Belfast, UK
trung.q.duong@qub.ac.uk

Abstract. The rapid increase in mobile users (MUs) and various video applications and services (VASs) poses a set of challenges to 5G networks. Although device-to-device (D2D) communications and resource reuse techniques can improve the streaming performance of VASs, they have not utilized the fact that there are many MUs with available sharing downlink resources, namely cellular users (CUs), and many videos cached in dense MUs to establish an efficient video streaming session. In this paper, we propose a downlink resource sharing and segment allocation (RSA) optimization problem and solve it for high video streaming performance over multi-hop multi-path (MHMP) in dense D2D 5G networks. Particularly, the RSA provides a maximum capacity for the CUs that share their downlink resources with the D2D hops in each path by finding the optimal downlink resource sharing-receiving pairs between the CUs and the D2D hops. Then, the segments of a video cached in different D2D helpers (DHs) are sent to the D2D requester (DR) over the MHMP. By finding which segments are allocated to which paths for sending, the reconstructed distortion of the received segments at the DR is minimized for high playback quality. Simulation results are shown to investigate the benefits of the proposed solution compared to other schemes without RSA.

Keywords: D2D caching and communications ·
Downlink resource sharing · Multi-hop · Multi-path ·
Ultra-dense 5G networks · Video streaming

© ICST Institute for Computer Sciences, Social Informatics and Telecommunications Engineering 2019
Published by Springer Nature Switzerland AG 2019. All Rights Reserved
T. Q. Duong et al. (Eds.): INISCOM 2019, LNICST 293, pp. 26–39, 2019.
https://doi.org/10.1007/978-3-030-30149-1_3

1   Introduction

Device-to-device (D2D) communications and spectrum reuse techniques have been defined as the emerging solutions to improve the system capacity and resource efficiency in 5G networks [1–4]. Applying D2D communications and spectrum reuse techniques to offloading data in close proximity can further relax the macro base stations (MB) and the small-cell base stations, which have been suffered from high workload due to the rapid increase in mobile users (MUs) and various video applications and services (VASs) [5,6]. However, current studies in the literature have not utilized the fact that there are many MUs with available sharing downlink resources, i.e., cellular users (CUs), and many videos cached in dense MUs, i.e., D2D helpers (DHs), to establish an efficient multi-hop multi-path (MHMP) video streaming session of VASs.

In consideration of D2D communications and resource allocation based VASs, other additional techniques such as caching and clustering [7–10], scheduling and mode selection [11–13], and transmission [14,15], have been proposed to enhance the video streaming performance, i.e., high resource efficiency and high playback quality. The problem of these techniques is that the VASs are deployed in the context of single-hop D2D communications. On the one hand, this in turn requires high D2D transmission powers causing high interference impact on the CUs that share their downlink resources. On the other hand, the videos, which have been cached in the DHs located more than one hop far away from the requesters, are not exploited for VASs. In fact, MHMP D2D communications have been studied to gain the benefits from dense D2D 5G networks [16–20]. However, the proposed solutions are not efficient enough, especially for VASs, because they do not utilize the videos cached in the HDs for multi-source streaming nor do they consider the characteristics of the videos (e.g., rate-distortion (RD) models) and the users' behavior (i.e., represented by the video access rate/popularity) for high video streaming quality.

In addition, few of the studies have focussed on VASs over multi-source and MHMP D2D communications [21–23]. Particularly, in [21], the paths and the video traffics are scheduled to balance the energy consumption amounts of the D2D links to prolong the cooperative streaming duration of D2D networks. The authors in [22] have proposed a cross-layer solution to optimally select the D2D modes, video coding modes, and transmission paths to enhance the average quality of received video under a given energy constraint. Especially, by packetizing the video into multiple descriptions at optimal encoding rates and allocating the optimal numbers of descriptions to different DHs for being transmitted, the video is received at the D2D requester (DR) with high playback quality and low quality fluctuation while consuming low energy [23]. It can be observed that the characteristics of videos and users [21,22] and the advantages of downlink resource sharing allocation and multi-hop D2D communications [23] are not considered for high video streaming performance.

In this paper, we take the advantages of available resources of CUs and videos cached in the DHs in consideration of the characteristics of both videos and users, to propose a downlink resource sharing and segment allocation (RSA)

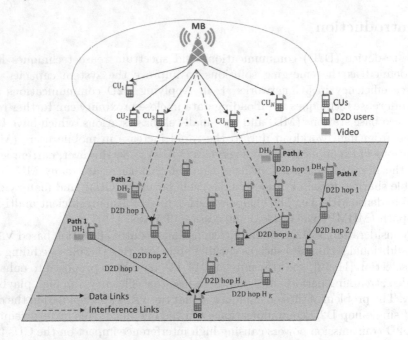

Fig. 1. VASs over MHMP in dense D2D 5G network.

optimization strategy for high performance of VASs over MHMP in dense D2D 5G networks. The RSA strategy consists of two optimization problems, namely resource sharing allocation (RA) optimization problem and segment allocation (SA) optimization problem. The RA optimization problem is solved for optimal resource sharing-receiving pairs between the CUs and the D2D hops for MHMP D2D communications. The objective is to maximize the capacity of the CUs that share their downlink resources with the D2D hops in each path. Based on the optimal results of the RA solution (all D2D hops with corresponding shared donwlink resources in each path from the DHs to the DR are identified), the SA optimization problem is then solved for optimally allocating which segments to which paths for sending. This way, the reconstructed distortion of the received video is minimized for high playback quality.

The rest of this paper is organized as follows. In Sect. 2, we introduce our system models including the VASs over MHMP in dense D2D 5G networks, downlink resource sharing based CUs capacity, and video segment transmission. The RSA optimization problems and solutions are presented in Sect. 3. We show the simulation results with detailed performance evaluation in Sect. 4. Finally, Sect. 5 is dedicated to concluding the paper.

2 System Models

2.1 VASs over MHMP in Dense D2D 5G Networks

In this paper, we consider the VASs over MHMP in a dense D2D 5G network as shown in Fig. 1. Thanks to dense D2D users in 5G networks, we assume that there are at least K DHs that have cached the requested video consisting of S segments ($S \leq K$). We also assume that there are K paths established from the K DHs to the DR by using [19, 24], here the path k has H_k D2D hops. The D2D users are randomly located and modeled as 2-D homogeneous Poisson point process (PPP) in a circular cell with intensity λ_D [25, 26]. In the system, there are N CUs ($N \geq K$) that share their downlink resources with H_k D2D hops in the path k. An arbitrary CU can share its downlink resource with only one D2D hop in a path, but with up to K D2D hops in K paths. The transmission over the D2D hop h_k in path k is done by reusing the downlink resource shared by the CU n optimally selected from N CUs such that the average capacity of the CUs is maximized. Finally, S segments of the requested video are optimally allocated to K paths for sending so as to minimize the reconstructed distortion of received video at the DR.

2.2 Downlink Resource Sharing Based CUs Capacity

In our aforementioned downlink sharing scheme, the CU n suffers interference from the transmitters of up to K D2D hops in K paths. Therefore, the signal to interference plus noise ratio (SINR) at the CU n is given by

$$SINR_C^n = \frac{P_M G_{M,C}^n}{N_0 + \sum_{k=1}^{K} P_D^{h_k} G_{D,C}^{n,h_k}}, \tag{1}$$

where P_M and $P_D^{h_k}$ are the transmission powers of the MB and the transmitter of the D2D hop h_k in the path k; N_0 is the power of additive white Gaussian noise (AWGN); $G_{M,C}^n$ and $G_{D,C}^{n,h_k}$ are the channel gains from the MB and the transmitter of the D2D hop h_k to the CU n, which are modelled as an exponential power fading coefficient with unit mean ($\sim\exp(1)$) and a standard power law path loss function with path loss exponent η [25, 26].

The problem is that how to allocate the downlink resource of the CU n to the D2D hop h_k, h_k=1, 2, ..., H_k, such that the total interference effect on the CU n is minimized, i.e., gaining the highest SINR at the CU n. To do so, we add a downlink resource sharing allocation (RA) index v_{n,h_k} to (1) to indicate that if the CU n agrees to share ($v_{n,h_k} = 1$) with the D2D hop h_k or not ($v_{n,h_k} = 0$). Therefore, Eq. (1) is re-written as

$$SINR_C^n = \frac{P_M G_{M,C}^n}{N_0 + \sum_{k=1}^{K} \sum_{h_k=1}^{H_k} v_{n,h_k} P_D^{h_k} G_{D,C}^{n,h_k}}. \tag{2}$$

Based on Shannon-like capacity, the capacity delivered to the CU n and the average capacity per each CU are respectively given by

$$C_C^n = W \log_2(1 + SINR_C^n) \tag{3}$$

and

$$\overline{C} = \frac{1}{N} \sum_{n=1}^{N} C_C^n, \tag{4}$$

where W is the system bandwidth.

In this paper, $\overline{C}$ in (4) is maximized by finding the optimal values of v_{n,h_k}. This so-called RA optimization problem and solution are introduced in the next section.

2.3 Video Segment Transmission

In the SA optimization problem, S segments of the requested video are allocated to K paths such that the reconstructed distortion of the received video at the DR is minimized. To do so, we take into account the RD based video packetization scheme at the DHs and the multi-hop lossy wireless environment analysis, which are presented below.

RD Based Video Packetization Scheme: Aiming at providing the highest data protection strategy and the robustness over the diverse characteristics of the MHMP wireless environment, we exploit the benefits of scalable extensions of high efficiency video coding (SHVC) to packetize each segment of the requested video into M descriptions for transmission by using layered multiple description coding with embedded forward error correction (LMDC-FEC) [23,27–34]. The most important characteristic of SHVC based LMDC-FEC is that the reconstructed distortion of the received segment is more reduced for higher playback quality if more descriptions are correctly received without considering the orders of the received descriptions.

By further applying the RD model given in [30,31] to SHVC, if the segment s has m out of M descriptions received correctly, this segment is decoded for playing back at the rate R_s^m corresponding to the reconstructed distortion $D_s(R_s^m)$. The relationship between rate R_s^m (measured in Kbps) and the reconstructed distortion $D_s(R_s^m)$ follows a decaying exponential function given by

$$D_s(R_s^m) = \gamma_s (R_s^m)^{\beta_s}, \tag{5}$$

where γ_s and β_s are the independent parameters found by analysing the experimental characteristic of the segment s.

Description Multi-hop Delay: A description of the segment s is lost over the path k if it does not arrive in time at the DR for playing back. Following [35,36] and ignoring the total processing delay, given a delay threshold $\tau_{s,k}$, the loss probability of a description of the segment s is expressed as

$$L_D^{s,k} = \int_{\tau_{s,k}}^{\infty} \frac{\mu_k}{\Gamma(H_k + 1)} (\mu_k t)^{H_k} e^{-\mu_k t} dt, \tag{6}$$

where the delay threshold $\tau_{s,k}$ and the average waiting delay per hop $1/\mu_k$ are respectively given by

$$\tau_{s,k} = \kappa \sum_{h_k=1}^{H_k} t_{h_k} + \frac{(s-1)F}{Sf} \tag{7}$$

and

$$\frac{1}{\mu_k} = \frac{\sum_{h_k=1}^{H_k} t_{h_k}}{H_k}, \tag{8}$$

where $\kappa \geq 1$ is the startup delay coefficient, t_{h_k} is the waiting delay of hop h_k, and F and f are the number of frames and the frame rate of the considered video, respectively.

Description Multi-hop Transmission Error: In addition, a description of a segment is lost over the path k due to transmission error. After finding which CUs share their downlink resources with which D2D hops in all K paths, let P_{h_k} be the outage probability of the hop h_k, if a description of a segment is sent over the path k, the loss probability of this description is expressed as

$$L_T^k = 1 - p_k \prod_{h_k=1}^{H_k} (1 - P_{h_k}), \tag{9}$$

where p_k is the probability that the video has been cached in the DH k. By following [25,26] in which N_0 is negligible, given a threshold of capacity (i.e, bit rate) C_{th} for reliable communications, and assuming that the statistical model of channel over the hop h_k is Rayleigh fading, P_{h_k} can be computed as

$$P_{h_k} = P\{C_D^{h_k} < C_{th}\} = 1 - \exp\left\{ -\xi_{h_k}\left[\lambda_M\left(\frac{P_M}{P_D^{h_k}}\right)^{\frac{2}{\eta}} + \lambda_D \right] \right\}, \tag{10}$$

where $\xi_{h_k} = \sum_{n=1}^{N} v_{n,h_k} \pi d_{h_k}^2 \Gamma(1+\frac{2}{\eta})\Gamma(1-\frac{2}{\eta})(2^{\frac{C_{th}}{W}} - 1)^{2/\eta}$, d_{h_k} is the distance from the transmitter to the receiver of the hop h_k, and $C_D^{h_k}$ is the capacity at the receiver of the D2D hop h_k coming from its corresponding signal-to-interference-plus-noise ratio $SINR_D^{h_k}$, expressed as

$$SINR_D^{h_k} = \frac{\sum_{n=1}^{N} v_{n,h_k} P_D^{h_k} G_{D,D}^{n,h_k}}{N_o + P_M G_{M,R}^n + \sum_{n=1}^{N} \sum_{l=1,l\neq k}^{K} \sum_{h_l=1}^{H_l} v_{n,h_l} P_D^{h_l} G_{D,R}^{n,h_l}} \tag{11}$$

and

$$C_D^{h_k} = W \log_2(1 + SINR_D^{h_k}).$$ (12)

Description Loss Probability: So far, a description of the segment s is lost if it arrives at the DR later than the threshold or it is not received correctly at the DR due to transmission error. The loss probability of a description of the segment s is computed as

$$L_{s,k} = L_T^k + (1 - L_T^k)L_D^{s,k}.$$ (13)

Finally, the probability that m out of M descriptions of the segment s are correctly received over the path k is given by

$$P_{M,m}^{s,k} = \binom{m}{M}(1 - L_{s,k})^m (L_{s,k})^{M-m}.$$ (14)

Reconstructed Distortion: Based on the aforementioned analysis of the RD based video packetization scheme and the multi-hop lossy wireless environment, we further take into account the segment allocation index $u_{s,k}$ to indicate that if the segment s is selected to be sent over the path k $(u_{s,k} = 1)$ or not $(u_{s,k} = 0)$ so as to minimize the average reconstructed distortion of received video at the DR. Consequently, the average reconstructed distortion of the received video at the DR is computed as

$$\overline{D} = \sum_{s=1}^{S} r_s \sum_{k=1}^{K} u_{s,k} \sum_{m=0}^{M} P_{M,m}^{s,k} D_s(R_s^m),$$ (15)

where r_s, which is the access rate (popularity) of the segment s representing the users' behavior, is modelled by following Zipf-like distribution [37], given by

$$r_s = \frac{s^{-\alpha}}{\sum_{s=1}^{S} s^{-\alpha}},$$ (16)

here α is the skewed access rate among different segments of the considered video. It means that if $\alpha = 0$, all segments have the same access rate, while the higher value of α yields the higher skewed access rate among different segments.

3   RSA Optimization Problems and Solutions

In the RA problem, it is to maximize the average capacity $\overline{C}$ per each CU that shares its resource for MHMP D2D communications. The optimal values of v_{n,h_k} are found such that the interference effect on the CUs caused by the transmitters of D2D hops is minimized. In addition, we further take into account the constraints of $\sum_{h_k=1}^{H_k} v_{n,h_k} \leq 1$ (each CU can share with up to one D2D hop in each path), $\sum_{n=1}^{N} v_{n,h_k} = 1$ (each D2D hop in a path can be shared by only one CU), and $\sum_{n=1}^{N} \sum_{h_k=1}^{H_k} v_{n,h_k} = H_k$ (all CUs can share up to H_k D2D hops in each path). The RA problem is formulated as follows:

$$\max_{v_{n,h_k}} \overline{C} \tag{17}$$

$$s.t. \begin{cases} \sum_{h_k=1}^{H_k} v_{n,h_k} \leq 1, n = 1, 2, ..., N, k = 1, 2, ..., K, \\ \sum_{n=1}^{N} v_{n,h_k} = 1, k = 1, 2, ..., K, \\ \sum_{n=1}^{N} \sum_{h_k=1}^{H_k} v_{n,h_k} = H_k, k = 1, 2, ..., K. \end{cases} \tag{18}$$

After solving (17) and (18), all the D2D hops in K paths are established for video streaming from the DHs to the DR. In the SA problem, the optimal values of $u_{s,k}$ are found to yield the optimal matching allocation between the access rate r_s and the RD of the segment s induced lossy characteristic of the path k such that the average reconstructed distortion $\overline{D}$ of the received video at the DR is minimized. In addition, because we have S segments, i.e., $S \leq K$, sent over K paths, there have some paths that are not used. We further take into account the constraints of $\sum_{s=1}^{S} u_{s,k} \leq 1$, $\sum_{k=1}^{K} u_{s,k} = 1$, and $\sum_{s=1}^{S} \sum_{k=1}^{K} u_{s,k} = S$ to ensure that each path can send up to one segment, each segment is sent over only one path, and K paths must send all S segments. The SA problem is formulated as follows:

$$\min_{u_{s,k}} \overline{D} \tag{19}$$

$$s.t. \begin{cases} \sum_{s=1}^{S} u_{s,k} \leq 1, k = 1, 2, ..., K, \\ \sum_{k=1}^{K} u_{s,k} = 1, s = 1, 2, ..., S, \\ \sum_{s=1}^{S} \sum_{k=1}^{K} u_{s,k} = S. \end{cases} \tag{20}$$

Both the RA and SA optimization problems can be solved by using exhaustive binary matrix search, in which finding optimal values of v_{n,h_k} and $u_{s,k}$ is actually finding the optimal matrices $\{V^*_{N \times H_1}; V^*_{N \times H_2}; ...; V^*_{N \times H_K}\}$ and $U^*_{S \times K}$ by searching the following binary matrix spaces:

$$\mathcal{V} = \{V^1_{N \times H_1}, V^2_{N \times H_1}, \quad, V^{2^{N \times H_1}}_{N \times H_1}; V^1_{N \times H_2}, V^2_{N \times H_2}, ..., V^{2^{N \times H_2}}_{N \times H_2}; \tag{21}$$
$$...; V^1_{N \times H_K}, V^2_{N \times H_K}, ..., V^{2^{N \times H_K}}_{N \times H_K} \}$$

and

$$\mathcal{U} = \{U^1_{S \times K}, U^2_{S \times K}, ..., U^{2^{S \times K}}_{S \times K}\}. \tag{22}$$

4   Performance Evaluation

The system parameters setting is given in Table 1. Furthermore, the distance of D2D hops, distance from D2D transmitters to the CUs, and distance from the MB to the CUs and the D2D receivers, are respectively in the ranges of $[1, 10]$ m, $[1, 50]$ m, and $[300, 1500]$ m. The system is covered by a circular area with radius of 1500 m, and thus $\lambda_M = 0.14147 \times 10^{-6}$ and $\lambda_D = 1.9806 \times 10^{-6}$. We evaluate the performance of the proposed **RSA** (i.e., **RA** and **SA**) by comparing it to the other two benchmarks without RSA, namely Average (**Ave**) and Minimum (**Min**). In **Ave**, the capacity of the CUs and the reconstructed distortion of received video are averaged over the number of feasible solution sets of $\mathcal{V}$ and the number of feasible solutions of $\mathcal{U}$ that satisfy (18) and (20). Meanwhile in **Min**, the capacity of the CUs and the reconstructed distortion of received video are computed by finding the worst feasible solution set of $\mathcal{V}$ and the worst feasible solution of $\mathcal{U}$ that cause the average capacity of CUs and the playback quality at the DR minimum.

Table 1. Parameters setting

Symbols	Specifications
S	3 segments
K	4 paths
H_k	$\{1, 2, 3, 4\}$ D2D hop(s)
N	5 CUs
W	10 MHz
P_M	10 W
p_k	$\{1, 1, 1, 1\}$
M	32 descriptions
C_{th}	1 bps
γ_s	$\{10, 20, 30\}$
β_s	$\{-0.5, -0.75, -1\}$
η	4
F	300 frames
f	24 frames/s
R_s^M	3000 Kbps (full rate of each segment)
κ	1
t_{h_k}	Randomly distributed in the range of $[0.0002, 0.002]$ s
$P_D^{h_k}$	Randomly distributed in the range of $[0.1, 0.5]$ W
N_0	10^{-13} W

We first evaluate the capacity performance of the **RA**, **Ave**, and **Min** versus the number of paths by changing K from 1 to 5. As shown in Fig. 2, if the number of paths increases, the CUs suffer from higher interference impacts caused by

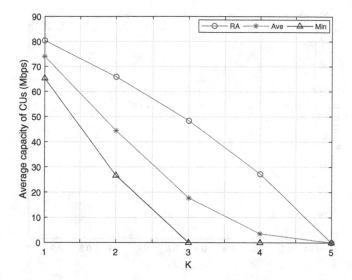

Fig. 2. Average capacity of CUs versus number of paths.

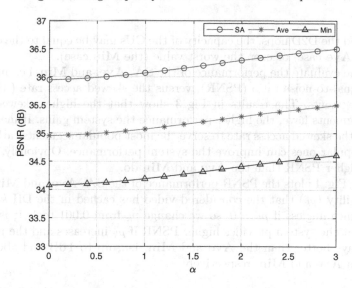

Fig. 3. Quality of received video at DR versus skewed access rate of segments.

dense D2D hops, and thus the average capacity of CUs decreases, even approximately to zero when $K = 5$. In comparison, the **RA** outperforms the **Ave** and **Min**. It is noted that the number of paths for MHMP D2D communications is carefully selected to guarantee the CUs high quality of service. In addition, if we do not optimally allocate the downlink resource sharing-receiving pairs between

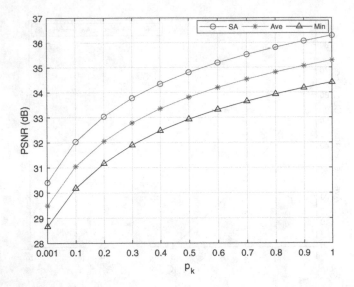

Fig. 4. Quality of received video at DR versus caching probability.

the CUs and the D2D hops, the capacity of the CUs may be equal to the common value (the **Ave** case) or even the worst value (the **Min** case).

Next, we evaluate the performance of the **SA**, **Ave**, and **Min**, i.e., measured in peak signal-to-noise ratio (PSNR), versus the skewed access rate (α) among different segments. The results in Fig. 3 show that the higher skewed access rate the segments have, the higher performance the system gains. It means that exploiting the skewed access rate to serve the most popular segments rather than the less popular ones can improve the system performance. Obviously, the **SA** provides higher PSNR than the **Ave** and **Min** do.

Finally, Fig. 4 plots the PSNR performance of the **SA**, **Ave**, and **Min** versus the probability (p_k) that the considered video has cached in the DH k. In this case, it is meaningless if $p_k = 0$, so we change p_k from 0.001 to 1. It is easy to observe that the system provides higher PSNR if p_k increases and the proposed **SA** is always better than the **Ave** and **Min**, i.e., up to 1 dB and about 2 dB higher than **Ave** and **Min**, respectively.

5   Conclusion

In this paper, we have proposed a downlink resource sharing and segment allo-cation (RSA) optimization solution for high performance of video streaming applications and services over multi-hop multi-path (MHMP) in dense D2D 5G networks. In particular, the RSA can exploit the dense characteristic of D2D users with cached video and the available downlink resources of the CUs to establish a MHMP video streaming session that can guarantee the CUs max-imum average capacity. In addition, by considering the characteristics of the video, the users' behavior, and the multi-hop lossy wireless environment, the RSA can provide the D2D requester with minimum reconstructed distortion of the received video for high playback quality.

Acknowledgement. This research is funded by Vietnam National Foundation for Science and Technology Development (NAFOSTED) under grant number 102.04-2018.308.

References

1. Mishra, P.K., Pandey, S., Biswash, S.K.: Efficient resource management by exploiting D2D communication for 5G networks. IEEE Access **4**, 9910–9922 (2016)
2. Ali, M., Qaisar, S., Naeem, M., Mumtaz, S.: Energy efficient resource allocation in D2D-assisted heterogeneous networks with relays. IEEE Access **4**, 4902–4911 (2016)
3. Nguyen, L.D.: Resource allocation for energy efficiency in 5G wireless networks. EAI Trans. Ind. Netw. Intell. Syst. **5**, 1–7 (2018)
4. Thanh, T.L., Hoang, T.M.: Cooperative spectrum-sharing with two-way AF relaying in the presence of direct communications. EAI Trans. Ind. Netw. Intell. Syst. **5**, 1–9 (2018)
5. Vo, N.-S., Duong, T.Q., Guizani, M., Kortun, A.: 5G optimized caching and downlink resource sharing for smart cities. IEEE Access **6**, 31457–31468 (2018)
6. Bui, M.-P., Vo, N.-S., Nguyen, T.-T., Tran, Q.-N., Tran, A.-T.: Social-aware caching and resource sharing optimization for video delivering in 5G networks. In: Duong, T.Q., Vo, N.-S., Phan, V.C. (eds.) Qshine 2018. LNICST, vol. 272, pp. 73–86. Springer, Cham (2019). https://doi.org/10.1007/978-3-030-14413-5_6
7. Zhu, H., Cao, Y., Hu, Q., Wang, W., Jiang, T., Zhang, Q.: Multi-bitrate video caching for D2D-enabled cellular networks. IEEE Trans. Multimed. **26**, 10–20 (2019)
8. Wu, D., Liu, Q., Wang, H., Yang, Q., Wang, R.: Cache less for more: exploiting cooperative video caching and delivery in D2D communications. IEEE Trans. Multimed. **21**, 1–12 (2018)
9. Shen, Y., Jiang, C., Quek, T.Q.S., Ren, Y.: Device-to-device-assisted communications in cellular networks: an energy efficient approach in downlink video sharing scenario. IEEE Wirel. Commun. **15**, 1575–1587 (2016)
10. Zhang, X., Wang, J.: Heterogeneous statistical QoS-driven resource allocation for D2D cluster-caching based 5G multimedia mobile wireless networks. In: Proceedings of IEEE International Conference on Communications (ICC), Kansas City, MO, pp. 1–6, May 2018
11. Liu, M., Yu, F.R., Teng, Y., Leung, V.C.M., Song, M.: Distributed resource allocation in blockchain-based video streaming systems with mobile edge computing. IEEE Wirel. Commun. **8**, 695–708 (2019)
12. Li, J., Feng, R., Sun, W., Chen, L., Xu, X., Li, Q.: Joint mode selection and resource allocation for scalable video multicast in hybrid cellular and D2D network. IEEE Access **8**, 64350–64358 (2018)
13. Ye, C., Gursoy, M.C., Velipasalar, S.: Power control and mode selection for VBR video streaming in D2D networks. In: Proceedings of IEEE Wireless Communications and Networking Conference (WCNC), Barcelona, Spain, pp. 1–6, April 2018

14. Ren, Q., Chen, J., Chen, B., Jin, L.: A video streaming transmission scheme based on frame priority in device-to-device multicast networks. IEEE Access **7**, 20187–20198 (2019)
15. Yun, J., Piran, M.J., Suh, D.Y.: QoE-driven resource allocation for live video streaming over D2D-underlaid 5G cellular networks. IEEE Access **6**, 72563–72580 (2018)
16. Ying, B., Nayak, A.: A power-efficient and social-aware relay selection method for multi-hop D2D communications. IEEE Commun. Lett. **22**, 1450–1453 (2018)
17. Gui, J., Deng, J.: Multi-hop relay-aided underlay D2D communications for improving cellular coverage quality. IEEE Access **8**, 14318–14338 (2018)
18. Lu, X., Zheng, J., Liu, C., Xiao, J.: A mobility and activeness aware relay selection algorithm for multi-hop D2D communication underlaying cellular networks. In: Proceedings of IEEE International Conference on Communications (ICC), Paris, France, pp. 1–6, May 2017
19. Yuan, H., Guo, W., Jin, Y., Wang, S., Ni, M.: Interference-aware multi-hop path selection for device-to-device communications in a cellular interference environment. IET Commun. **11**, 1741–1750 (2017)
20. Wei, L., Hu, R.Q., Qian, Y., Wu, G.: Energy efficiency and spectrum efficiency of multihop device-to-device communications underlaying cellular networks. IEEE Trans. Veh. Technol. **65**, 367–380 (2016)
21. Liu, B., Cao, Y., Wang, W., Jiang, T.: Energy budget aware device-to-device cooperation for mobile videos. In: Proceedings of IEEE Global Communications Conference (GLOBECOM), San Diego, CA, pp. 1–7, December 2015
22. Wang, Q., Wang, W., Jin, S., Zhu, H., Zhang, N.T.: Joint coding mode and multi-path selection for video transmission in D2D-underlaid cellular network with shared relays. In: Proceedings of IEEE Global Communications Conference (GLOBE-COM), San Diego, CA, pp. 1–6, December 2015
23. Vo, N.-S., Duong, T.Q., Tuan, H.D., Kortun, A.: Optimal video streaming in dense 5G networks with D2D communications. IEEE Access **6**, 209–223 (2017)
24. Hung, D.T., Duy, T.T., Trinh, D.Q.: Security-reliability analysis of multi-hop LEACH protocol with fountain codes and cooperative jamming. EAI Trans. Ind. Netw. Intell. Syst. **6**, 1–7 (2019)
25. Bhardwaj, A., Agnihotri, S.: Channel allocation for multiple D2D-multicasts in underlay cellular networks using outage probability minimization. In: Proceedings of National Conference on Communications (NCC), Hyderabad, India, pp. 1–6, February 2018
26. Bhardwaj, A., Agnihotri, S.: Energy- and spectral-efficiency trade-off for D2D-multicasts in underlay cellular networks. IEEE Wirel. Commun. Lett. **7**, 546–549 (2018)
27. Boyce, J.M., Ye, Y., Chen, J., Ramasubramonian, A.K.: Overview of SHVC: scalable extensions of the high efficiency video coding standard. IEEE Trans. Circ. Syst. Video Technol. **26**, 20–34 (2016)
28. Sullivan, G.J., Ohm, J.-R., Han, W.-J., Wiegand, T.: Overview of the high efficiency video coding (HEVC) standard. IEEE Trans. Circ. Syst. Video Technol. **22**, 1649–1668 (2012)
29. Vanne, J., Viitanen, M., Hamalainen, T.D., Hallapuro, A.: Comparative rate-distortion-complexity analysis of HEVC and AVC video codecs. IEEE Trans. Circ. Syst. Video Technol. **22**, 1885–1898 (2012)
30. Xiang, W., Zhu, C., Siew, C.K., Xu, Y., Liu, M.: Forward error correction-based 2-D layered multiple description coding for errorresilient H.264 SVC video transmission. IEEE Trans. Circ. Syst. Video Technol. **19**, 1730–1738 (2009)

31. Jurca, D., Frossard, P., Jovanovic, A.: Forward error correction for multipath media streaming. IEEE Trans. Circ. Syst. Video Technol. **19**, 1315–1326 (2009)
32. Chou, P.A., Wang, H.J., Padmanabhan, V.: Layered multiple description coding. In: Proceedings of Packet Video Workshop, Nantes, France, pp. 1–7, April 2003
33. Puri, R., Ramchandran, K.: Multiple description source coding using forward error correction codes. In: Proceedings of Asilomar Conference on Signals, Systems and Computers, Pacific Grove, CA, pp. 342–346, October 1999
34. Albanese, A., Blomer, J., Edmonds, J., Luby, M., Sudan, M.: Priority encoding transmission. IEEE Trans. Inf. Theory **42**, 1737–1744 (1996)
35. Vo, N.-S., Duong, T.Q., Shu, L.: Bit allocation for multi-source multi-path P2P video streaming in VoD systems over wireless mesh networks. In: Proceedings of IEEE International Conference on Communications (ICC), Ottawa, ON, Canada, pp. 5360–5364, June 2012
36. Chou, P., Miao, Z.: Rate-distortion optimized streaming of packetized media. IEEE Trans. Multimed. **8**, 390–404 (2006)
37. Breslau, L., Cao, P., Fan, L., Phillips, G., Shenker, S.: Web caching and zipf-like distributions: evidence and implications. In: Proceedings of IEEE International Conference on Computer Communications (INFOCOM), New York, NY, pp. 126–134, March 1999

QoE-Aware Video Streaming over HTTP and Software Defined Networking

Pham Hong Thinh[1,2], Nguyen Thanh Dat[1], Pham Ngoc Nam[3],
Nguyen Huu Thanh[1], and Truong Thu Huong[1(✉)]

[1] Hanoi University of Science and Technology, Hanoi, Vietnam
huong.truongthu@hust.edu.vn
[2] Quy Nhon University, Binhdinh, Vietnam
[3] Vin University, Hanoi, Vietnam

Abstract. Due to the increase in video streaming traffic over the Internet, more innovative methods are in demand for improving both Quality of Experience (QoE) of users and Quality of Service (QoS) of providers. In recent years, HTTP Adaptive Streaming (HAS) has received significant attention from both industry and academia based on its impacts in the enhancement of media streaming services. However, HAS-alone cannot guarantee a seamless viewing experience, since this highly relies on the Network Operators' infrastructure and evolving network conditions. Along with the development of future Internet infrastructure, Software-Defined Networking (SDN) has been researched and newly implemented as a promising solution in improving services of different Internet layers. In order to enhance quality of video delivery, we try to combine the above two technologies, which has not been well-studied in academia. In this paper, we present a novel architecture incorporating bitrate adaptation and dynamic route allocation. At the client side, adaptation logic of VBR videos streaming is built based on the MPEG-DASH standard. On the network side, a SDN controller is implemented with several routing strategies on top of the OpenFlow protocol. Our experimental results show that the proposed solution enhances at least 38% up to 185% in term of average bitrate in comparison with some existing solutions as well as achieves smoother viewing experience than the traditional Internet.

Keywords: Dynamic routing · Adaptive streaming · SDN · DASH

1 Introduction

The last decade has witnessed a tremendous escalation of media content consumption, especially high-definition videos over the Internet. Cisco forecasts that the global Internet traffic in 2021 will equivalent to 127 times of that of the year 2005. In 2017, among the services over the Internet such as web, email, file sharing, etc., video streaming takes part in more than 74% of the global Internet traffic and will continue to rise over 81% by 2021 [1]. Those numbers figures

© ICST Institute for Computer Sciences, Social Informatics and Telecommunications Engineering 2019
Published by Springer Nature Switzerland AG 2019. All Rights Reserved
T. Q. Duong et al. (Eds.): INISCOM 2019, LNICST 293, pp. 40–52, 2019.
https://doi.org/10.1007/978-3-030-30149-1_4

show the necessity of developing methods in optimization of video streaming over the Internet that satisfy both efficiency for service providers and the quality for users. In that context, one technology has become the de facto standard for Internet streaming: HTTP-Based Adaptive Streaming (HAS) [2]. Using HTTP HAS is leveraging an ubiquitous and highly optimized delivery infrastructure, originally created for the web traffic, which includes, e.g., Content Delivery Networks (CDNs), caches, and proxies.

One of the enablers of the success of HAS was the open standard MPEG-DASH (Dynamic Adaptive Streaming over HTTP) [2,3]. The fundamental principle of DASH is encoding video content into multiple versions at different discrete bitrates. And clients can request segment's version actively which avoids the overhead computation at the server when numerous clients connect to the server simultaneously. Rate adaptation algorithms of HAS clients were often categorized into three categories, namely throughput based, buffer based and hybrid method throughput-buffer based.

From another aspect, Software-Defined Networking [4] is a new network architecture, that centralizes network intelligence in one network component by disassociating the forwarding process of network packets (data plane) from the routing process (control plane). In SDN, the common logical architecture in all switches, routers, and other network devices are managed by an SDN controller that allows networks policies be dynamically designed to support each single specific application.

In this study, we develop new methods to unite the advantages of dynamic adaptive streaming over HTTP and Software-Defined Networking. At the client side, the proposed adaptation algorithm based on DASH that frequently requires quality levels of video segments (a bitrate level of video segment, a high quality level means a high video quality) up or down depending on the bandwidth and client's buffer status. Within the transportation network, in order to improve the bandwidth received at client, two routing policies are proposed, called periodically routing and on-demand routing. Experimental results prove that our approach out-performs the existing state-of-the-art streaming solutions.

The remainder of this paper is organized as follows. Section 2 provides the related work on several existing adaptation method and bandwidth allocation schemes. Section 3 introduces our bitrate adaptation algorithm and SDN-based dynamic routing solutions. The experiment setup and performance evaluation of the proposed solution is also presented in this section. Conclusion and possible future extensions are presented in the last section.

2   Related Work

The DASH standard is designed to cope with highly varying delivery conditions. Over the past few years, many bitrate adaptation algorithms have been introduced in order to improve user's Quality of Experience (QoE). Their difference is mainly the required input information, ranging from network characteristics to application-layer parameters such as the playback buffer or the download speed.

In DASH, all algorithms decide the bitrate of next download segment based on the throughput variation or buffer level at the client side. These algorithms can be roughly divided into three main types such as the throughput-based group [5,6]; the buffer-based group [7,8] and the mixed type of rate adaptation algorithms, [9,10].

In throughput-based methods, bitrate is decided based on the estimated throughput without considering buffer. These schemes mainly aim at dynamically adapting a video bitrate to an available bandwidth, which usually leads to a low bandwidth utilization and cannot reach the maximum quality allowed by the available bandwidth. This is because, in such schemes, video bitrates higher than available bandwidth are never allowed to be selected to avoid playback interruptions. The key differences between these methods are the ways to estimate and use the throughput. As discussed in [5,6], authors proposed a rate adaptation algorithm based on the estimated throughput (called aggressive method) where the last segment throughput is simply used as the estimated throughput. It is currently the most responsive method to capture the dynamic changes of throughput. In the aggressive method, the bitrate is decided as the highest bitrate that is lower than the estimated throughput.

Buffer-based schemes basically employ buffer thresholds to decide the changes of bitrate. Compared to the throughput-based methods, buffer-based methods provide smoother video bitrate curves in on-demand streaming. Usually, during a certain range of buffer level, a client will try to maintain the current bitrate, resulting in a stable bitrate curve and a rather unstable buffer level. However, when bandwidth is drastically reduced, the buffer-based methods may cause sudden change of bitrate, still. This is mainly because there is a trade-off between the stability of buffer occupancy and the smoothness of video bitrate due to the time-varying bandwidth. In [7,8] Huang et al. proposed a class of buffer-based bitrate adaptation algorithm for HTTP video streaming, called BBA, that is based only on the current playback buffer occupancy as bitrate is selected heuristically without throughput estimation. The objective of BBA is to maximize the average video quality by selecting the available highest bitrate level that the network can support and avoid stalling events.

There are some ABR (Adaptive Bitrate Selection) algorithms in the literature that consider also the path bandwidth combining with the current buffer occupancy to select the most suitable video version for the next segment. In, [9], the authors proposed a rate adaptation algorithm for VBR video which based on buffer thresholds to request the next video quality level. The authors in [10] proposed a segment-aware rate adaptation (SARA) algorithm that considers the segment size variation in addition to the estimated path bandwidth and the current buffer occupancy to accurately predict the time required to download the next segment. This ensures that the best possible representation of the video is downloaded while avoiding video buer starvation.

Overall, current adaptation algorithms for HTTP Adaptive Streaming adjust the quality version of video segments to achieve the highest bandwidth utilization, smooth playback as well as avoid buffer underflow or overflow. However, the algorithms almost focus on improving the adaptation policy at the client side without considering the available resources in the networks.

There are some studies proposed in the literature for HAS over SDN. In [11], the authors proposed an adaptive HTTP video streaming framework utilizing the flow route selection capability of SDN networks. In this method the SDN controller reroutes DASH flows in each segment period after the client has completely downloaded. This leads to the controller being overloaded when great number of clients accesses at the same time or when network load increases rapidly. In [12], the authors proposed an SDN architecture to monitor network conditions of streaming ow in real time and dynamically change routing paths using multi-protocol label switching to provide reliable video watching experience. In SDNHAS [13], Bentaleb et al. relies on an SDN-based management and resource allocation architecture with the goal to estimates optimal QoE policies for groups of users and requests a bandwidth constraint slice allocation, while providing encoding recommendations to HAS players. However, these studies only present the general adaptation mechanism.

3   Problem Formulation

In this section, we propose a HTTP adaptive streaming solution included adaptation algorithm at client and network routing policies for video contents over Software-Defined Networking platform.

3.1   Network Context of HTTP Streaming over SDN

As Fig. 1. depicts, HTTP video streaming is carried out over an Openflow/SDN transportation network where Openflow switches are controlled by the SDN controller. In our design, this controller can reinforce routing policies based on

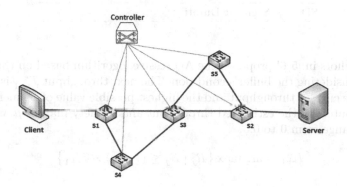

Fig. 1. HTTP streaming over SDN

network conditions, while at the client side, a bitrate adaptation algorithm can be implemented in order to acquire a seamless playback as well as a best possible video quality.

3.2 Adaptation Problems

Many adaptation algorithms have been proposed to attain a good experience in poor bandwidth conditions (highly variable throughput or low bandwidth etc.). With a proper method, a client can avoid video freezing caused by sudden severe bandwidth drop and achieve an acceptable video's quality. Before presenting the proposal, three adaptation algorithms corresponding to the three well known mechanism as mentioned before will be introduced first. Table 1 shows the symbols used in this study.

Table 1. Symbols using in the paper.

Symbol	Description
i	Segment index
j	Version index
q	The number of quality level
B_i	The buffer level at segment i
I_i	The representation's index of segment i
D_i	The download rate of segment i
R_j	The bitrate of version j
B_{i+1}^e	The estimate buffer for segment $i+1$
B^{Th}	The buffer threshold
D^{Th}	The bitrate threshold
T_i	The measured throughput at segment i
T_{i+1}^e	The estimate throughput for segment $i+1$
RTT	Round Trip Time
SD	Segment Duration

The authors in [5,6] proposed the Aggressive algorithm based on throughput without considering the buffer's condition. The next throughput T_{i+1}^e is assigned equal to the current throughput and the highest possible value of segment bitrate R_j is computed by the estimated throughput and a safety margin μ as in Eq. 1 with μ is range from 0 to 0.5.

$$I_{i+1} = \arg\max_{j} \left\{ R_j \mid R_j \leq (1 - \mu) \times T_{i+1}^e \right\} \tag{1}$$

BBA is a very well-known buffer-based adaptation algorithm. According to BBA [7,8], the average segment size of each corresponding bitrate is mapped to an instantaneous buffer level, in a linear manner with two fixed points for the lowest and highest bit-rate. BBA is too conservative during startup. The network can sustain a much higher video rate, but the algorithm is just not aware of it yet. There is a trade-off between buffer occupancy and video quality in BBA.

The Segment Aware Rate Adaptation - SARA [10] estimated next weighted Harmonic Mean throughput based on all measured throughputs in the past as Eq. 2, where w_i and d_i are the volume and the download rate of segment number i, respectively. SARA selects the most suitable representation for the next segment to be downloaded based on H_n the buffer occupancy at any given time, B_{curr}. The rate adaptation is done in four stages including fast start, addictive increase, aggressive switching, delayed download corresponding to B_{curr} value.

$$H_n = \frac{\sum_{i=1}^{n} w_i}{\sum_{i=1}^{n} \frac{w_i}{d_i}} \tag{2}$$

3.3 Our Proposed Adaptation Algorithm – MUNTH (iMpede sUspend and AttaiN PoTential PatH)

In our solution, next throughput is estimated based on two previous measured throughput at client as described in Eq. 3 [9], where γ is a dynamic constant range from [0,1]. This enables a client to adapt well with the highly bandwidth fluctuations especially with highly variable bandwidth which is the main cause of video freezing. The γ is usually set to be 0.5, when $\gamma = 1.0$, and the estimation is exactly the same as aggressive method.

$$T_{i+1}^e = \gamma \times T_i + (1 - \gamma) \times T_{i-1} \tag{3}$$

To avoid video freezing, we propose a new quality selection mechanism based on the estimate buffer level. Equation 4 [14] is used to estimate the next buffer level if client selects quality level j for Segment number i where SD is segment duration. Our objective is to choose a suitable quality level to keep the buffer greater than a threshold B^{Th} to prevent stalling events. The details of our method are shown in Algorithm 1.

$$B_{i+1}^e = B_i + SD - RTT - \frac{SD \times R_i}{T_{i+1}^e} \tag{4}$$

Algorithm 1. Bitrate Adaptation Algorithm - MUNTH

Input: T_i, R_n, B_i, B^{Th}, D_i, RTT, D^{Th}, SD
Output: I_{i+1}

1 $T_{i+1}^e \leftarrow \gamma \times T_i + (1 - \gamma) \times T_{i-1}$;   // Estimate throughput.
2 $I_{i+1} \leftarrow 0$
3 **if** $D_i \leq D^{Th}$ **then**
4 | Request for a new path;
5 **else**
6 | **for** $j \leftarrow q - 1$ **to** 0 **do**
7 | | $B_{i+1}^e \leftarrow B_i + SD - RTT - \frac{SD \times R_j}{T_{i+1}^e}$;   // Estimate next buffer
      | | level.
8 | | **if** $B_{i+1}^e \geq B^{Th}$ **then**
9 | | | $I_{i+1} = j$;
10 | | **end**
11 | **end**
12 **end**

In the proposed algorithm, when the download rate D_i is smaller than $D^{Th} = 1000$ kbps, client is going to send a message to the network controller to request an optimal path which is sufficiency to transport video data. The detail policies at network controller will be discussed in the next subsection.

3.4 Routing Policies

With the OpenFlow/SDN-based architecture [15] which can control and manage all types of data flows in the control layer, the SDN platform provides ability to flexibly perform any routing rules in the network. Furthermore, the centralized scheme of SDN has the ability to entirely monitor the network topology and routing status, and timely modify the path selection according to the changes of network states.

Periodical Routing. We proposed a periodical routing mechanism to select the optimal path every T seconds based on the stability and availability of each path. Every path from a client to the server will be represented by $Repre^p$ as shown in Eq. 5.

$$Repre^p = (1 - \omega^p) \times BW_{inst}^p \tag{5}$$

$Repre^p$ is calculated from the stability w^p and the availability BW_{inst}^p which are:

BW_{inst}^p: The instant bandwidth of each path measured by the SDN controller, p represents for path number p in n available paths (indexed from 0 to n−1) from a client to the server. If a path includes multiple links, BW_{inst}^p is assigned to the bandwidth of the bottleneck link.

w^p: The stability of a path p, calculated by the Eq. 6. Where: BW_i^p is the measured bandwidth on path p at $i \times T$ seconds before. w^p is in range $[0,1]$ and the higher w^p is, the less stable path p is.

$$\omega^p = \frac{\sqrt{\frac{\sum_1^m (BW_i^p - \overline{BW^p})^2}{m}}}{\sum_{p=0}^{n-1} \sqrt{\frac{\sum_1^m (BW_i^p - \overline{BW^p})^2}{m}}} \qquad (6)$$

The Controller selects the most stable and available path which has the highest $Repre^p$ as shown in Eq. 7 where $Indx$ is the selected path index.

$$Indx = indexOf(\max_{p=0..n-1}\{Repre^p\}) \qquad (7)$$

Adaptive Routing - MUNTH. We propose an active mechanism from client namely MUNTH to request for the optimal path when the current bandwidth is not satisfying their demand. A client can send a 'reroute' message to the controller when download rate is lower than 1000 kbps and the controller then selects the optimal path having highest BW_{inst}^p to serve the stream as shown in Eq. 8. the controller only has to seek a new path when poor network conditions occur otherwise the path will be kept during the stream session. This mechanism has two advantages: firstly, it reduces computation at the controller compared to the periodical routing scheme. Secondly, a client knows best about its perceived network condition as well as its ability therefore giving client an ability to control the adaptive rates is meaningful.

$$Indx = indexOf(\max_{p=0..n-1}\{BW_{inst}^p\}) \qquad (8)$$

3.5 Experiment Setting

In this section we present the experiment setting as well as evaluate the performance of our proposed scheme. The setup video streaming system includes three parts: DASH client, SDN network and Video Server. The client is installed by libdash library [16] on linux to serve the DASH standards. Client's media player is implemented by Qtsampleplayer [16] with buffer size of 50 s. The emulated server stores a VBR video clip named "Elephants Dream" [17] with length of 10 min 52 s, 2 s segment, 24 fps and Full-HD resolution (1920 × 1080). Every segment is encoded to 12 different versions corresponding to 12 QP values (13 to 46) of the H264 standard. In this paper, safety margin μ is set equal to 0.1. We test our solution in two experiments as follows:

Experiment 1: Buffer threshold (B^{Th}) optimization
- Client runs the MUNTH adaptation algorithm with B^{Th} of 10 s, 15 s, 20 s, 25 s with the adaptive routing policy at the controller ($m = 5$)

Experiment 2: To evaluate the performance of MUNTH, we investigate the related existing schemes by setting up as follows:

- Scheme "AGG_DF": Client runs the aggressive algorithm, the controller is active with the conventional routing policy (e.g. Dijkstra algorithm).
- Scheme "SARA": Client runs SARA algorithm controller have the same configuration as scheme "AGG_DF".
- Scheme "BBA": Client runs BBA algorithm (BBA parameter), controller have the same configuration as scheme "AGG_DF".
- Scheme "AGG_RR": Client runs Aggressive algorithm, controller now uses periodical routing policy.
- And our scheme "MUNTH": Client runs MUNTH algorithm with the optimal B^{Th} selected from Experiment 1.

3.6   Performance Evaluation

B^{Th} **Optimization for MUNTH.** The Table 2 shows the quality metric resulted from MUNTH. As we can see, with the lowest buffer thershold, with $B^{Th} = 10$ s the stalling duration gets highest at 43.1 s. It also has the highest rate of buffer level lower than 10 s as well as lowest average buffer. In contrast, experiment with $B^{Th} = 20$ s shows the best results. Stalling duration is zero, percentage of buffer ≤ 10 s is only 1.83% while keeping average buffer in a comparable level than $B^{Th} = 25$ s. However, a higher B^{Th} can harm the video quality because clients concentrate in avoiding stalling events more than video representation. As we can see, when $B^{Th} = 25$ s both average bitrate and percentage of bitrate ≥ 8000 kbps is smaller than the rest.

Table 2. MUNTH performance with different B^{Th} values

Criteria	$B^{Th} = 10$ s	$B^{Th} = 15$ s	$B^{Th} = 20$ s	$B^{Th} = 25$ s
Stalling event	6	2	0	3
Stalling duration	43.1	38.9	0	21.45
Percentage buffer $\leq$ **10 s** (%)	21.95	6.4	1.83	4.27
Average birate	10613.64	10496.54	10764.62	9575.02
Bitrate $\geq$ **8000 kbps**	46.34	44.21	46.34	40.55
Switching down version	42	33	39	44

In conclusion, the experiment results show that our MUNTH algorithm with $B^{Th} = 20$ s enables a client to occupy enough buffer to avoid freezing event while achieving a comparable video quality compared with the others. The optimal value for B^{Th} was chosen to be 20 s.

Performance Comparison. In this part, the MUNTH algorithm with optimized $B^{Th} = 20$ s will be compared with the other algorithms as listed in Experiment 2.

Figure 2 shows the occupied buffer for 5 different schemes respectively. The periodical routing mechanism enables AGG_RR to delivery segments on a stable path and a client can achieve a smoother streaming compared to AGG_DF. The result for AGG_RR shows that in only the duration from segment 50 to 100 stalling events occurs. MUNTH successfully avoids stalling events (0 times) by smooth throughput estimation and selection of suitable video rate to preserve buffer on an optimal path.

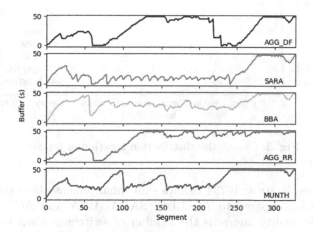

Fig. 2. Buffer level of all methods

Figure 3 shows the Cumulative Distribution Function (CDF) of bitrate during the playing time. The first thing to notice from the bitrate CDF figure is that the percentage of segments downloaded at the high bitrate of MUNTH is greater than the rest. Specifically, around 40% of MUNTH's segments has bitrate greater than 10000 kbps. The CDF value at 10000 kbps for AGG_DF and SARA is lower about 20% while the figure for BBA is lowest at 10%. The periodical routing policy of AGG_RR enable about 30% of segments are downloaded with bitrate rate $\geq$ 10000 kbps.

Table 3 shows the precise numerical metrics accumulated during our test. From the table, we can see that AGG_DF has the average quality level compared to the other methods while having the highest freeze frequency and freeze duration. SARA is the mixed throughput - buffer algorithm which can reduce stalling events compared to AGG_DF but the video's average bitrate is slightly lower. BBA successfully avoids stalling events. However, the average bitrate (in kpbs) of BBA is lowest at only 3754.68 kbps. By using the periodical rerouting policy, AGG_RR acquires a higher bitrate than AGG_DF and can avoid freezing by choosing the most stable path every T second. MUNTH shows a superior result compared to the other methods. Firstly, buffer statistics show that MUNTH can avoid stalling events while keeping a high buffer occupancy (Avg Buffer = 35.24 s). Secondly, by the adaptive routing policy, MUNTH achieves the

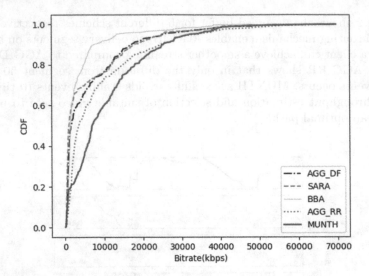

Fig. 3. Cumulative distribution function of bitrate

highest Average bitrate at 10734.18 kbps - an enhancement about 69.5%, 80.3%, 185.9%, 21.79% compared to AGG_DF, SARA, BBA and AGG_RR respectively. Another quality metric is the number of switching-down version event, users are sensitive with the changing in the video quality, especially with suddenly drop of bitrate versions. As we can see, MUNTH obtains the comparable switch-down version with BBA and SARA. Whereas achieving such better performance, MUNTH consumes much less computation at the controller compared with AGG_RR by a politic switching path mechanism.

Table 3. Results' statistics of the methods

Criteria	AGG_DF	SARA	BBA	AGG_RR	MUNTH
Stalling duration	125.11	75.32	0	81.3	0
Number of stalling events	26	8	0	13	0
Average buffer (s)	29.27	17.61	32.55	35.24	31.50
Average bitrate (kbps)	6331.33	5953.73	3754.68	8800.26	10734.18
Number of version switch-downs	79	21	25	105	39
Switch path	N/A	N/A	N/A	23	2

4 Conclusion

In this article we proposed a combined solution both from the client and network perspective to enhance users' experience while using HTTP Adaptive Streaming applications over SDN network. From the client side, our proposed adaptation algorithm - MUNTH uses smooth throughput estimation and buffer occupancy estimation mechanism to select a segment representation which helps the client to deal with bandwidth fluctuation therefore able to request a suitable bitrate version without harming buffer level leading to stalling event. From the network side, we proposed two routing policies, periodical routing and adaptive routing – MUNTH. Two metrics named stability w^p and availability BW_{inst}^p of a path are used to select the optimal path from the client to the sever in periodical routing manner. MUNTH routing policy is a client active scheme, where the client can actively request a new path that best satisfies its requirement. The experiment results show that our proposals is superior to the predecessors.

For future work, we will solve a problem where topology is more complicated and multiple clients connect to multiple servers. Therefore, the problem is not only about rate adaptation and path selection but also about fairness, stability and resource utilization among clients.

Acknowledgement. This work was supported by the University project grant, ID [T2018-PC-065]. The grant is funded by Hanoi University of Science and Technology.

References

1. Cisco Visual Networking Index: Forecast and methodology, 2016–2021, white paper, San Jose, CA, USA 1 (2016). https://www.cisco.com/c/en/us/solutions/col lateral/service-provider/visual-networking-index-vni/complete-white-paper-c11-4 81360.html
2. Stockhammer, T.: Dynamic adaptive streaming over HTTP-: standards and design principles, pp. 133–144 (2011)
3. Sodagar, I.: The MPEG-DASH standard for multimedia streaming over the internet. IEEE Multimedia **18**(4), 62–67 (2011)
4. Open networking Foundation: Open networking Foundation. https://www.opennetworking.org/. Accessed 14 Apr 2019
5. Romero, L.R.: A dynamic adaptive HTTP streaming video service for Google android. Master's Degree Project, The Royal Institute of Technology (2011)
6. Thang, T.C., Ho, Q.D., Kang, J.W., Pham, A.T.: Adaptive streaming of audio-visual content using MPEG DASH. IEEE Trans. Consum. Electron. **58**(1), 78–85 (2012)
7. Huang, T.Y., Johari, R., McKeown, N., Trunnell, M., Watson, M.: Using the buffer to avoid rebuffers: evidence from a large video streaming service. arXiv preprint arXiv:1401.2209 (2014)
8. Huang, T.Y., Johari, R., McKeown, N., Trunnell, M., Watson, M.: A buffer-based approach to rate adaptation: evidence from a large video streaming service. ACM SIGCOMM Comput. Commun. Rev **44**(4), 187–198 (2015)
9. Nguyen, H.N., Vu, T., Le, H.T., Ngoc, N.P., Thang, T.C.: Smooth quality adaptation method for VBR video streaming over HTTP. In: 2015 International Conference on Communications, Management and Telecommunications (ComManTel), pp. 184–188. IEEE (2015)
10. Juluri, P., Tamarapalli, V., Medhi, D.: Sara: segment aware rate adaptation algorithm for dynamic adaptive streaming over HTTP, pp. 1765–1770 (2015)

11. Cetinkaya C, K.E.: SDN for segment based flow routing of dash. In: 2014 IEEE Fourth International Conference on Consumer Electronics Berlin (ICCE-Berlin), pp. 74–77 (2014)
12. Nam, H., Kim, K.H., Kim, J.Y., Schulzrinne, H.: Towards QoE-aware video streaming using SDN. In: 2014 IEEE Global Communications Conference, pp. 1317–1322. IEEE (2014)
13. Bentaleb, A., Begen, A.C., Zimmermann, R.: SDNDASH: improving QoE of HTTP adaptive streaming using software defined networking. In: Proceedings of the 24th ACM International Conference on Multimedia, pp. 1296–1305. ACM (2016)
14. Vu, T., Le, H.T., Nguyen, D.V., Ngoc, N.P., Thang, T.C.: Future buffer based adaptation for VBR video streaming over HTTP. In: 2015 IEEE 17th International Workshop on Multimedia Signal Processing (MMSP), pp. 1–5. IEEE (2015)
15. Owens II, H., Durresi, A.: Video over software-defined networking (VSDN). Comput. Netw. **92**, 341–356 (2015)
16. Bitmovin: Libdash - bitmovin. https://github.com/bitmovin/libdash. Accessed 14 Apr 2019
17. Orange Open Movie Project Studio: Elephants dream (2009). https://orange.blender.org/. Accessed 3 Dec 2018

An Optimal Power Allocation for D2D Communications over Multi-user Cellular Uplink Channels

Adeola Omorinoye$^{(\boxtimes)}$, Quoc-Tuan Vien, Tuan Anh Le, and Purav Shah

Middlesex University, The Burroughs, London NW4 4BT, UK
AO991@live.mdx.ac.uk, {Q.Vien,T.Le,P.Shah}@mdx.ac.uk
https://www.mdx.ac.uk

Abstract. Device-to-Device (D2D) communications has emerged as a promising technology for optimizing spectral efficiency, reducing latency, improving data rate and increasing system capacity in cellular networks. Power allocation in D2D communication to maintain Quality-of-Service (QoS) remains as a challenging task. In this paper, we investigate the power allocation in D2D underlaying cellular networks with multi-user cellular uplink channel reuse. Specifically, this paper aims at minimizing the total transmit power of D2D users and cellular users (CUs) subject to QoS requirement at each user in terms of the required signal-to-interference-plus-noise ratio (SINR) at D2D users and base station (BS) over uplink channel as well as their limited transmit power. We first derive expressions of SINR at the D2D users and BS based on which an optimization framework for power allocation is developed. We then propose an optimal power allocation algorithm for all D2D users and CUs by taking into account the property of non-negative inverse of a Z-matrix. The proposed algorithm is validated through simulation results which show the impacts of noise power, distance between D2D users, the number of D2D pairs and the number of CUs on the power allocation in the D2D underlaying cellular networks.

Keywords: Device-to-device communication · Uplink · Power allocation

1 Introduction

One of the fundamental motivation behind using Device-to-Device (D2D) communication underlaying cellular networks is to enable direct connection between a pair of proximity devices without involvement of Base Station (BS). Although in the current cellular systems, D2D communications along with the development of small cells can cover a large area providing an enhanced Quality-of-Service (QoS), this may require a considerably increased operating expense [1–5].

© ICST Institute for Computer Sciences, Social Informatics and Telecommunications Engineering 2019
Published by Springer Nature Switzerland AG 2019. All Rights Reserved
T. Q. Duong et al. (Eds.): INISCOM 2019, LNICST 293, pp. 53–64, 2019.
https://doi.org/10.1007/978-3-030-30149-1_5

In spite of the benefits of D2D communication in cellular networks, energy efficiency and interference management have become the fundamental requirements [6] to control the interference caused by the D2D users, while simultaneously extending the battery lifetime of the User Equipment (UE). Cellular links only suffer from cross-tier interference from D2D transmitter, whereas D2D links not only deal with inter-D2D interference, but also with cross-tier interference from cellular transmission. Channel allocation and power allocation have been bestowed in the literature as strategies to diminish interference in cellular networks. In [7], close-loop and open-loop power control schemes used in LTE were investigated with an optimization based approach aimed at reducing total power consumption and increasing spectrum efficiency for D2D communications.

Additionally, green communication has been proposed attracting a number of research works with various power control and resource allocation approaches to enhance energy efficiency (EE) of D2D-aided heterogeneous network. In [1], the aim of controlling and limiting the interference of a D2D communication to the cellular network was investigated. There are basically two extensive categories of power control in D2D underlaying cellular networks which include distributed [8,9] and centralized approaches [10,11]. In the distributed approach resource allocation and power control are performed independently by the UEs, whereas they are both carried out at the BS in the centralized approach.

Considering an interference limited environment, resource allocation for D2D communications has been investigated in various research works, e.g. [12]. In this paper, we investigate the resource allocation in D2D underlaying cellular networks where the D2D users exploit multi-user cellular uplink channels.[1]

We first develop an optimization problem to find the optimal power for D2D users and CUs so as to minimize the total power consumption of the system subject to per-user QoS constraints in terms of the required signal-to-interference-plus-noise ratio (SINR) and limited transmit power at each user. In order to solve the developed problem, the property of Z-matrix is exploited to find the optimal power allocation at all users. The impact of the number of CUs and D2D users, noise power and the distance between D2D users are investigated and validated through the simulation. The proposed algorithm is shown to be able to allocate power to all D2D users and CUs achieving the minimum total power subject to various QoS constraints, while not affecting the performance of the CUs. Given a low QoS requirement, it is shown that a considerable transmit power of the D2D users can be saved for an increased energy efficiency of the overall system. In particular, the number of CUs is shown to have a significant impact on the average transmit power of the D2D users due to the interference from the CUs.

---

[1] This paper is different from [12] which considered only a CU in the uplink channel.

2   System Model

2.1   System Description

Figure 1 illustrates the system model of a D2D underlaying cellular network where we focus on a multi-user cellular uplink channels for D2D communication consisting of a BS, K CUs $\{CU_1, CU_2, \ldots, CU_K\}$ and N pairs of D2D users. The D2D communication exploits the uplink resource of cellular networks, i.e. K CUs operating together with N D2D pairs. Specifically, N D2D transmitters, i.e. $\{DT_1, DT_2, \ldots, DT_N\}$ send their data to N desired D2D receivers, i.e. $\{DR_1, DR_2, \ldots, DR_N\}$. The D2D receivers suffer the interference from not only other D2D transmitters, but also the CUs. Similarly, over the uplink channels, the BS receives unwanted signals from the D2D transmitters in addition to those from other CUs in the network.

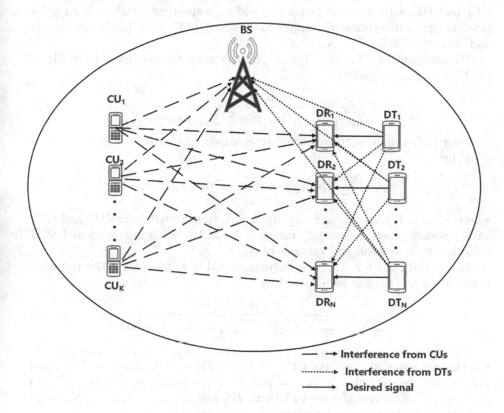

Fig. 1. System model.

Let d_{b,c_k}, $d_{i,j}$, d_{j,c_k}, $d_{b,i}$, $\{i,j\} \in \{1, 2, \ldots, N\}$ and $k = 1, 2, \ldots, K$, denote the distances between CU_k and BS, between DR_i and DT_j, between CU_k and DR_j, and between DT_i and BS, respectively. The links $CU_k \to$ BS, $DT_j \to$

DR_i, $CU_k \to DR_j$, and $DT_i \to \mathrm{BS}$, $\{i,j\} \in \{1,2,\ldots,N\}$, $k = 1,2,\ldots,K$, are assumed to experience Rayleigh flat fading channels having channel coefficients h_{b,c_k}, $h_{i,j}$, h_{j,c_k}, and $h_{b,i}$, respectively, with $E[|h_{b,c_k}|^2] = 1/d_{b,c_k}^\alpha$, $E[|h_{i,j}|^2] = 1/d_{i,j}^\alpha$, $E[|h_{j,c_k}|^2] = 1/d_{j,c_k}^\alpha$, and $E[|h_{b,i}|^2] = 1/d_{b,i}^\alpha$. Here, $E[\cdot]$ and α denote the expectation operator and pathloss exponent, respectively.

2.2   Channel Model

Over shared uplink channels, the channels dedicated for CUs can be reused by D2D transmitters. The received signal at BS is thus given by

$$y_b = \sum_{k=1}^{K} \sqrt{p_{c_k}} h_{b,c_k} x_{c_k} + \sum_{i=1}^{N} \sqrt{p_i} h_{b,i} x_i + n_b, \tag{1}$$

where $x_{c_k}, k = 1,2,\ldots,K$ and x_i, $i = 1,2,\ldots,N$, are signals transmitted from CU_k and DT_i with transmit power p_{ck} and p_i, respectively, and n_b is an independent circularly symmetric complex Gaussian (CSCG) noise having zero mean and variance of $E[|n_b|^2] = N_0$.

With respect to CU_k, $k = 1,2,\ldots,K$, the instantaneous received SINR at the BS can be obtained by

$$\gamma_{b_k} = \frac{p_{c_k} |h_{b,c_k}|^2}{\sum_{j=1,j\neq k}^{K} p_{c_j} |h_{b,c_j}|^2 + \sum_{i=1}^{N} p_i |h_{b,i}|^2 + N_0}. \tag{2}$$

Over D2D channels, the expected received signal at DR_i, $i = 1,2,\ldots,N$, is given by

$$y_i = \sum_{j=1}^{N} \sqrt{p_j} h_{i,j} x_j + \sum_{k=1}^{K} \sqrt{p_{c_k}} h_{i,c_k} x_{c_k} + n_i, \tag{3}$$

where x_j, $j = 1,2,\ldots,N$, and x_{c_k} are signals transmitted from DT_j and CU_k with transmit power p_j and p_{c_k}, respectively, and n_i is an independent CSCG noise having zero mean and variance of N_0.

In (3), DR_i, $i = 1,2,\ldots,N$, is only interested in x_i from DT_i. The instantaneous SINR at DR_i can be obtained by

$$\gamma_i = \frac{p_i |h_{i,i}|^2}{\sum_{j=1,j\neq i}^{N} p_j |h_{i,j}|^2 + \sum_{k=1}^{K} p_{c_k} |h_{i,c_k}|^2 + N_0}. \tag{4}$$

Let g_{b,c_k}, $g_{i,j}$, g_{j,c_k} and $g_{b,i}$, $\{i,j\} \in \{1,2,\ldots,N\}$, $k = 1,2,\ldots,K$, denote the channel gains of the links $CU_k \to \mathrm{BS}$, $DT_j \to DR_i$, $CU_k \to DR_j$, and $DT_i \to \mathrm{BS}$, respectively, i.e. $g_{b,c_k} = |h_{b,c_k}|^2$, $g_{i,j} = |h_{i,j}|^2$, $g_{j,c_k} = |h_{j,c_k}|^2$, and $g_{b,i} = |h_{b,i}|^2$. The instantaneous SINR at BS and DR_i, $i = 1,2,\ldots,N$, in (2) and (4) can be accordingly rewritten as

$$\gamma_{b_k} = \frac{p_c g_{b,c_k}}{\sum_{i=1}^{K} p_i g_{b,i} + \sum_{j=1,j\neq}^{k} p_{c_j} g_{b,c_j} + N_0}, \tag{5}$$

$$\gamma_i = \frac{p_i g_{i,i}}{\sum_{j=1,j\neq i}^{N} p_j g_{i,j} + \sum_{i=1}^{N} p_{c_k} g_{i,c_k} + N_0}. \tag{6}$$

3   Proposed Optimization Problem for both D2D and Cellular Communications

In this section, we first formulate the optimization problem to minimize the total transmit power of all users in a D2D underlaying cellular network as illustrated in Fig. 1. A QoS-driven power allocation scheme is then developed for all users subject to constraints of the required SINR and limited transmit power at these users in the network.

(i) D2D communications: Given constraints of SINR and transmit power, the optimization problem to minimize the total transmit power in D2D communications can be formulated as

$$
\min_{p_i} \quad \sum_{i=1}^{N} p_i,
$$
$$
\text{s. t.} \quad \gamma_i \geqslant \bar{\gamma}_i, \forall i = 1, \ldots, N,
$$
$$
p_i \leq p_i^{\max}, \forall i = 1, \ldots, N,
$$
(7)

where γ_i is given by (6), $p_i^{\max}$ is the maximum transmit power at DT_i, $\bar{\gamma}_i$ is the required SINR level at DR_i.

(ii) Cellular communications: The optimal problem for uplink cellular communications can be expressed as

$$
\min_{p_{c_k}} \quad \sum_{k=1}^{K} p_{c_k},
$$
$$
\text{s. t.} \quad \gamma_{b_k} \geqslant \bar{\gamma}_{b_k}, \forall k = 1, 2, \ldots, K,
$$
$$
p_{c_k} \leq p_{c_k}^{\max}, \forall k = 1, 2, \ldots, K,,
$$
(8)

where γ_{b_k} is given by (5), $p_{c_k}^{\max}$ is the maximum transmit power at the CU_k, and $\bar{\gamma}_{b_k}$ is the required SINR level at the BS for the uplink channel from CU_k.

Considering an overall system, the problems in (7) and (8) can be combined as in the following form:

$$
\min_{p_i} \quad \sum_{i=1}^{N+K} p_i,
$$
$$
\text{s. t.} \quad \frac{p_i g_{i,i}}{\sum_{j=1, j \neq i}^{N+K} p_j g_{i,j} + N_0} \geqslant \bar{\gamma}_i, \forall i = 1, \ldots, N+K,
$$
$$
p_i < p_i^{\max}, \forall i = 1, \ldots, N + K,
$$
(9)

Here the notation is slightly abused by using the index $N + K$ to represent both D2D and cellular communications, i.e. $p_{N+k} = p_{c_k}$, $g_{i,N+k} = g_{i,c_k}$, $g_{N+k,j} = g_{b,j}$, $g_{N+k,N+k} = g_{b,c_k}$, and $\bar{\gamma}_{N+k} = \bar{\gamma}_{b_k}$.

The optimization problem in (9) can be rewritten by rearranging the SINR constraints in (9) in the following equivalent form:

$$
\min_{p_i} \sum_{i=1}^{N+K} p_i,
$$

$$
\text{s. t.} \quad p_i g_{i,i} - \bar{\gamma}_i \sum_{j=1,j\neq i}^{N+K} p_j g_{i,j} \geqslant N_0 \bar{\gamma}_i, \forall i = 1, 2, \ldots, N+K, \tag{10}
$$

$$
p_i \leq p_i^{\max}, \forall i = 1, 2, \ldots, N+K,
$$

The scalar form of (10) can be further simplified by introducing the following vectors:

$$
\mathbf{G} \triangleq \begin{bmatrix}
g_{1,1} & -\bar{\gamma}_1 g_{1,2} & \cdots & -\bar{\gamma}_1 g_{1,N} & -\bar{\gamma}_1 g_{1,c_1} & -\bar{\gamma}_1 g_{1,c_2} & \cdots & -\bar{\gamma}_1 g_{1,c_K} \\
-\bar{\gamma}_2 g_{2,1} & g_{2,2} & \cdots & -\bar{\gamma}_2 g_{2,N} & -\bar{\gamma}_2 g_{2,c_1} & -\bar{\gamma}_2 g_{2,c_2} & \cdots & -\bar{\gamma}_1 g_{2,c_K} \\
\vdots & \vdots & \ddots & \vdots & \vdots & \vdots & \ddots & \vdots \\
-\bar{\gamma}_N g_{N,1} & -\bar{\gamma}_N g_{N,2} & \cdots & g_{N,N} & -\bar{\gamma}_N g_{N,c_1} & -\bar{\gamma}_N g_{N,c_2} & \cdots & -\bar{\gamma}_N g_{N,c_K} \\
-\bar{\gamma}_{b_1} g_{b,1} & -\bar{\gamma}_{b_1} g_{b,2} & \cdots & -\bar{\gamma}_{b_1} g_{b,N} & g_{b,c_1} & -\bar{\gamma}_{b_1} g_{1,c_2} & \cdots & -\bar{\gamma}_{b_1} g_{1,c_K} \\
\vdots & \vdots & \ddots & \vdots & \vdots & \vdots & \ddots & \vdots \\
-\bar{\gamma}_{b_K} g_{b,1} & -\bar{\gamma}_{b_K} g_{b,2} & \cdots & -\bar{\gamma}_{b_K} g_{b,N} & -\bar{\gamma}_{b_K} g_{b,c_1} & g_{b,c_K} & \cdots & -\bar{\gamma}_{b_K} g_{b,c_K}
\end{bmatrix},
$$
$$\tag{11}$$

$$
\mathbf{p} \triangleq \begin{bmatrix} p_1 \ p_2 \ \cdots \ p_N \ p_{c_1} \ p_{c_2} \ \cdots \ p_{c_K} \end{bmatrix}^T, \tag{12}
$$

$$
\mathbf{n} \triangleq \begin{bmatrix} N_0 \bar{\gamma}_1 \ N_0 \bar{\gamma}_2 \ \cdots \ N_0 \bar{\gamma}_N \ N_0 \bar{\gamma}_{b_1} \ N_0 \bar{\gamma}_{b_2} \ \cdots \ N_0 \bar{\gamma}_{b_K} \end{bmatrix}^T, \tag{13}
$$

$$
\mathbf{p}_{\max} \triangleq \begin{bmatrix} p_1^{\max} \ p_2^{\max} \ \cdots \ p_N^{\max} \ p_{c_1}^{\max} \ p_{c_2}^{\max} \ \cdots \ p_{c_K}^{\max} \end{bmatrix}^T. \tag{14}
$$

Hence, the problem (10) can be rewritten as

$$
\min_{p_i} \sum_{i=1}^{N+K} p_i,
$$

$$
\text{s. t.} \quad \mathbf{Gp} \succeq \mathbf{n}, \tag{15}
$$

$$
\mathbf{p} \preceq \mathbf{p}_{\max}.
$$

where $\succeq$ and $\preceq$ denote the element-wise greater and less operators, respectively.

The optimal solution to problem (15) can be found by using the following lemma

Lemma 1. *If matrix* $\mathbf{G}$ *defined in* (11) *satisfies*

$$
g_{i,i} > \bar{\gamma}_i \sum_{j=1,j\neq i}^{N+K} g_{i,j}, \forall i = 1, 2, \ldots, N+K, \tag{16}
$$

then there exists a unique lower bound for the power allocation for problem (15) *as*

$$\mathbf{p}_{min} = \mathbf{G}^{-1}\mathbf{n}. \tag{17}$$

Proof. The proof follows the same approach as in [12] where the basic idea is to treat (11) as a Z-*matrix* [15,16]. By observing (11), one can conclude that all the off-diagonal elements of matrix $\mathbf{G}$ are non-positive. Hence, as shown in [13,14], matrix $\mathbf{G}$ is called a Z-*matrix*. If $\mathbf{G}$ satisfies the condition in (16), then $\mathbf{G}$ is strictly diagonally dominant matrix. According to [13, Chap. 6, Theorem 2.3], all principal minors of $\mathbf{G}$ are positive. Since $\mathbf{G}$ is a Z-*matrix*, according to [14, theorem 3.11.10], $\mathbf{G}^{-1}$ exist and all of its elements are non-negative. In addition, all the elements of vector $\mathbf{n}$ in (13) are non-negative, and thus $\mathbf{p}$ is lower bounded by $\mathbf{p}_{min} = \mathbf{G}^{-1}\mathbf{n} \succeq 0$. The proof is complete.

Remark 1. Notice that if $\mathbf{p}_{min}$ defined in (17) satisfies $p_{min} \preceq \mathbf{p}_{max}$ then $\mathbf{p}_{min}$ is the optimal solution to the optimization problem (15).

4   Simulation Results

In this section, we provide numerical results for the D2D underlaying cellular network. We implement the Monte Carlo method in MATLAB to evaluate the

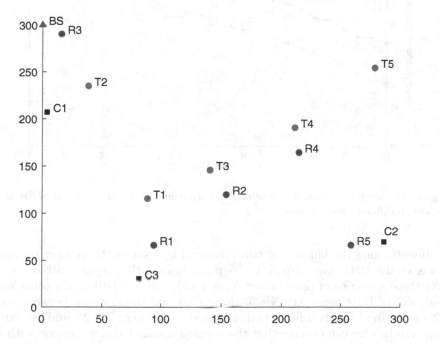

Fig. 2. A typical example of simulation model for a D2D underlaying cellular network consisting of a BS, 3 CUs, and 5 pairs of D2D users within an area of $300\,\text{m} \times 300\,\text{m}$.

performance of the proposed power allocation algorithm. In the simulation, the nodes are located within an area of 300 m × 300 m, the pathloss exponent is set as $\alpha = 2$, the required SINR of all D2D users and CUs are equally set and varies as $\bar{\gamma}_i = \bar{\gamma}_{b_k} \in [-20, 5]$ dB, $\forall i = 1, 2, \ldots, N$, $\forall k = 1, 2, \ldots, K$, and the maximum transmit power is $p_i^{\max} = p_c^{\max} = 30$ dBm. It is assumed that BS is at the top left corner, i.e. $\{x_{BS}, y_{BS}\} = \{0, 300\}$ m, while the locations of other nodes, i.e. CUs and D2D users, are uniformly distributed in the range $[0, 300]$ m. Due to the requirement that mobile devices should be in short range for D2D communications, the distance between the D2D transmitter and D2D receiver are limited in $[d_{\min}, d_{\max}]$, where 10 m $\leqslant d_{\min} < d_{\max} \leqslant 50$ m. An illustration of the simulation settings is shown in Fig. 2 where 5 pairs of D2D users and 3 CUs are plotted with $d_{\min} = 10$ m and $d_{\max} = 25$ m.

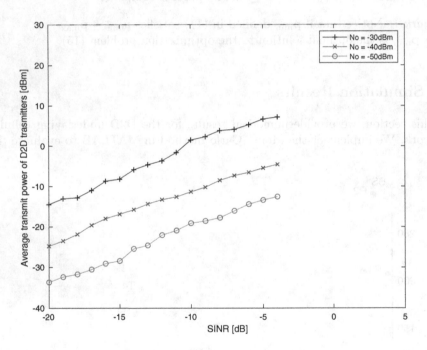

Fig. 3. The average transmit power of D2D transmitters versus required SINR with respect to different noise power.

Investigating the impacts of noise power, Fig. 3 shows the average transmit power of the D2D transmitters, i.e. $E[\mathbf{p}_{\min}]$ against the required SINR, i.e. $\bar{\gamma}$, with three scenarios of noise power $N_0 \in \{-30, -40, -50\}$ dBm. We considered five pairs of D2D users, i.e. $N = 5$, three CUs and the distance between each D2D pairs and CUs are uniformly distributed in the range $[10, 25]$ and $[30, 90]$ m respectively. One can observe that the average transmit power increase with an increase in the SINR requirement, which follows the intuition that increased in noise power contribute to a higher transmit power. For instance, when the required SINR is -5 dBm, the average transmit power required is -12 dBm with noise power of -50 dBm compared to when the noise power -30 dBm with an average transmit power of 8 dBm. The fluctuating in the graph is due to the fact that the instantaneous SINR is considered over different fading generations, among which some cause the matrix $\mathbf{G}$ defined in (11) is not convertible, i.e. does not satisfy the condition in Lemma 1.

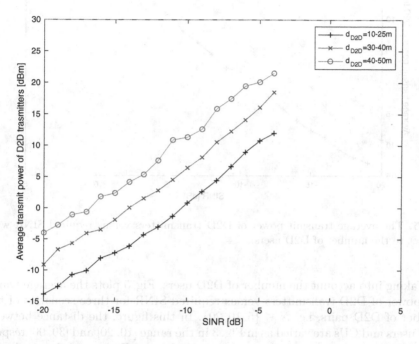

Fig. 4. The average transmit power of D2D transmitters versus required SINR with respect to distance between D2D users.

The impacts of the distance between D2D users are investigated in Fig. 4 where the average transmit power of D2D transmitters is plotted over the required SINR with respect to three cases of distance between D2D users, i.e. $\{[d_{\min}, d_{\max}]\} \in \{[10, 25], [30, 40], [40, 50]\}$ m. The noise power is fixed as $N_0 = -30$ dBm. There are five pairs of D2D users and their locations are similarly set as in Fig. 3. It can be observed that a higher transmit power is required with D2D users having distance within 40 m to 50 m compared to those with D2D users with shorter distance such as 10 m, -25 m and 30 m to 40 m. This means that, the distance between the D2D users has a considerable impact on the average power at the D2D users.

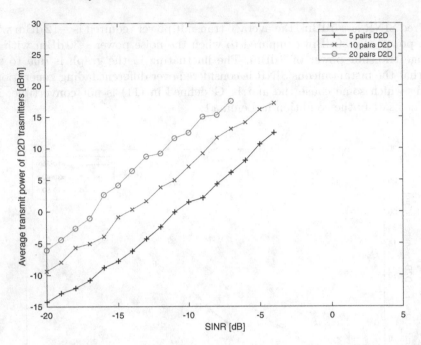

Fig. 5. The average transmit power of D2D transmitters versus required SINR with respect to the number of D2D users.

Taking into account the number of D2D users, Fig. 5 plots the average transmit power of D2D transmitters versus required SINR for three scenarios of the number of D2D pairs, i.e. $N \in \{5, 10, 20\}$. In this figure, the distance between D2D users and CUs are varied as in Fig. 3 in the range $[10, 20]$ and $[30, 90]$ respectively. The noise power parameter is set similarly as in Fig. 4, and the number of D2D pairs ranging from $[5, 10, 20]$. It can be observed that the proposed power allocation demonstrates that the number of D2D pairs has a significant impact on the transmit power due to the interference from both other D2D transmitters and other CUs.

Taking into account the impact of the number of CUs, Fig. 6 plots the average transmit power of the D2D transmitter against the required SINR with respect to four scenarios of CUs (ranging from 1 to 4) with distance range $[0, 70]$ m. The parameter is set as in Fig. 3 i.e. the distance between D2D users $[10, 25]$ m, and the number of D2D pairs $N = 5$, the noise power is set similar to Fig. 4 $N_0 = -30$ dBm. It can be observed that there is much difference in the average transmit power required with four CUs compared to one CU due to the interference from other CUs and D2D transmitters.

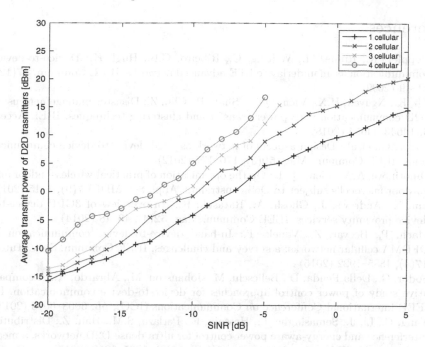

Fig. 6. The average transmit power of D2D transmitters versus required SINR with respect to the number of CUs.

5 Conclusion

In this paper, we have introduced a power allocation approach for D2D users over uplink in D2D underlaying cellular networks with multiple CUs. We have developed an optimization problem by incorporating the SINR constraints while maintaining the QoS of the system. An optimal power allocation has been proposed by taking into account the property of non-negative inverse of a Z-matrix. Moreover, the impact of the number of the CUs, noise power, the distance between D2D users and the number of D2D pairs have been evaluated for the considered system. Our simulation results have shown that deploying more number of either the CUs or D2D users has a significant impact on the transmit power of the D2D users due to the interference caused by both D2D and cellular communications. Also, the transmit power of the D2D users has shown to be dependent of not only their location but also the location of the CUs. For future work, we will consider a power control approach for an ultra-dense network with statistical modelling for multiple cells.

References

1. Doppler, K., Rinne, M., Wijting, C., Ribeiro, C.B., Hugl, K.: Device-to-device communication as an underlay to LTE-advanced networks. IEEE Commun. **47**(12), 42–49 (2009)
2. Ali, K., Nguyen, H.X., Vien, Q.-T., Shah, P., Chu, Z.: Disaster management using D2D communication with power transfer and clustering techniques. IEEE Access **6**, 14643–14654 (2018)
3. Fodor, G., et al.: Design aspects of network assisted device-to-device communications. IEEE Commun. Mag. **50**(3), 170–177 (2012)
4. Omorinoye, A.A., Vien, Q.-T.: On the optimisation of practical wireless indoor and outdoor microcells subject to QoS constraints. Appl. Sci. MDPI **7**(9), 1–15 (2017)
5. Lin, X., Andrews, J., Ghosh, A., Ratasuk, R.: An overview of 3GPP device-to-device proximity services. IEEE Commun. Mag. **52**(4), 40–48 (2014)
6. Mach, P., Becvar, Z., Vanek, T.: In-band device-to-device communication in OFDMA cellular networks: a survey and challenges. IEEE Commun. Surv. Tutor. **17**(4), 1885–1922 (2015)
7. Fodor, G., Della Penda, D., Belleschi, M., Johansson, M., Abrardo, A.: A comparative study of power control approaches for device-to-device communication. In: IEEE International Conference on Communications (ICC), pp. 6008–6013 (2013)
8. Yang, C., Li, J., Semasinghe, P., Hossain, E., Perlaza, S.M., Han, Z.: Distributed interference and energy-aware power control for ultra-dense D2D networks: a mean field game. IEEE Int. Conf. Commun. (ICC) **16**(2), 1205–1217 (2017)
9. Zhang, G., Hu, J., Heng, W., Li, X., Wang, G.: Distributed power control for D2D communications underlaying cellular network using Stackelberg game. In: IEEE Wireless Communications and Networking Conference (WCNC), pp. 1–6 (2017)
10. Huang, J., Xiong, Z., Li, J., Chen, Q., Duan, Q., Zhao, Y.: A priority-based access control model for device-to-device communications underlaying cellular network using network calculus. In: Cai, Z., Wang, C., Cheng, S., Wang, H., Gao, H. (eds.) WASA 2014. LNCS, vol. 8491, pp. 613–623. Springer, Cham (2014). https://doi.org/10.1007/978-3-319-07782-6_55
11. ElSawy, H., Hossain, E., Alouini, M.S.: Analytical modeling of mode selection and power control for underlay D2D communication in cellular networks. IEEE Trans. Commun. **62**(11), 4147–4161 (2014)
12. Omorinoye, A., Vien, Q., Le, T.A., Shah, P.: On the resource allocation for D2D underlaying uplink cellular networks. In: IEEE ICT Proceedings (2019)
13. Berman, A., Plemmons, R.: Nonnegetive Matrices in the Mathematical Science. Society for Industrial and Applied Mathematics. SIAM, Philadelphia (1994)
14. Cottle, R., Pang, J., Stone, R.: The Linear Complementarity Problem. Society for Industry and Applied Mathematics. SIAM, Philadelphia (2009)
15. Attar, A., Nakhai, M.R., Aghvami, A.H.: Cognitive radio game for secondary spectrum access problem. IEEE Trans. Wirel. Commun. **8**(4), 2121–2131 (2009)
16. Le, T.A., Nakhai, M.R.: Downlink optimization with interference pricing and statistical CSI. IEEE Trans. Commun. **61**(6), 2339–2349 (2013)

Secrecy Performance Enhancement Using Path Selection over Cluster-Based Cognitive Radio Networks

Pham Minh Nam[1], Phan Van Ca[1], Tran Trung Duy[2(✉)], and Khoa N. Le[3]

[1] University of Technology and Education, Ho Chi Minh City 700000, Vietnam
1727002@student.hcmute.edu.vn, capv@hcmute.edu.vn
[2] Posts and Telecommunications Institute of Technology, Ho Chi Minh City, Vietnam
trantrungduy@ptithcm.edu.vn
[3] Western Sydney University, Penrith South, NSW, Australia
lenkhoa@gmail.com

Abstract. In this paper, we propose three path selection methods for cluster-based cognitive radio (CR) networks for secrecy enhancement by formulating the probability of non-zero secrecy capacity (PNSC). In the proposed work, it is assumed that uniform transmit power for the secondary transmitters and jammers must be adjusted to guarantee quality of service (QoS) of the primary network, follows a simple and efficient power allocation strategy. To improve the channel capacity, the best receiver is selected at each cluster to relay the source data to the next hop. Additionally, a jammer is randomly chosen at each cluster to generate noises on an eavesdropper, and to reduce the quality of the eavesdropping links. Three methods are studied in this paper. First, we propose the BEST path selection method (BEST) to maximize the end-to-end instantaneous secrecy capacity. Second, the path obtaining the MAXimum Value for the average end-to-end PNSC (MAXV) is selected for data transmission. Third, we also propose a RAND method in which a RANDom path is employed. For performance evaluation and comparison, we derive exact closed-form expressions for the end-to-end PNSC of the BEST, MAXV and RAND methods over Rayleigh fading channel. Monte Carlo simulations are then performed to verify the derived theoretical results.

Keywords: Physical-layer security · Cognitive radio · Cluster networks · Path selection · Secrecy capacity

1 Introduction

Physical-layer security (PLS) [1,2] has recently emerged as an efficient method to provide security for wireless sensor networks (WSNs) and Internet of Things (IoT) networks. Under PLS context, secure communication can be obtained when

T. Q. Duong et al. (Eds.): INISCOM 2019, LNICST 293, pp. 65–80, 2019.
https://doi.org/10.1007/978-3-030-30149-1_6

channel capacity of a data link is higher than that of an eavesdropping link. Therefore, diversity-based transmit/receive methods [3–5] and cooperative relaying transmission [6–8] have been widely used to enhance secrecy performance, in terms of average secrecy capacity (ASC), secrecy outage probability (SOP), and probability of non-zero secrecy capacity (PNSC). Additionally, the methods reported in [3–8] can be combined with cooperative jamming techniques reported in [9,10], i.e., jammers can practically collude with authorized receivers so that generated artificial noise can only interfere on the eavesdropper nodes.

Until now, there have been only several published works for performance evaluation of multi-hop transmission in PLS [8,11–14]. In [11], authors proposed a cluster-based secure communication with relay selection methods at each hop. In addition, the eavesdropper in [11] can use maximal ratio combining (MRC) to decode the data received over multiple hops. In [12], authors developed a system model by combining a randomize-and-forward (RF) method and cooperative jamming techniques. Particularly, the transmitters randomly generate codebooks when forwarding the source data, while the selected receivers and jammers collaborate to remove interferences in the received signals. In [13,14], full-duplex relaying methods for enhancing security over multi-hop relaying systems were proposed and evaluated. Being different with [11–14], authors of [8] considered a multi-hop amplify-and-forward (AF) relaying model in PLS, employing compress sensing.

Recently, PLS in cognitive radio (CR) has gained much attention from researchers. The authors of [15,16] proposed cooperative cognitive protocols using cooperative jamming to enhance secrecy performance for secondary networks. In [17,18], radio frequency energy harvesting (RF-EH) based secure communication protocols employing overlay and underlay spectrum sharing approaches were investigated. Authors in [19] focused on designing a routing protocol for cooperative jamming multi-hop multi-antenna secondary networks in the presence of random eavesdroppers. In [20], a cooperative routing scheme was proposed to enhance secrecy performance for multi-hop relaying secondary networks. In [21], authors considered transmit antenna selection (TAS)/selection combining (SC) and harvest-to-transmit based secure multi-hop transmission over underlay CR environments in the presence of multi-antenna eavesdroppers, and hardware imperfection.

To the best of our knowledge, PLS in cluster-based multi-hop multi-path over underlay CR networks has not yet been studied. This has motivated us to propose and evaluate secrecy performance of CR networks. This paper thus focuses on end-to-end PNSC performance of secondary networks, where a secondary source communicates with a secondary destination using a multi-hop multi-path relaying approach. For the underlay spectrum sharing, the secondary transmitters including source and relays must adjust their transmit power to satisfy QoS of the primary network. Under this power constraint, we propose the best receiver selection at each cluster to improve data transmission reliability. On the other hand, to lessen the severity of eavesdropping channels, a jammer at each cluster is randomly selected to realize the effectiveness of the cooperative jamming process. The main contributions of this paper are summarized as follows:

- We propose a simple power allocation strategy for the secondary transmitters and jammers to satisfy the required primary QoS.
- We propose three efficient path selection methods, BEST, MAXV and RAND. For the BEST method, the path with the highest end-to-end instantaneous secrecy capacity is chosen for data transmission. For the MAXV method, the system selects the path obtaining maximum value of the average end-to-end PNSC. For the RAND method, a random path is used to transmit source data to the destination.
- We derive exact closed-form expressions for the end-to-end PNSC of the BEST, MAXV and RAND methods over Rayleigh fading channels, which are then verified by Monte Carlo simulation.

The remainder of this paper is organized as follows. The system model of the considered methods is described in Sect. 2. In Sect. 3, expressions for the end-to-end PNSC of the BEST, MAXV and RAND methods are derived. Simulation results are shown in Sect. 4 to verify the derived theoretical results. Finally, Sect. 5 concludes the main findings and outlines possible future work.

2   System Model

Figure 1 presents the system model for the proposed underlay CR network, where the primary network shares licensed bands to the secondary network. For the primary network, the primary transmitter (PT) sends its data to the primary receiver (PR). In the secondary network, the source S attempts to transmit its data to the destination D using the multi-hop relaying approach, in the presence of the eavesdropper E, who illegally listens to the source data. Assuming that there are M available disjoint paths that are established at the network layer, and the source would select one path for the data transmission. On the m-th path, there are N_m clusters between S and D, denoted by $CL_{m,1}$, $CL_{m,2}$, ..., CL_{m,N_m}, where $m = 1, 2, ..., M$ and $N_m \geq 1$. We also denote $CL_{m,0}$ and CL_{m,N_m+1} as the clusters including the source and the destination, respectively. In addition, let us denote $L_{m,u}$ as the number of nodes belonging to the cluster $CL_{m,u}$, and $\{R_{m,u,1}, R_{m,u,2}, ..., R_{m,u,L_{m,u}}\}$ as the set of these nodes, where $u = 1, 2, ..., N_m$. Considering the cluster CL_{m,N_m+1}, except the destination, the remaining cluster nodes are named as $R_{m,N_m+1,2}, ..., R_{m,N_m+1,L_{m,N_m+1}}$, where $L_{m,N_m+1} = L_D (\forall m)$. Figure 1 presents the system model for the proposed underlay CR network, where the primary network shares licensed bands to the secondary network. For the primary network, the primary transmitter (PT) sends its data to the primary receiver (PR). In the secondary network, the source S attempts to transmit its data to the destination D using the multi-hop relaying approach, in the presence of the eavesdropper E, who illegally listens to the source data. Assuming that there are M available disjoint paths that are established at the network layer, and the source would select one path for the data transmission. On the m-th path, there are N_m clusters between S and D, denoted by $\mathrm{CL}_{m,1}$, $\mathrm{CL}_{m,2}$, ..., CL_{m,N_m}, where $m = 1, 2, ..., M$ and $N_m \geq 1$. We also denote $\mathrm{CL}_{m,0}$ and CL_{m,N_m+1} as the clusters including the source and the destination, respectively. In addition, let us denote $\mathrm{L}_{m,u}$ as the number of nodes belonging to the cluster $\mathrm{CL}_{m,u}$, and $\{\mathrm{R}_{m,u,1}, \mathrm{R}_{m,u,2}, ..., \mathrm{R}_{m,u,L_{m,u}}\}$ as the set of these nodes, where $u = 1, 2, ..., N_m$. Considering the cluster CL_{m,N_m+1}, except the destination, the remaining cluster nodes are named as $\mathrm{R}_{m,N_m+1,2}, ..., \mathrm{R}_{m,N_m+1,L_{m,N_m+1}}$, where $\mathrm{L}_{m,N_m+1} = \mathrm{L}_\mathrm{D}\,(\forall m)$.

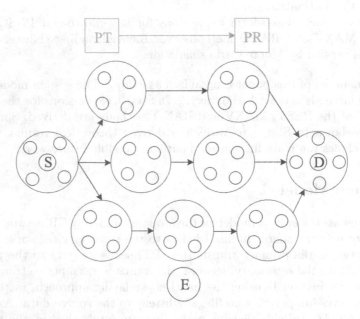

Fig. 1. System model of the proposed methods.

It is assumed that (i) all of the nodes are equipped with single antennas, and operate on the half-duplex mode; (ii) all the channels are block slow Rayleigh fading, which remains coherently constant over the length of a code word. Hence, the channel gain $\gamma_{\mathrm{X,Y}}$ of the X $\rightarrow$ Y link is an exponential random variable (RV), where $(\mathrm{X, Y}) \in \{\mathrm{PT, PR}, \mathrm{R}_{m,u,v}, \mathrm{E}\}$, $m = 1, 2, ..., M$, $u = 0, 1, ..., N_m + 1$, $v = 1, 2, ..., L_{m,u}$, and $\mathrm{R}_{m,0,v} \equiv \mathrm{S}$, $\mathrm{R}_{m,N_m+1,1} \equiv \mathrm{D}\,(\forall m, v)$. Then, the cumulative distribution function (CDF) and probability density function (PDF) of $\gamma_{\mathrm{X,Y}}$ can be expressed, respectively as

$$F_{\gamma_{\mathrm{X,Y}}}(z) = 1 - \exp\left(-\lambda_{\mathrm{X,Y}}z\right), \; f_{\gamma_{\mathrm{X,Y}}}(z) = \lambda_{\mathrm{X,Y}}\exp\left(-\lambda_{\mathrm{X,Y}}z\right), \quad (1)$$

where $\lambda_{\mathrm{X,Y}} = d_{\mathrm{X,Y}}^\beta$ [22], $d_{\mathrm{X,Y}}$ is the link distance between X and Y, and β is the path-loss exponent. Since the nodes in the cluster $\mathrm{CL}_{m,u}$ are closely spaced, we can assume the distances $d_{X,R_{m,u,v}}$ are statistically unchanged, i.e., $d_{X,R_{m,u,v}} = d_{X,R_{m,u,t}}$ $(\forall v,t)$.

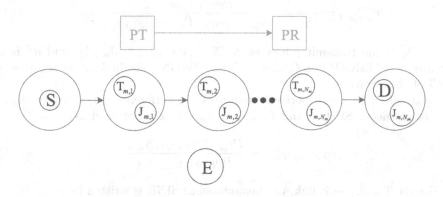

Fig. 2. Data transmission on the m-th path.

Assume that the m-th path is selected (see Fig. 2). Because of the half-duplex constraint, the transmit data stream is split into $N_m + 1$ orthogonal time slots. At the first time slot, the source sends the data to the best node belonging to the cluster $CL_{m,1}$, which is chosen to maximize the channel gain of the first hop as [11]:

$$T_{m,1} : \arg \max_{v=1,2,...,L_{m,1}} \left(\gamma_{S,R_{m,1,v}} \right), \qquad (2)$$

where $T_{m,1}$ denotes the selected node.

Moreover, during the $S \to T_{m,1}$ transmission, one of the remaining nodes of the cluster $CL_{m,1}$, named $J_{m,1}$, is randomly chosen to generate jamming noise on the eavesdropper E. It is worth noting that $T_{m,1}$ can remove the interference from $J_{m,1}$ via exchanging secure messages with $T_{m,1}$ [12]. Then, after decoding the source data, $T_{m,1}$ re-encodes and forwards it to the next cluster in the second time slot.

Generally, the best receiver at the u-th time slot, named $T_{m,u}$, is selected by

$$T_{m,u} : \arg \max_{v=1,2,...,L_{m,u}} \left(\gamma_{T_{m,u-1},R_{m,u,v}} \right), \qquad (3)$$

where $u = 1, 2, ..., N_m$, and $T_{m,0} \equiv S$.

Let us consider the last time slot, the transmitter T_{m,N_m} transmits the source data to the destination D, while the jammer J_{m,N_m+1}, a member of the cluster CL_{m,N_m+1}, is employed to realize the cooperative jamming operation. Because the eavesdropper can overhear multiple hops, the transmitters $T_{m,u}$ employ the randomize-and-forward method [12] so that it cannot combine the received signals using MRC.

Considering the data transmission between PT and PR in the u-th time slot of the m-th path, because of the co-channel interference caused by $T_{m,u-1}$ and

$J_{m,u}$, the instantaneous signal-to-interference-plus-noise ratio (SINR) obtained at PR can be formulated by

$$\psi_{\text{PT,PR}}^{(m,u)} = \frac{P_{\text{PT}}\gamma_{\text{PT,PR}}}{P_{T_{m,u-1}}\gamma_{T_{m,u-1},\text{PR}} + P_{J_{m,u}}\gamma_{J_{m,u},\text{PR}} + \sigma^2},$$ (4)

where P_X is the transmit power of X ($X \in \{\text{PT}, T_{m,u-1}, J_{m,u}\}$), and σ^2 is a variance of Additive White Gaussian Noise (AWGN). For brevity, we set $\sigma^2 = 1$ at all of the receivers.

Next, since $T_{m,u}$ can perfectly remove the interference generated by $J_{m,u}$, the instantaneous SINR of the $T_{m,u-1} \to T_{m,u}$ link is obtained as

$$\psi_{T_{m,u-1},T_{m,u}} = \frac{P_{T_{m,u-1}}\gamma_{T_{m,u-1},T_{m,u}}}{P_{\text{PT}}\gamma_{\text{PT},T_{m,u}} + 1}.$$ (5)

For the $T_{m,u-1} \to E$ link, the instantaneous SINR is written by

$$\psi_{T_{m,u-1},E} = \frac{P_{T_{m,u-1}}\gamma_{T_{m,u-1},E}}{P_{\text{PT}}\gamma_{\text{PT},E} + P_{J_{m,u}}\gamma_{J_{m,u},E} + 1}.$$ (6)

From (4), (5) and (6), we can express the instantaneous channel capacity for the primary link, the secondary data link and the secondary eavesdropping link respectively as given in (7)–(9)

$$C_{\text{PT,PR}}^{(m,u)} = \frac{1}{N_m + 1}\log_2\left(1 + \psi_{\text{PT,PR}}^{(m,u)}\right),$$ (7)

$$C_{T_{m,u-1},T_{m,u}} = \frac{1}{N_m + 1}\log_2\left(1 + \psi_{T_{m,u-1},T_{m,u}}\right),$$ (8)

$$C_{T_{m,u-1},E} = \frac{1}{N_m + 1}\log_2\left(1 + \psi_{T_{m,u-1},E}\right),$$ (9)

where the factor $1/(1 + N_m)$ indicates that the data transmission of the secondary network is split into $(1 + N_m)$ orthogonal time slots.

To guarantee QoS of the primary network at any time slot, we focus on the lowest channel capacity of the PT $\to$ PR link, i.e.,

$$C_{\text{PT,PR}}^{m,\min} = \min_{u=1,2,\ldots,N_m+1}\left(C_{\text{PT,PR}}^{(m,u)}\right).$$ (10)

Using (10), the outage probability (OP) of the primary network can be defined by

$$\text{OP}_m = \text{Pr}\left(C_{\text{PT,PR}}^{m,\min} < R_P\right),$$ (11)

where R_P is a target rate of the primary network.

Considering the secondary network, the secrecy capacity at the u-th time slot is given as

$$SC_{m,u} = \left[C_{T_{m,u-1},T_{m,u}} - C_{T_{m,u-1},E} \right]^+, \tag{12}$$

where $[x]^+ = \max(0, x)$.

Due to the randomize-and-forward strategy, the end-to-end secrecy capacity obtained on the m-th path is expressed as

$$SC_m^{e2e} = \min_{u=1,2,\ldots,N_m+1} (SC_{m,u}). \tag{13}$$

Then, the end-to-end PNSC on the m-th path is defined by

$$PNSC_m = \Pr\left(SC_m^{e2e} > 0\right). \tag{14}$$

3 Performance Evaluation

3.1 OP of Primary Network

In this sub-section, we calculate OP of the primary network when the m-th path is selected. Combining (7), (10) and (11), we can write

$$OP_m = 1 - \prod_{u=1}^{N_m+1} \left(1 - \Pr\left(\psi_{PT,PR}^{(m,u)} < \rho\right)\right), \tag{15}$$

where $\rho = 2^{((N_m+1)R_P)} - 1$. Substituting (4) into $\Pr\left(\psi_{PT,PR}^{(m,u)} < \rho\right)$ in (15), which yields

$$\Pr\left(\psi_{PT,PR}^{(m,u)} < \rho\right) = \Pr\left(\gamma_{PT,PR} < \frac{P_{T_{m,u-1}}\rho}{P_{PT}}\gamma_{T_{m,u-1},PR} + \frac{P_{J_{m,u}}\rho}{P_{PT}}\gamma_{J_{m,u},PR} + \frac{\rho}{P_{PT}}\right)$$

$$= \int_0^{+\infty} \int_0^{+\infty} \left[F_{\gamma_{PT,PR}}\left(\frac{P_{T_{m,u-1}}\rho}{P_{PT}}x + \frac{P_{J_{m,u}}\rho}{P_{PT}}y + \frac{\rho}{P_{PT}}\right) \right] dxdy. \tag{16}$$

$$\phantom{=\int_0^{+\infty} \int_0^{+\infty}} \left[f_{\gamma_{T_{m,u-1},PR}}(x) f_{\gamma_{J_{m,u},PR}}(y) \right]$$

Substituting CDF and PDF given in (1) into (16), after some manipulation, we can obtain

$$\Pr\left(\psi_{PT,PR}^{(m,u)} < \rho\right) = 1 - \frac{\lambda_{T_{m,u-1},PR}P_{PT}}{\lambda_{T_{m,u-1},PR}P_{PT} + \lambda_{PT,PR}P_{T_{m,u-1}}\rho}$$

$$\times \frac{\lambda_{J_{m,u},PR}P_{PT}}{\lambda_{J_{m,u},PR}P_{PT} + \lambda_{PT,PR}P_{J_{m,u}}\rho} \exp\left(-\frac{\lambda_{PT,PR}\rho}{P_{PT}}\right). \tag{17}$$

Substituting (17) into (15), we obtain the exact closed-form expression for OP_m as follows:

$$OP_m = 1 - \exp\left(-\frac{(N_m+1)\lambda_{PT,PR}\rho}{P_{PT}}\right)$$

$$\times \prod_{u=1}^{N_m+1} \left(\frac{\lambda_{T_{m,u-1},PR}P_{PT}}{\lambda_{T_{m,u-1},PR}P_{PT} + \lambda_{PT,PR}P_{T_{m,u-1}}\rho} \frac{\lambda_{J_{m,u},PR}P_{PT}}{\lambda_{J_{m,u},PR}P_{PT} + \lambda_{PT,PR}P_{J_{m,u}}\rho}\right). \tag{18}$$

To guarantee the QoS of the primary network, the following condition should be satisfied:

$$OP_m \leq \varepsilon_{OP}, \tag{19}$$

where ε_{OP} is the target OP of the PT $-$ PR link.

3.2 Power Allocation of Secondary Transmitters

This sub-section proposes a simple power allocation for the secondary transmitter to satisfy the required primary QoS. Firstly, we assume that all nodes in the cluster $CL_{m,u}$ have the same transmit power, i.e., $P_{T_{m,u}} = P_{J_{m,u}} = Q_{m,u}$ for all m, u. Secondly, if the distance between $T_{m,u}$ and PR is longer than that between $T_{m,v}$ and PR, $Q_{m,u}$ should be higher than $Q_{m,v}$, and vice verse, where $(u, v) \in \{0, 1, ..., N_m + 1\}$ and $u \neq v$. Therefore, the ratio between $Q_{m,u}$ and $Q_{m,v}$ can be formulated as a function of the link distance as

$$\frac{Q_{m,u}}{Q_{m,v}} = \frac{d^{\beta}_{T_{m,u},PR}}{d^{\beta}_{T_{m,v},PR}} = \frac{\lambda_{T_{m,u},PR}}{\lambda_{T_{m,v},PR}} \Leftrightarrow \frac{Q_{m,u}}{\lambda_{T_{m,u},PR}} = \frac{Q_{m,v}}{\lambda_{T_{m,v},PR}} = \chi. \tag{20}$$

From (20), substituting $P_{T_{m,u}} = P_{J_{m,u}} = Q_{m,u} = \chi\lambda_{T_{m,u},PR}$ into (18), we arrive at

$$OP_m = 1 - \left(\frac{P_{PT}}{P_{PT} + \lambda_{PT,PR}\chi\rho}\right)^{2(N_m+1)} \exp\left(-\frac{(N_m + 1)\lambda_{PT,PR}\rho}{P_{PT}}\right). \tag{21}$$

Now, solving $OP_m = \varepsilon_{OP}$, we can express χ as

$$\chi = \frac{P_{PT}}{\lambda_{PT,PR}\rho}\left[\left((1 - \varepsilon_{OP})\exp\left(\frac{(N_m + 1)\lambda_{PT,PR}\rho}{P_{PT}}\right)\right)^{-\frac{1}{2(N_m+1)}} - 1\right]. \tag{22}$$

Because the transmit power $Q_{m,u}$ is not negative, we can provide an exact closed-form formula of $Q_{m,u}$ as follows:

$$Q_{m,u} = \frac{\lambda_{T_{m,u},PR}P_{PT}}{\lambda_{PT,PR}\rho}\left[\left((1-\varepsilon_{OP})\exp\left(\frac{(N_m+1)\lambda_{PT,PR}\rho}{P_{PT}}\right)\right)^{-\frac{1}{2(N_m+1)}} - 1\right]^{+}. \tag{23}$$

3.3 End-to-End PNSC of m-th Path

For the m-th path, the end-to-end PNSC can be formulated as

$$PNSC_m = \prod_{u=1}^{N_m+1} Pr\left(\frac{Q_{m,u-1}\gamma_{T_{m,u-1},T_{m,u}}}{P_{PT}\gamma_{PT,T_{m,u}} + 1} > \frac{Q_{m,u-1}\gamma_{T_{m,u-1},E}}{P_{PT}\gamma_{PT,E} + Q_{m,u}\gamma_{J_{m,u},E} + 1}\right)$$

$$= \prod_{u=1}^{N_m+1} Pr\left(Z_u^D > Z_u^E\right), \tag{24}$$

where

$$Z_u^D = \frac{\gamma_{T_{m,u-1},T_{m,u}}}{P_{PT}\gamma_{PT,T_{m,u}} + 1}, \ Z_u^E = \frac{\gamma_{T_{m,u-1},E}}{P_{PT}\gamma_{PT,E} + Q_{m,u}\gamma_{J_{m,u},E} + 1}. \quad (25)$$

Furthermore, we can rewrite $\Pr\left(Z_u^D > Z_u^E\right)$ in (25) as

$$\Pr\left(Z_u^D > Z_u^E\right) = \int_0^{+\infty} \left(1 - F_{Z_u^D}(x)\right) f_{Z_u^E}(x)\, dx. \quad (26)$$

Next, we find the CDF $F_{Z_u^D}(x)$ which can be formulated by

$$F_{Z_u^D}(x) = \Pr\left(Z_u^D < x\right) = \int_0^{+\infty} F_{\gamma_{T_{m,u-1},T_{m,u}}}(P_{PT}xy + x) f_{\gamma_{PT,T_{m,u}}}(y)\, dy. \quad (27)$$

Since $\gamma_{T_{m,u-1},T_{m,u}} = \max\limits_{v=1,2,\ldots,L_{m,u}}\left(\gamma_{T_{m,u-1},R_{m,u,v}}\right)$, we can obtain the CDF $F_{\gamma_{T_{m,u-1},T_{m,u}}}(P_{PT}xy + x)$ as

$$F_{\gamma_{T_{m,u-1},T_{m,u}}}(P_{PT}xy + x) = \left(1 - \exp\left(-\lambda_{T_{m,u-1},T_{m,u}}(P_{PT}xy + x)\right)\right)^{L_{m,u}}$$

$$= 1 + \sum_{v=1}^{L_{m,u}} (-)^v C_{L_{m,u}}^v \exp\left(-v\lambda_{T_{m,u-1},T_{m,u}}x\right)\exp\left(-v\lambda_{T_{m,u-1},T_{m,u}}P_{PT}xy\right). \quad (28)$$

It is noted that when $u = N_m + 1$, then $L_{m,N_m+1} = 1$ and

$$F_{\gamma_{T_{m,N_m},T_{m,N_m+1}}}(P_{PT}xy + x) = 1 - \exp\left(-\lambda_{T_{m,N_m},T_{m,N_m+1}}x\right)\exp\left(-\lambda_{T_{m,N_m},T_{m,N_m+1}}P_{PT}xy\right). \quad (29)$$

Substituting (28) and $f_{\gamma_{PT,T_{m,u}}}(y) = \lambda_{PT,T_{m,u}}\exp\left(-\lambda_{PT,T_{m,u}}y\right)$ into (27), after algebraic simplifications, we obtain

$$F_{Z_u^D}(x) = 1 + \sum_{v=1}^{L_{m,u}} \frac{(-)^v C_{L_{m,u}}^v \lambda_{PT,T_{m,u}}}{\lambda_{PT,T_{m,u}} + v\lambda_{T_{m,u-1},T_{m,u}}P_{PT}x} \exp\left(-v\lambda_{T_{m,u-1},T_{m,u}}x\right)$$

$$= 1 + \sum_{v=1}^{L_{m,u}} (-)^v C_{L_{m,u}}^v \frac{\omega_{0,u,v}}{\omega_{0,u,v} + x} \exp\left(-v\lambda_{T_{m,u-1},T_{m,u}}x\right), \quad (30)$$

where

$$\omega_{0,u,v} = \frac{\lambda_{PT,T_{m,u}}}{v\lambda_{T_{m,u-1},T_{m,u}}P_{PT}}. \quad (31)$$

Similarly, the CDF of Z_u^E can be obtained as

$$F_{Z_u^E}(x) = \int_0^{+\infty} \begin{bmatrix} F_{\gamma_{T_{m,u-1},E}}(P_{PT}xy + Q_{m,u}xz + x) \\ f_{\gamma_{PT,E}}(y) f_{\gamma_{J_{m,u},E}}(z) \end{bmatrix} dy\, dz$$

$$= 1 - \frac{\lambda_{PT,E}}{\lambda_{PT,E} + \lambda_{T_{m,u-1},E}P_{PT}x} \frac{\lambda_{T_{m,u},E}}{\lambda_{T_{m,u},E} + \lambda_{T_{m,u-1},E}Q_{m,u}x} \exp\left(-\lambda_{T_{m,u-1},E}x\right)$$

$$= 1 - \frac{\omega_{1,u}}{\omega_{1,u} + x} \frac{\omega_{2,u}}{\omega_{2,u} + x} \exp\left(-\lambda_{T_{m,u-1},E}x\right), \quad (32)$$

where $\lambda_{J_{m,u},E} = \lambda_{T_{m,u},E}$, $\omega_{1,u} = \frac{\lambda_{PT,E}}{\lambda_{T_{m,u-1},E}P_{PT}}$ and $\omega_{2,u} = \frac{\lambda_{T_{m,u},E}}{\lambda_{T_{m,u-1},E}Q_{m,u}}$.

Assume that $\omega_{1,u} \neq \omega_{2,u}$, the CDF $F_{Z_u^E}(x)$ can be rewritten as

$$F_{Z_u^E}(x) = 1 - \frac{\kappa_u}{\omega_{1,u} + x} \exp\left(-\lambda_{T_{m,u-1},E}x\right)$$
$$+ \frac{\kappa_u}{\omega_{2,u} + x} \exp\left(-\lambda_{T_{m,u-1},E}x\right), \tag{33}$$

where $\kappa_u = \frac{\omega_{1,u}\omega_{2,u}}{\omega_{2,u} - \omega_{1,u}}$. Then, the PDF $f_{Z_u^E}(x)$ is obtained as

$$f_{Z_u^E}(x) = \frac{\kappa_u}{(\omega_{1,u} + x)^2} \exp\left(-\lambda_{T_{m,u-1},E}x\right) + \frac{\kappa_u\lambda_{T_{m,u-1},E}}{\omega_{1,u} + x} \exp\left(-\lambda_{T_{m,u-1},E}x\right)$$
$$- \frac{\kappa_u}{(\omega_{2,u} + x)^2} \exp\left(-\lambda_{T_{m,u-1},E}x\right) - \frac{\kappa_u\lambda_{T_{m,u-1},E}}{\omega_{2,u} + x} \exp\left(-\lambda_{T_{m,u-1},E}x\right). \tag{34}$$

Using (26), (30) and (34), we have

$$\Pr\left(Z_u^D > Z_u^E\right) = \sum_{v=1}^{L_{m,u}} (-)^{v+1}C_{L_{m,u}}^v \omega_{0,u,v}\kappa_u$$
$$\times \left(I_{1,u,v} - I_{2,u,v} + I_{3,u,v} - I_{4,u,v}\right), \tag{35}$$

where $\xi_{u,v} = v\lambda_{T_{m,u-1},T_{m,u}} + \lambda_{T_{m,u-1},E}$, and

$$I_{1,u,v} = \int_0^{+\infty} \frac{1}{(\omega_{0,u,v} + x)(\omega_{1,u} + x)^2} \exp\left(-\xi_{u,v}x\right)dx,$$
$$I_{2,u,v} = \int_0^{+\infty} \frac{1}{(\omega_{0,u,v} + x)(\omega_{2,u} + x)^2} \exp\left(-\xi_{u,v}x\right)dx,$$
$$I_{3,u,v} = \int_0^{+\infty} \frac{\lambda_{T_{m,u-1},E}}{(\omega_{0,u,v} + x)(\omega_{1,u} + x)} \exp\left(-\xi_{u,v}x\right)dx,$$
$$I_{4,u,v} = \int_0^{+\infty} \frac{\lambda_{T_{m,u-1},E}}{(\omega_{0,u,v} + x)(\omega_{2,u} + x)} \exp\left(-\xi_{u,v}x\right)dx. \tag{36}$$

Now, we can rewrite the integral $I_{1,u,v}$ as

$$I_{1,u,v} = \frac{1}{(\omega_{0,u,v} - \omega_{1,u})^2} \int_0^{+\infty} \frac{1}{\omega_{0,u,v} + x} \exp\left(-\xi_{u,v}x\right)dx$$
$$- \frac{1}{(\omega_{0,u,v} - \omega_{1,u})^2} \int_0^{+\infty} \frac{1}{\omega_{1,u} + x} \exp\left(-\xi_{u,v}x\right)dx$$
$$+ \frac{1}{\omega_{0,u,v} - \omega_{1,u}} \int_0^{+\infty} \frac{1}{(\omega_{1,u} + x)^2} \exp\left(-\xi_{u,v}x\right)dx, \tag{37}$$

where $\omega_{0,u,v} \neq \omega_{1,u}$. After algebraic simplifications, we obtain

$$
\begin{aligned}
I_{1,u,v} = {} & \frac{1}{(\omega_{0,u,v} - \omega_{1,u})^2} \exp(\omega_{0,u,v}\xi_{u,v}) E_1(\omega_{0,u,v}\xi_{u,v}) \\
& - \frac{1}{(\omega_{0,u,v} - \omega_{1,u})^2} \exp(\omega_{1,u}\xi_{u,v}) E_1(\omega_{1,u}\xi_{u,v}) \\
& + \frac{1}{\omega_{0,u,v} - \omega_{1,u}} \left(\frac{1}{\omega_{1,u}} - \xi_{u,v} \exp(\omega_{1,u}\xi_{u,v}) E_1(\omega_{1,u}\xi_{u,v}) \right),
\end{aligned}
\tag{38}
$$

where $E_1(.)$ is exponential integral [23]. Similarly, we have

$$
\begin{aligned}
I_{2,u,v} = {} & \frac{1}{(\omega_{0,u,v} - \omega_{2,u})^2} \exp(\omega_{0,u,v}\xi_{u,v}) E_1(\omega_{0,u,v}\xi_{u,v}) \\
& - \frac{1}{(\omega_{0,u,v} - \omega_{2,u})^2} \exp(\omega_{2,u}\xi_{u,v}) E_1(\omega_{2,u}\xi_{u,v}) \\
& + \frac{1}{\omega_{0,u,v} - \omega_{2,u}} \left(\frac{1}{\omega_{2,u}} - \xi_{u,v} \exp(\omega_{2,u}\xi_{u,v}) E_1(\omega_{2,u}\xi_{u,v}) \right).
\end{aligned}
\tag{39}
$$

Next, the integrals $I_{3,u,v}$ and $I_{4,u,v}$ can be calculated, respectively as

$$
\begin{aligned}
I_{3,u,v} = {} & \frac{\lambda_{T_{m,u-1},E}}{\omega_{1,u} - \omega_{0,u,v}} \\
& \times [\exp(\omega_{0,u,v}\xi_{u,v}) E_1(\omega_{0,u,v}\xi_{u,v}) - \exp(\omega_{1,u}\xi_{u,v}) E_1(\omega_{1,u}\xi_{u,v})],
\end{aligned}
\tag{40}
$$

$$
\begin{aligned}
I_{4,u,v} = {} & \frac{\lambda_{T_{m,u-1},E}}{\omega_{2,u} - \omega_{0,u,v}} \\
& \times [\exp(\omega_{0,u,v}\xi_{u,v}) E_1(\omega_{0,u,v}\xi_{u,v}) - \exp(\omega_{2,u}\xi_{u,v}) E_1(\omega_{2,u}\xi_{u,v})].
\end{aligned}
\tag{41}
$$

Substituting (38)–(41) into (35), we obtain the exact closed-form expression for $\Pr(Z_u^D > Z_u^E)$, which is then substituted into (24) to obtain PNSC_m.

3.4 Path-Selection Methods

For the BEST method, the best path is selected by the following strategy:

$$
\text{Path } a: \ SC_a^{e2e} = \max_{m=1,2,\ldots,M}(SC_m^{e2e}).
\tag{42}
$$

From (42), the end-to-end PNSC of the BEST method is computed by

$$
\mathrm{PNSC}_{\mathrm{BEST}} = \Pr(SC_a^{e2e} > 0) = 1 - \prod_{m=1}^{M}(1 - \Pr(SC_m^{e2e} > 0))
$$

$$
= 1 - \prod_{m=1}^{M}(1 - \mathrm{PNSC}_m).
\tag{43}
$$

Then, substituting the expression for derived in Subsect. 3.3 into (43), we obtain an exact closed-form expression for the end-to-end PNSC using the BEST method.

However, the implementation of the BEST method which requires the instantane-ous channel state information (CSIs) of all of the links appears complex. In practice, the statistical CSIs, i.e., average channel gain, can be readily obtained. Therefore, we propose the MAXV method, where the path providing the maximum average end-to-end PNSC is selected for the data transmission. Mathematically, we thus can write

$$\text{PNSC}_{\text{MAXV}} = \max_{m=1,2,...,M} (\text{PNSC}_m).\tag{44}$$

Finally, if the instantaneous and statistical CSIs of the data and/or eaves-dropping and/or interference links are unknown because of the complexity, delay time constraint or the random presence of the eavesdropper, the random path selection (RAND) will be an appropriate solution. In this method, the end-to-end PNSC is ex-pressed by

$$\text{PNSC}_{\text{RAND}} = \frac{1}{M} \sum_{m=1}^{M} \text{PNSC}_m.\tag{45}$$

In (45), due to the random selection, the probability that the m-th path is selected for the data transmission equals to $1/M$ for all values of m.

4   Simulation Results

In this section, we perform Monte Carlo simulations to verify the proposed the-oretical results obtained in Sect. 3. For illustration purposes only, in all of the simulations, the path loss exponent (β) is fixed by 3, the target rate of the pri-mary network (R_P) is set at 0.25, and the required QoS of the primary network (ε_{OP}) is assumed to 0.05. We also assume that there are three available paths between the source and the destination $(M = 3)$, the number of clusters are 2, 3, 4 $(N_1 = 2, N_2 = 3, N_3 = 4)$, and the number of nodes in each cluster is set at 3 $(L_{m,u} = 3, \forall u, v)$.

For the simulation environment, we consider a two-dimensional plane Oxy, in which the secondary source S and the secondary destination D are placed at (0,0) and (1,0), respectively. In addition, the cluster nodes $R_{m,u,v}$ have the same location at $(u/(N_m + 1), 0)$, and the position of the eavesdropper E is (x_E, y_E), where $m = 1, 2, 3$, $u = 1, 2, ..., N_m$. For the primary network, the primary trans-mitter (PT) and the primary receiver (PR) have been located at (x_{PT}, y_{PT}) and (x_{PR}, y_{PR}), respectively.

In Fig. 3, we present the transmit power of the secondary transmitters of the first path $(N_1 = 2)$, including the source (S or $T_{1,0}$), the selected receivers and jammers ($T_{1,1}$, $T_{1,2}$, $T_{1,3}$), as a function of the transmit power of PT (P_{PT}). As

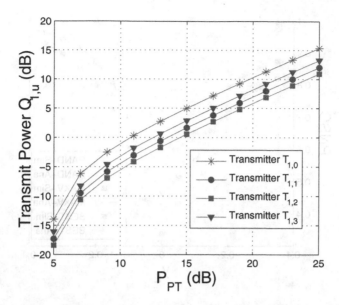

Fig. 3. Transmit power of the secondary transmitters on the first path as a function of P_{PT} in dB when $x_{PT} = 0.5$, $y_{PT} = 1$, $x_{PR} = 0.6$, and $y_{PR} = 0.6$.

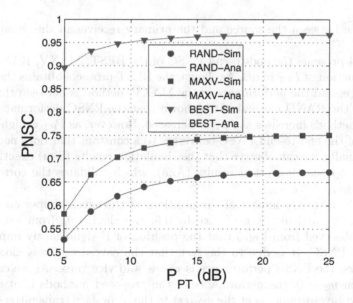

Fig. 4. End-to-end PNSC as a function of P_{PT} in dB when $x_{PT} = 0.5$, $y_{PT} = 1$, $x_{PR} = 0.5$, $y_{PR} = 0.75$, $x_E = 0.5$, and $y_E = -0.15$.

we can see, the transmit power $Q_{1,u}$ ($u = 0, 1, 2, 3$) increases as P_{PT} is increased. Indeed, as given in (23), $Q_{1,u}$ is an increasing function of (P_{PT}). Moreover, we can see in Fig. 3 that the transmit power of the source ($Q_{1,0}$) is highest because

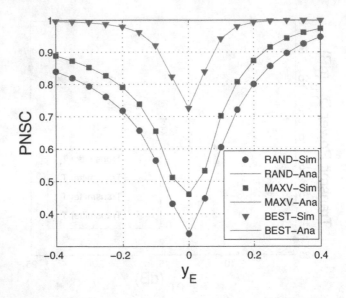

Fig. 5. End-to-end PNSC as a function of y_E when $P_{PT} = 15\,\mathrm{dB}$, $x_{PT} = 0.5$, $y_{PT} = 1$, $x_{PR} = 0.5$, $y_{PR} = 0.75$, $x_E = 0.5$, and $x_E = 0.5$.

the distance between the source and the primary receiver in this simulation is longest.

Figure 4 presents the end-to-end PNSC of the BEST, MAXV, RAND methods as a function of P_{PT} in dB. As shown, the BEST protocol obtains the highest performance, and the performance of the MAXV method is between that of the BEST and the RAND methods. It is shown that the PNSC performance of the proposed methods increases as P_{PT} is increased. However, as P_{PT} is high enough, the value of the end-to-end PNSC converges to a constant that does not depend on P_{PT}. Finally, we can observe that the simulation results (Sim) exactly match with the proposed theoretical results (Ana), which validates the correction of our derivations.

In Fig. 5, we study impact of the positions of the eavesdropper on the end-to-end PNSC. Particularly, we fix x_E by 0.5, and change y_E from -0.4 to 0.4. It can be observed from Fig. 5 that the position of E significantly impacts the end-to-end PNSC. It is due to the fact that the eavesdropper is close to the transmitters, the PNSC performance is worse, and vice verse. As we can see in Fig. 5, when $y_E = 0$, the performance of the proposed methods is at its worst because the eavesdropper is at the nearest to the secondary transmitters. Again, consistent matching for the simulation and theoretical results has been achieved, which verifies our proposed theoretical analyzes.

5   Conclusion

In this paper, we have proposed and evaluated secrecy performance of three path-selection methods over cluster-based underlay CR networks by computing their end-to-end PNSC. Our results have shown that the BEST method attained the best performance, while that of the RAND method has appeared to be the worst. However, the implementation of the RAND and MAXV methods has been much simpler than the BEST method, which thus has presented trade-off between method complexity and secrecy performance. Moreover, the position of the eavesdropper and the transmit power of the primary transmitter significantly has impacted the end-to-end PNSC performance of the proposed methods. Future work on analyzing the proposed methods under more advanced fading environments will also be presented in a separate publication.

Acknowledgment. This research is funded by Vietnam National Foundation for Science and Technology Development (NAFOSTED) under grant number 102.04-2017.317.

References

1. Gopala, P.K., Lai, L., Gamal, H.E.: On the secrecy capacity of fading channels. IEEE Trans. Inf. Theory **54**(10), 4687–4698 (2008)
2. Zhang, J., Duong, T.Q., Woods, R., Marshall, A.: Securing wireless communications of the internet of things from the physical layer. Overv. Entropy **19**(8), 420 (2017)
3. Yang, N., Yeoh, P.L., Elkashlan, M., Schober, R., Collings, I.B.: Transmit antenna selection for security enhancement in MIMO wiretap channels. IEEE Trans. Commun. **61**(1), 144–154 (2013)
4. Xiong, J., Tang, Y., Ma, D., Xiao, P., Wong, K.-K.: Secrecy performance analysis for TAS-MRC system with imperfect feedback. IEEE Trans. Inf. Forensics Secur. **10**(8), 1617–1629 (2015)
5. Zhao, R., Lin, H., He, Y.-C., Chen, D.-H., Huang, Y., Yang, L.: Secrecy performance of transmit antenna selection for MIMO relay systems with outdated CSI. IEEE Trans. Commun. **66**(2), 546–559 (2018)
6. Krikidis, I.: Opportunistic relay selection for cooperative networks with secrecy constraints. IET Commun. **4**(15), 1787–1791 (2010)
7. Yang, M., Guo, D., Huang, Y., Duong, T.Q., Zhang, B.: Secure multiuser scheduling in downlink dual-hop regenerative relay networks over Nakagami-m fading channels. IEEE Trans. Wirel. Commun. **15**(12), 8009–8024 (2016)
8. Qing, L., Guangyao, H., Xiaomei, F.: Physical layer security in multi-hop AF relay network based on compressed sensing. IEEE Commun. Lett. **22**(9), 1882–1885 (2018)
9. Hoang, T.M., Duong, T.Q., Vo, N.-S., Kundu, C.: Physical layer security in cooperative energy harvesting networks with a friendly jammer. IEEE Wirel. Commun. Lett. **6**(2), 174–177 (2017)
10. Ma, H., Cheng, J., Wang, X., Ma, P.: Robust MISO beamforming with cooperative jamming for secure transmission from perspectives of QoS and secrecy rate. IEEE Trans. Commun. **66**(2), 767–780 (2018)

11. Duy, T.T., Kong, H.Y.: Secrecy performance analysis of multihop transmission protocols in cluster networks. Wirel. Pers. Commun. **82**(4), 2505–2518 (2015)
12. Tin, P.T., Duy, T.T., Phuong, T.T., Voznak, M.: Secrecy performance of joint relay and jammer selection methods in cluster networks: with and without hardware noises. In: International Conference on Advanced Engineering - Theory and Applications, Busan, pp. 769–779 (2016)
13. Lee, J.-H.: Full-duplex relay for enhancing physical layer security in multi-hop relaying systems. IEEE Commun. Lett. **19**(4), 525–528 (2015)
14. Atapattu, S., Ross, N., Jing, Y., He, Y., Evans, J.S.: Physical-layer security in full-duplex multi-hop multi-user wireless network with relay selection. IEEE Trans. Wirel. Commun. **18**(2), 1216–1232 (2019)
15. Liu, Y., Wang, L., Duy, T.T., Elkashlan, M., Duong, T.Q.: Relay selection for security enhancement in cognitive relay networks. IEEE Wirel. Commun. Lett. **4**(1), 46–49 (2015)
16. Nguyen, V.D., Duong, T.Q., Dobre, O.A., Shin, O.-S.: Joint information and jamming beamforming for secrecy rate maximization in cognitive radio networks. IEEE Trans. Inf. Forensics Secur. **11**(11), 2609–2623 (2016)
17. Liu, Y., Wang, L., Zaidi, S.A.R., Elkashlan, M., Duong, T.Q.: Secure D2D communica-tion in large-scale cognitive cellular networks: a wireless power transfer model. IEEE Trans. Commun. **64**(1), 329–342 (2016)
18. Li, M., Yin, H., Huang, Y., Wang, Y., Yu, R.: Physical layer security in overlay cognitive radio networks with energy harvesting. IEEE Trans. Veh. Technol. **67**(11), 11274–11279 (2018)
19. Xu, Q., Ran, P., He, H., Xu, D.: Security-aware routing for artificial-noise-aided multi-hop secondary communications. In: IEEE Wireless Communications and Networking Conference, pp. 1–6. IEEE, Barcelona (2018)
20. Tin, P.T., Hung, D.T., Tan, N.N., Duy, T.T., Voznak, M.: Secrecy performance en-hancement for underlay cognitive radio networks employing cooperative multi-hop transmission with and without presence of hardware impairments. Entropy **221**(2), 217 (2019)
21. Tin, P.T., Nam, P.M., Duy, T.T., Phuong, T.T., Voznak, M.: Secrecy performance en-hancement for underlay cognitive radio networks employing cooperative multi-hop transmission with and without presence of hardware impairments. Sensors **19**(5), 1160 (2019)
22. Laneman, J.N., Tse, D.N., Wornell, G.W.: Cooperative diversity in wireless networks: efficient protocols and outage behavior. IEEE Trans. Inf. Theory **50**(12), 3062–3080 (2004)
23. Gradshteyn, I.S., Ryzhik, I.M.: Table of Integrals, Series, and Products, 7th edn. Elsevier Inc., San Diego (2007)

Physical Layer Secrecy Enhancement for Non-orthogonal Multiple Access Cooperative Network with Artificial Noise

Van-Long Nguyen[1](✉) and Dac-Binh Ha[2]

[1] Graduate School, Duy Tan University, Da Nang, Vietnam
vanlong.itqn@gmail.com
[2] Faculty of Electrical and Electronics Engineering, Duy Tan University,
Da Nang, Vietnam
hadacbinh@duytan.edu.vn

Abstract. In this paper, the physical layer secrecy performance of non-orthogonal multiple access (NOMA) in a downlink cooperative network is studied. This considered system consists of one source, multiple legitimate user pairs and presenting an eavesdropper. In each pair, the better user forwards the information from the source to the worse user by using the decode-and-forward (DF) scheme and assuming that the eavesdropper attempts to extract the worse user's message. We propose the artificial noise - cooperative transmission scheme, namely ANCO-TRAS, to improve the secrecy performance of this considered system. In order to evaluate the effectiveness of this proposed scheme, the lower bound and exact closed-form expressions of secrecy outage probability are derived by using statistical characteristics of signal-to-noise ratio (SNR) and signal-to-interference-plus-noise ratio (SINR). Moreover, we investigate the secrecy performance of this considered system according to key parameters, such as power allocation ratio, average transmit power and number of user pair to verify our proposed scheme. Finally, Monte- Carlo simulation results are provided to confirm the correctness of the analytical results.

Keywords: Non-orthogonal multiple access · Cooperative network · Decode and forward · Artificial noise · Secrecy outage probability

1 Introduction

Nowadays, wireless devices, i.e., smartphones, smart control devices, and so on, are indispensable things in human life. Due to the mobility and convenience of wireless communication devices, the wireless system and devices are booming, i.e., now toward the fifth generation network (5G). The multiple access techniques applied in 4G and earlier generation networks belong to conventional orthogonal multiple access (OMA) type, such as FDMA, TDMA, and CDMA.

© ICST Institute for Computer Sciences, Social Informatics and Telecommunications Engineering 2019
Published by Springer Nature Switzerland AG 2019. All Rights Reserved
T. Q. Duong et al. (Eds.): INISCOM 2019, LNICST 293, pp. 81–98, 2019.
https://doi.org/10.1007/978-3-030-30149-1_7

Due to the significantly growing number of users and wireless devices, the next generation networks, i.e., 5G networks, are required to support the demand for low latency, low-cost and diversified services at higher quality and data rate. The conventional OMA techniques can not satisfy these demands anymore due to the limited channel resource and the spectral efficiency loss. The most prominent candidate that can meet these requirements 5G is the non-orthogonal multiple access (NOMA) technique [1–5]. Due to the advantage of the power domain to serve multiple users at the same time/frequency/code, the use of NOMA can ensure a significant spectral efficiency. In addition, compared with conventional OMA, NOMA offers better user fairness. The cooperative relaying technique can improve the performance and extend the coverage of wireless networks [6,7]. Naturally, this technique can be applied in NOMA networks and attracts a number of researchers to focus on this topic [8–12]. The work [8] proposed a cooperative NOMA transmission scheme that fully exploits prior information available in NOMA systems, in which the users with better channel conditions decode the messages of the others. Prior information is used as relays to improve the reception reliability for users with poor connections. The authors of [9] investigated the NOMA cooperative relaying system and proposed the best relay selection (BRS) scheme. The result has shown that the NOMA-based BRS obtains more rate gain than the conventional BRS when a number of relays becomes large. The works of [10,11] studied the NOMA cooperative relaying system with energy harvesting. Another communication protocol for cooperative NOMA system was proposed in [12]. The authors concluded that this proposed protocol can overcome the problem of the direct link between the paired users unavailable due to weak transmission conditions.

In the information age, security of information is the most essential issue to ensure that confidential information cannot be used by illegitimate users. However, due to the broadcast nature of wireless communications, the information transmission between transceivers can be eavesdropped by wiretappers which are difficult to detect. Although there are a number of solutions to solve this problem, such as RSA (Rivest Shamir Adleman) and DES (Data Encryption Standard), most of them are applied at higher layers, i.e., application or network layers. Moreover, the conditions at such higher layers assume that the link between the transmitter and receiver (physical layer) is error-free and that eavesdroppers have restricted computational power and lack efficient algorithms [13]. In order to enhance the security of wireless networks, the physical layer secrecy (PLS) is proposed to achieve secure transmission by exploiting the dynamic characteristics of the transmission channels [14–18]. This approach can be applied to NOMA relaying networks to improve the secrecy ability of NOMA communication networks. However, the employment of successive interference cancellation (SIC) in NOMA technique makes the secrecy performance analysis of the physical layer secrecy of NOMA different from that of conventional OMA technique. Recently, there are a number of works focusing on the physical layer secrecy of NOMA relaying being networks published [19–23]. The PLS of a downlink NOMA system was investigated in [19], in which users' quality of service (QoS) requirements to perform NOMA and all channels are assumed to undergo Nakagami-m fading. The results have shown that better secrecy performance of overall communication process can be obtained when there is not much difference in the level of priority between legitimate users. The work in [20] presented the secrecy performance analysis of a two-user downlink NOMA system. The authors considered two schemes those are single-input and single-output and multiple-input and single-output systems with different transmit antenna selection (TAS). The exact and approximated closedform expressions of the secrecy outage performance (SOP) for different TAS schemes were derived. The results have shown that when increasing the transmit power, the SOP for the far user with fixed power allocation scheme degraded as the transmit power beyond the threshold and then reaches a floor as the interference from the near user increases. The authors in [21] investigated the PLS of simple NOMA model of large-scale networks through the SOP. In this model, stochastic geometry approaches were used to model the locations of NOMA users and eavesdroppers. In [22], the artifical noise is generated at the base station that can improve the security of a beamforming-aided multiple-antenna system. The secrecy beamforming scheme for multiple-input single-output non-orthogonal multiple access (MISO-NOMA) systems is proposed. In this proposed scheme, the artificial noise is used to protect the confidential information of two NOMA assisted legitimate users, such that the system secrecy performance improved. The authors of [23] studied the PLS for cooperative NOMA systems, in which the amplify-and-forward (AF) and decode-and-forward (DF) protocols are considered. They concluded that AF and DF schemes almost achieve the same secrecy performance and this secrecy performance is independent of the channel conditions between the relay and the poor user.

Unlike the above works, in this work, we consider the PLS of a downlink NOMA cooperative relay communication system combining with artifical noise. In other words, we study the PLS performance of the NOMA system in which the base station (BS) simultaneously transmits information to user pairs based on power-domain, and then users with better channel conditions (better user) relay information to the worse user assuming that the eavesdropper only tries to listen to the worse users information. In order to ensure the PLS performance, artificial noise is used in BS to confuse the eavesdropper. Given the considered networks, our work provides the following contributions:

1. The artificial noise with cooperative transmission scheme, namely ANCO-TRAS, is proposed.
2. The lower bound and exact closed-form expressions of secrecy outage probability for each user and overall system are derived.
3. By means of secrecy outage probability expressions, we carry out evaluating the PLS performance.
4. The behavior of the considered system is assessed with respect to different key parameters, such as power allocation ratio, average transmit power and number of user pair.

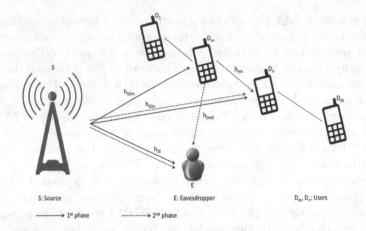

Fig. 1. System model for secured cooperative NOMA

The rest of this paper is organized as follows. The system model is presented in Sect. 2. Secrecy performance of the considered system is analyzed in Sect. 3. The numerical results are shown in Sect. 4. Finally, Sect. 5 draws the conclusion of our paper

2   Network and Channel Models

A cooperative communication system for downlink NOMA is considered as the Fig. 1. One source S, i.e., base station, intends to transmit information to M mobile users denoted as D_i, $(i = 1 \leq m < n \leq M)$ in the presence of an eavesdropper E. In this considered system, we can divided M users into multiple pairs, such as $\{D_m, D_n\}$, $m < n$, to perform NOMA [8] and these two paired users help to forward information signals to each other. It means that the mth user and the nth user are paired to perform cooperative NOMA. The better user D_m forward the information of the worse user D_n after use applying successive interference cancellation (SIC) to detect the D_m's signal.

The two-phase ANCOTRAS protocol of this considered system is proposed as follows

- In the first phase: S transmits information signal $x = \sqrt{a_m}s_m + \sqrt{a_n}s_n$ with power P_S to user pair $\{D_m, D_n\}$ in the time of αT $(0 < \alpha < 1, T$ is block time), where s_m and s_n are the message for the mth user D_m and the nth user D_n, respectively; a_m and a_n are the power allocation coefficients that satisfy the conditions: $0 < a_m < a_n$ and $a_m + a_n = 1$ by following the NOMA scheme.
- In the second phase: Applying NOMA, D_m uses SIC to detect message s_n and subtracts this component from the received signal to obtain its own message s_m, then decodes and forwards s_n to D_n in the time of $(1 - \alpha T)$. By this time, the S cooperates with D_m to broadcast an artificial noise to confuse

the eavesdropper. Finally, D_n combines two received signals, i.e., the direct signal from the S and the relaying signal from D_m, to decode its own message by using selection combining (SC) technique.

Meanwhile, the eavesdropper tries to extract the poor user's message s_n from the links $S - D_n$ and $D_m - D_n$.

Without loss of generality, assume that all the channel gains between S and users follow the order of $|h_{SD_1}|^2 \leq |h_{SD_2}|^2 \leq \cdots \leq |h_{SD_m}|^2 \leq |h_{SD_n}|^2 \leq \cdots \leq |h_{SD_M}|^2$ are decoded as the ordered channel gains of the mth user and the nth user. Denote that $|h_{mn}|^2$ is the channel gains of the links between the mth user and the nth user; $|h_{SE}|^2$ and $|h_{D_mE}|^2$ are channel gains of the links of $S-E$ and $D_m - E$, respectively. We assume that all the nodes are single-antenna devices and operate in a half-duplex mode. All wireless links are assumed to undergo independent frequency non-selective Rayleigh block fading and additive white Gaussian noise (AWGN) with zero mean and the same variance σ^2. We denote $d_{SD_m}, d_{SD_n}, d_{mn}, d_{SE}, d_{D_mE}$ as the Euclidean distances of $S-D_m, S-D_n, D_m-D_n, S-E, D_m-E$, respectively and θ denote the path-loss exponent.

2.1 The First Phase

In this phase, the source S broadcasts information to the users. The received signals at D_m and at D_n are

$$y_{SD_m} = \sqrt{\frac{P_S}{d_{SD_m}^\theta}}(\sqrt{a_m}s_m + \sqrt{a_n}s_n)h_{SD_m} + n_{SD_m}, \tag{1}$$

$$y_{SD_n} = \sqrt{\frac{P_S}{d_{SD_n}^\theta}}(\sqrt{a_m}s_m + \sqrt{a_n}s_n)h_{SD_n} + n_{SD_n}, \tag{2}$$

respectively, where n_{SD_m} and n_{SD_n} are the AWGN with zero mean and variance σ^2.

Let $X_1 \triangleq |h_{SD_m}|^2, Y_1 \triangleq |h_{SD_n}|^2, X_2 \triangleq |h_{mn}|^2, Z_1 \triangleq |h_{SE}|^2$, and $Z_2 \triangleq |h_{D_mE}|^2$. The instantaneous SINR at the nth user to detect s_n transmitted from S can be given by

$$\gamma_{SD_n} = \frac{a_n \bar{\gamma}|h_{SD_n}|^2}{a_m \bar{\gamma}|h_{SD_n}|^2 + d_{SD_n}^\theta} = \frac{b_2 Y_1}{b_1 Y_1 + 1}, \tag{3}$$

where $\bar{\gamma} = \frac{P_S}{\sigma^2}$ is denoted as the average transmit SNR of $S - D_n$ link, $b_1 = \frac{a_m \bar{\gamma}}{d_{SD_n}^\theta}, b_2 = \frac{a_n \bar{\gamma}}{d_{SD_n}^\theta}$.

Similarly, the instantaneous SINR at the mth user to detect s_n transmitted from S can be written as

$$\gamma_{SD_m}^{s_n} = \frac{a_m \bar{\gamma}|h_{SD_m}|^2}{a_n \bar{\gamma}|h_{SD_m}|^2 + d_{SD_m}^\theta} = \frac{b_4 X_1}{b_3 X_1 + 1}, \tag{4}$$

where $b_3 = \frac{a_m \bar{\gamma}}{d_{SD_m}^\theta}, b_4 = \frac{a_n \bar{\gamma}}{d_{SD_m}^\theta}$.

At the same time, the received signal at E is given as follows

$$y_{SE} = \sqrt{\frac{P_S}{d_{SE}^\theta}}(\sqrt{a_m}s_m + \sqrt{a_n}s_n)h_{SE} + n_{SE}, \tag{5}$$

where n_{SE} is the AWGN with zero mean and variance σ_E^2. Due to assuming that the eavesdropper only tries to detect s_n, therefore the instantaneous SINR at E is given by

$$\gamma_{SE} = \frac{a_n\bar{\gamma}_E|h_{SE}|^2}{a_m\bar{\gamma}_E|h_{SE}|^2 + d_{SE}^\theta} = \frac{b_6 Z_1}{b_5 Z_1 + 1}, \tag{6}$$

where $b_5 = \frac{a_m\bar{\gamma}_E}{d_{SE}^\theta}, b_6 = \frac{a_n\bar{\gamma}_E}{d_{SE}^\theta}$.

2.2 The Second Phase

In this phase, D_m uses the power P_{Dm} to forward s_n to D_n and S simultaneously uses the power P_S to broadcast an AN to users and eavesdrooper. The instantaneous SINR at D_n in the second phase is as follows

$$\gamma_{mn} = U_1 = \frac{c_1 X_2}{c_3 Y_1 + 1}, \tag{7}$$

where $c_1 = \frac{\bar{\gamma}_{Dm}}{d_{mn}^\theta}, c_3 = \frac{\bar{\gamma}}{d_{SDn}^\theta}, \bar{\gamma}_{Dm} = \frac{P_{Dm}}{\sigma^2}$.

Similarly, the instantaneous SINR at E in this phase is given by

$$\gamma_{DmE} = U_2 = \frac{c_2 Z_2}{c_4 Z_1 + 1}, \tag{8}$$

where $c_2 = \frac{\bar{\gamma}_{DmE}}{d_{DmE}^\theta}, c_4 = \frac{\bar{\gamma}_E}{d_{SE}^\theta}, \bar{\gamma}_{Dm E} = \frac{P_{Dm}}{\sigma_E^2}$.

Considering i.i.d. Rayleigh channels, the channel gains $|h_{SDm}|^2, |h_{SDn}|^2, |h_{SE}|^2$ and $|h_{mn}|^2$ follow exponential distributions with parameters $\lambda_{SDm}, \lambda_{SDn}, \lambda_{SE}$ and λ_{mn}, respectively. In order statistics, the probability density function (PDF) and the cumulative distribution function (CDF) of U, where $U \in \{X_1, Y_1\}$, are respectively given by [24]

$$f_U(x) = \frac{M!}{(M-i)!(i-1)!}\frac{1}{\lambda_{SDi}}\sum_{k=0}^{i-1}C_k^{i-1}(-1)^k e^{\frac{-x(M-i+k+1)}{\lambda_{SDi}}} \tag{9}$$

$$F_U(x) = \frac{M!}{(M-i)!(i-1)!}\sum_{k=0}^{i-1}C_k^{i-1}(-1)^k\frac{1}{M-i+k+1}\left[1 - e^{\frac{-x(M-i+k+1)}{\lambda_{SDi}}}\right] \tag{10}$$

where $i \in \{m, n\}$.

The PDF and CDF of V, where $V \in (X_2, Z_1, Z_2)$ are respectively expressed as

$$f_V(x) = \frac{1}{\lambda}e^{-\frac{x}{\lambda}} \tag{11}$$

$$F_V(x) = 1 - e^{-\frac{x}{\lambda}} \tag{12}$$

where $\lambda \in \{\lambda_{mn}, \lambda_{SE}, \lambda_{DmE}\}$.

For further calculation, we derive CDFs and PDFs of $\gamma_{SD_m}^{s_n}, \gamma_{SD_n}, \gamma_{SE}$, $\gamma_{mn}, \gamma_{D_mE}$. From above results, we calculate the CDF of γ_{SDn} as follows

$$F_{\gamma_{SDm}^{s_n}}(x) = Pr\left(\frac{b_4 X_1}{b_3 X_1 + 1} < x\right) \overset{(a)}{=} F_{X_1}\left(\frac{x}{b_4 - b_3 x}\right)$$

$$= \frac{M!}{(M-m)!(m-1)!}\sum_{k=0}^{m-1} C_k^{m-1}(-1)^k$$

$$\times \frac{1}{M-m+k+1}\left[1 - e^{\frac{-(M-m+k+1)x}{\lambda_{SDm}}}\right], \tag{13}$$

$$F_{\gamma_{SDn}}(x) = Pr\left(\frac{b_2 Y_1}{b_1 Y_1 + 1} < x\right) \overset{(b)}{=} F_{Y_1}\left(\frac{x}{b_2 - b_1 x}\right)$$

$$= \frac{M!}{(M-n)!(n-1)!}\sum_{k=0}^{n-1} C_k^{n-1}(-1)^k$$

$$\times \frac{1}{M-n+k+1}\left[1 - e^{\frac{-(M-n+k+1)x}{\lambda_{SDn}}}\right]. \tag{14}$$

Note that step (a) and (b) are obtained by assuming the following condition holds $x < \frac{b_2}{b_1}$, $x < \frac{b_4}{b_3}$, respectively.

Similarly, we respectively derive the CDF and PDF of γ_{SE} as follows

$$F_{\gamma_{SE}}(x) = Pr\left(\frac{b_6 Z_1}{b_5 Z_1 + 1} < x\right) \overset{(c)}{=} F_{Z_1}\left(\frac{x}{b_6 - b_5 x}\right)$$

$$= 1 - e^{-\frac{x}{\lambda_{SE}(b_6 - b_5 x)}}, \tag{15}$$

$$f_{\gamma_{SE}}(x) = \frac{b_6}{\lambda_{SE}(b_6 - b_5 x)^2}e^{-\frac{x}{\lambda_{SE}(b_6 - b_5 x)}}. \tag{16}$$

Note that step (c) is obtained by assuming the following condition holds $x < \frac{b_6}{b_5}$.

The CDF of the γ_{mn} is calculated as follows

$$F_{\gamma_{mn}}(x) = Pr\left(\frac{c_1 X_2}{c_3 Y_1 + 1} < x\right)$$

$$= \int_0^\infty F_{X_2}\left(\frac{x(c_3 y + 1)}{c_1}\right) f_{Y_1}(y) dy$$

$$= 1 - \frac{M!}{(M-n)!(n-1)!}\frac{1}{\lambda_{SDn}}\sum_{k=0}^{n-1} C_k^{n-1}(-1)^k \int_0^\infty e^{-\frac{x(c_3 y+1)}{c_1 \lambda_{mn}} - \frac{y(M-n+k+1)}{\lambda_{SDn}}} dy$$

$$= 1 - \frac{M!}{(M-n)!(n-1)!}\sum_{k=0}^{n-1} C_k^{n-1}(-1)^k$$

$$\times \frac{c_1 \lambda_{mn}}{c_3 \lambda_{SDn}x + c_1 \lambda_{mn}(M-n+k+1)}e^{-\frac{x}{c_1 \lambda_{mn}}}. \tag{17}$$

Similarly the CDF of the $\gamma_{D_m E}$ is given by

$$
\begin{aligned}
F_{\gamma_{D_m E}}(x) &= \Pr\left(\frac{c_2 Z_2}{c_4 Z_1 + 1} < x\right) \\
&= \int_0^\infty F_{Z_2}\left(\frac{x(c_4 y + 1)}{c_2}\right) f_{Z_1}(y) dy \\
&= 1 - \frac{1}{\lambda_{SE}} \int_0^\infty e^{-\frac{x(c_4 y + 1)}{c_2 \lambda_{DmE}} - \frac{y}{\lambda_{SE}}} dy \\
&= 1 - \frac{c_2 \lambda_{DmE}}{c_4 \lambda_{SE} x + c_2 \lambda_{DmE}} e^{-\frac{x}{c_2 \lambda_{DmE}}}.
\end{aligned}
\tag{18}
$$

The PDF of the $\gamma_{D_m E}$ is expressed as follows

$$
f_{\gamma_{D_m E}}(x) = \left(\frac{c_2 c_4 \lambda_{DmE} \lambda_{SE}}{(c_4 \lambda_{SE} x + c_2 \lambda_{DmE})^2} + \frac{1}{(c_4 \lambda_{SE} x + c_2 \lambda_{DmE})}\right) e^{-\frac{x}{c_2 \lambda_{DmE}}}.
\tag{19}
$$

3   Secrecy Performance Analysis

In this section, secrecy performance is analized in terms of secrecy outage probability. Secrecy outage probability (SOP) is an important performance metric that is usually used to characterize the secrecy performance of a wireless communication system. Here, we analyze the secrecy performance in terms of SOP at S and at D_m with assuming that E tries to extract the message of D_n.

3.1   Preliminaries

The instantaneous capacities of legitimate channel and eavesdropper channel can be respectively defined by

$$
C_M = B \log(1 + \gamma_M),
\tag{20}
$$
$$
C_N = B \log(1 + \gamma_N),
\tag{21}
$$

where B is bandwith (Hertz), $\gamma_M \in \{\gamma_{SD_m}, \gamma_{SD_n}, \gamma_{mn}\}$, $\gamma_N \in \{\gamma_{SE}, \gamma_{D_m E}\}$.

The instantaneous secrecy capacity for $S - D_m$, $S - D_n$ and $D_m D_n$ are given by

$$
\begin{aligned}
C_{S_1} &= \left[C_{\gamma_{SD_m}^{s_n}} - C_{\gamma_{SE}}\right]^+ \\
&= \begin{cases} \alpha \log_2\left(\dfrac{1 + \gamma_{SD_m}^{s_n}}{1 + \gamma_{SE}}\right), & \gamma_{SD_m}^{s_n} > \gamma_{SE} \\ 0, & \gamma_{SD_m}^{s_n} \leq \gamma_{SE} \end{cases} ,
\end{aligned}
\tag{22}
$$

$$
\begin{aligned}
C_{S_2} &= \left[C_{\gamma_{SD_n}} - C_{\gamma_{SE}}\right]^+ \\
&= \begin{cases} \alpha \log_2\left(\dfrac{1 + \gamma_{SD_n}}{1 + \gamma_{SE}}\right), & \gamma_{SD_n} > \gamma_{SE} \\ 0, & \gamma_{SD_n} \leq \gamma_{SE} \end{cases}
\end{aligned}
\tag{23}
$$

$$C_{S_3} = \left[C_{\gamma_{mn}} - C_{\gamma_{D_m E}}\right]^+$$

$$= \begin{cases} (1-\alpha)\log_2\left(\dfrac{1+\gamma_{mn}}{1+\gamma_{D_m E}}\right), & \gamma_{mn} > \gamma_{D_m E} \\ 0, & \gamma_{mn} \leq \gamma_{D_m E} \end{cases} \tag{24}$$

respectively. Here, for simplicity we assume $B = 1\,\mathrm{Hz}$.

SOP is defined as the probability that the instantaneous secrecy capacity falls below a predetermined secrecy rate threshold $R_S > 0$, given by $SOP = Pr(C_S < R_S)$. Notice that, in this considered system we only consider the case of the eavesdropper attempting to hear the message of D_n at S and at D_m. In the next subsection, we present the calculation the SOPs at S and at D_m.

3.2 Secrecy Outage Probability at S

The secrecy outage probability at S of $S - D_n$ link (SOP_1) can be calculated as follows

$$SOP_1 = Pr(C_{S_1} < R_S) = Pr\left(\frac{1+\gamma_{SD_n}}{1+\gamma_{SE}} < 2^{R_S/\alpha}\right)$$

$$= 1 - Pr\left(\gamma_{SE} < \frac{1+\gamma_{SD_n} - 2^{R_S/\alpha}}{2^{R_S/\alpha}}\right). \tag{25}$$

Because Eq. (25) is intractable to obtain the closed-form expression, here we only obtain the lower bound of SOP_1.

From the Eq. (3), we have $\gamma_{SD_n} < \frac{a_n}{a_m}$, therefore the lower bound of SOP_1 is calculated as follows

$$SOP_1 > SOP_{1_{lower}} = 1 - CDF_{\gamma_{SE}}\left(\frac{1+\frac{a_n}{a_m}}{2^{R_S/\alpha}} - 1\right)$$

$$= e^{-\frac{\beta}{\lambda_{SE}(b_6 - b_5\beta)}}, \tag{26}$$

where $\beta = \frac{1+\frac{a_n}{a_m}}{2^{R_S/\alpha}} - 1$.

Similarly, for SOP at S of $S - D_m$ link ($SOP_2 = Pr(C_{S_2} < R_S)$), the lower bound $SOP_{2_{lower}}$ is the same as $SOP_{1_{lower}}$:

$$SOP_2 > SOP_{2_{lower}} = e^{-\frac{\beta}{\lambda_{SE}(b_6 - b_5\beta)}}. \tag{27}$$

3.3 Secrecy Outage Probability at D_m

In this scenario, the secrecy outage event occurs when D_m cannot detect s_n or D_m can detect s_n but the secrecy capacity is below the secrecy threshold. Therefore, the secrecy outage probability at the D_m is calculated as follows

$$SOP_3 = Pr(\gamma_{SD_m}^{s_n} < \gamma_t) + Pr(\gamma_{SD_m}^{s_n} > \gamma_t, C_{S_3} < R_S)$$

$$= Pr\left(\frac{b_4 X_1}{b_3 X_1 + 1} < \gamma_t\right)$$

$$+ \left[1 - Pr\left(\frac{b_4 X_1}{b_3 X_1 + 1} < \gamma_t\right)\right] Pr\left(\frac{1+\gamma_{mn}}{1+\gamma_{D_m E}} < 2^{\frac{R_S}{1-\alpha}}\right). \tag{28}$$

Proposition 1. *Under Rayleigh fading, the SOP of the link $D_m - D_n$ with AN is given by*

$$SOP_3 = \Phi_1 + (1 - \Phi_1)(1 - \frac{M!}{(M-n)!(n-1)!}\sum_{k=0}^{n-1}C_k^{n-1}(-1)^k(\Phi_2 + \Phi_3)), \quad (29)$$

where

$$\Phi_1 = \begin{cases} \frac{M!}{(M-m)!(m-1)!}\sum_{k=0}^{m-1}(-1)^k C_k^{m-1}\frac{1}{M-m+k+1}\left[1 - e^{-\frac{\gamma_t(M-m+k+1)}{(b_4-b_3\gamma_t)\lambda_{SDm}}}\right], & \gamma_t < \frac{a_n}{a_m}, \\ 1, & \gamma_t > \frac{a_n}{a_m}, \end{cases}$$

$$\Phi_2 = c_1 c_2 c_4 \lambda_{DmE}\lambda_{SE}\lambda_{mn}e^{-\frac{2^{\frac{R_S}{(1-\alpha)}}-1}{c_1\lambda_{mn}}}\left[\frac{ce^{\frac{d\mu}{c}}\Gamma(0, \frac{d\mu}{c})}{(ad-bc)^2} - \frac{ce^{\frac{b\mu}{a}}\Gamma(0, \frac{b\mu}{a})}{(ad-bc)^2} + \frac{\mu e^{\frac{b\mu}{a}}\Gamma(-1, \frac{b\mu}{a})}{a(ad-bc)^2}\right],$$

$$\Phi_3 = c_1\lambda_{mn}e^{-\frac{2^{R_s}-1}{c_1\lambda_{mn}}}\left[\frac{1}{ad-bc}e^{\frac{b}{a}\mu}\Gamma(0, \frac{b}{a}\mu) - \frac{1}{ad-bc}e^{\frac{d}{c}\mu}\Gamma(0, \frac{d}{c}\mu)\right].$$

Denoted that, $a = c_4\lambda_{SE}, b = c_2\lambda_{DmE}, c = c_3\lambda_{SDn}2^{R_s/(1-\alpha)}, d = c_3\lambda_{SDn} (2^{R_s/(1-\alpha)} - 1) + c_1\lambda_{mn}(M-n+k+1), \mu = \frac{2^{R_s/(1-\alpha)}}{c_1\lambda_{mn}} + \frac{1}{c_2\lambda_{DmE}}.$

Proof. See Appendix A.

4   Numerical Results and Discussion

In this section, we investigate the physical layer secrecy performance of ANCO-TRAS protocol for this considered cooperative NOMA system by numerical results. Monte-Carlo simulation results are also provides to verify our analytical results. In this simulation, it is assumed that the power allocation coefficients of NOMA are $a_m = 0.3, a_n = 0.7$. The targeted data rates for the selected NOMA user pair are assumed to be $R_s = 0.5$ bit per channel use.

4.1   Secrecy Outage Probability at S

Figures 2 and 3 describe the result of SOP of the system at S. These figures show then when we crease $d_{SE}, \bar{\gamma}$ then PLS at the S decreases. This mean that the PLS capability of this model increases. Besides, in the two results, when we increase $\bar{\gamma}_E$, the SOP at S of the system also increases.

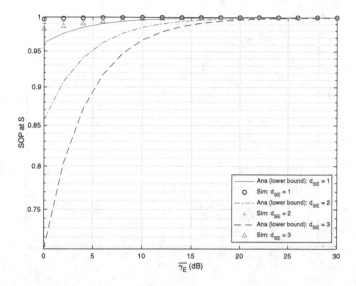

Fig. 2. The SOP at the Base Station versus $\overline{\gamma_E}$ with the change of d_{SE} and $d_{SDm} = 1$, $d_{SDn} = 2$, $\alpha = 0.3$, $M = 4$, $m = 2$, $n = 3$

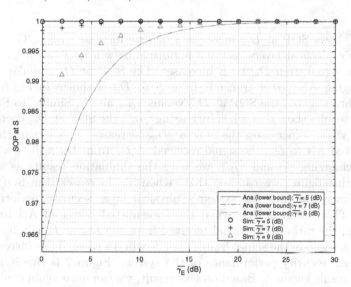

Fig. 3. The SOP at the Base Station versus $\overline{\gamma}_E$ with the change of $\overline{\gamma}$ and $d_{SDm} = 1$, $d_{SDn} = 2$, $\alpha = 0.3$, $M = 4$, $m = 2$, $n = 3$

4.2 Secrecy Outage Probability at D_m

Figure 4, shows the results of the SOP at D_m versus $\gamma_{\bar{D}m}$. In this simulation result, we can see that when the $\gamma_{\bar{D}m}$ increases, the SOP at D_m decreases, meaning that the physical layers performance increases. This simulation

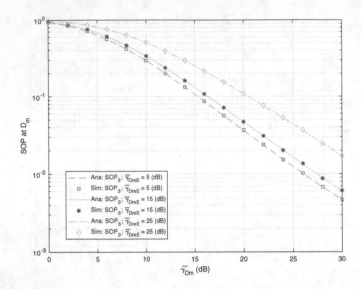

Fig. 4. The SOP at the Base Station versus $\gamma_{\bar{D}m}$ with the change of $\gamma_{\bar{D}mE}$ and $d_{SDm} = 1, d_{SDn} = 2, \alpha = 0.3, M = 4, m = 2, n = 3$

result also shows SOP at D_m increases when we increase $\gamma_{\bar{D}mE}$. That is when the system's physical layers security performance will decrease if the average SNR of the signal from D_m to E increases. The SOP at D_m simulation result when changing AN power transmit from S to D_n is shown in Fig. 5. In this figure, we investigated the SOP at D_m versus $\gamma_{\bar{D}m}$ and $\bar{\gamma}$. Similar to Fig. 4, the simulation results show that when increasing $\gamma_{\bar{D}m}$, the SOP at D_m decreases. In particular, when $\bar{\gamma}$ increases, the SOP at D_m increases. This means that when S transmitted AN to D_n, the signal received at D_n from D_m will be affected.

When changing $\bar{\gamma}_E$ and $\gamma_{\bar{D}m}$, we have the simulation results as shown in Fig. 6. In this figure, we can see that, when $\bar{\gamma}_E$ increases, the SOP at D_m decreases, meaning that the system's physical layer security performance is improved. The reason is that when the transmitted power of AN from S to E increases, the ratio of the signal received from the D_m at E will decrease, so the ability to decode the signal from D_n of E will decrease. Thus increasing the physical layer security performance of the system. Figure 7 is the SOP at D_m simulation result versus $\bar{\gamma}$. Based on this result, we can once again confirm that using AN will help to improve the system's physical layer security performance. Specifically, when $\bar{\gamma}_E$ increases, the SOP at D_m decreases. However, this result also shows that, when $\bar{\gamma}$ increases, the SOP at D_m will increase.

The system has many users, Fig. 8 is the SOP at D_m simulation result when there is a change in the number of users. Based on the simulation results, we see that when the number of users increases, the SOP at D_m will decrease, meaning that the system's physical layer security performance increases.

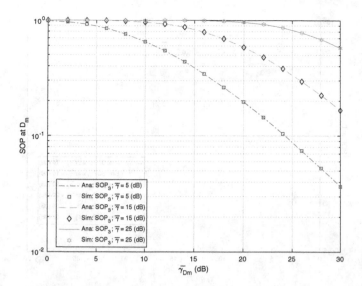

Fig. 5. The *SOP* at the Base Station versus $\gamma_{\bar{D}m}$ with the change of $\bar{\gamma}$ and $d_{SDm} = 1, d_{SDn} = 2, \alpha = 0.3, M = 4, m = 2, n = 3$

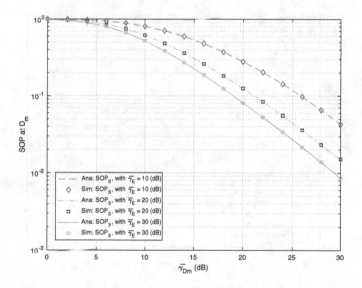

Fig. 6. The *SOP* at the Base Station versus $\gamma_{\bar{D}m}$ with the change of $\bar{\gamma}_{\bar{E}}$ and $d_{SDm} = 1, d_{SDn} = 2, \alpha = 0.3, M = 4, m = 2, n = 3$

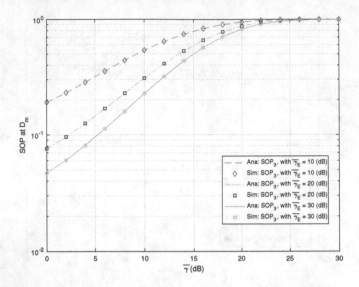

Fig. 7. The SOP at the Base Station versus $\bar{\gamma}$ with the change of $\bar{\gamma}_E$ and $d_{SDm} = 1, d_{SDn} = 2, \alpha = 0.3, M = 4, m = 2, n = 3$

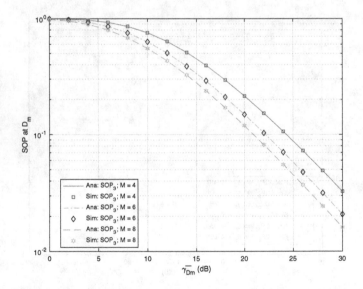

Fig. 8. The SOP at the Base Station versus $\gamma_{\bar{D}m}$ with the change of M and $d_{SDm} = 1, d_{SDn} = 2, \alpha = 0.3, m = 2, n = 3$

5 Conclusion

In this paper, the secrecy performance of (NOMA) in a downlink cooperative network with simultaneous wireless information and artificial noise was examined. Specifically, we used a method to broadcast artificial noise from the base station after signaling to prevent devices from eavesdropping on exchanged messages. In addition, new analytical expressions were derived in terms of the secrecy outage probability to determine the system secrecy performance. Meanwhile, the numerical results were presented to validate the analyses. Based on the analyses and simulations, we can conclude that the security performance of the network model proposed in this paper depends on the average transmit SNR: from (1) base station to better user; worse users; eavesdropping device; (2) from good users to bad users, eavesdropping devices and (3) number of users of the system. Using artificial noise can improve the security performance of the system.

Appendix A

Here, we derive the expression of SOP at D_m in the case of with AN.

$$
\begin{aligned}
SOP_3 &= \Pr\left(\frac{b_4 X_1}{b_3 X_1 + 1} < \gamma_t\right) + \left(1 - \Pr\left(\frac{b_4 X_1}{b_3 X_1 + 1} < \gamma_t\right)\right) \Pr\left(\frac{1 + \gamma_{mn}}{1 + \gamma_{D_m E}} < 2^{R_S/(1-\alpha)}\right) \\
&= \Phi_1 + (1 - \Phi_1) \int_0^\infty F_{U_1}(2^{R_S/(1-\alpha)}(1+y) - 1) f_{U_2}(y) dy \\
&= \Phi_1 + (1 - \Phi_1)\left[1 - \frac{M!}{(M-n)!(n-1)!} \sum_{k=0}^{n-1} C_k^{n-1}(-1)^k \right. \\
&\quad \times \int_0^\infty \frac{c_1 \lambda_{mn}}{c_3 \lambda_{SDn}(2^{R_S/(1-\alpha)}(1+y) - 1) + c_1 \lambda_{mn}(M-n+k+1)} e^{-\frac{(2^{R_S/(1-\alpha)}(1+y)-1)}{c_1 \lambda_{mn}}} \\
&\quad \left. \times \left(\frac{c_2 c_4 \lambda_{D_m E} \lambda_{SE}}{(c_4 \lambda_{SE} y + c_2 \lambda_{D_m E})^2} + \frac{1}{(c_4 \lambda_{SE} y + c_2 \lambda_{D_m E})}\right) e^{-\frac{y}{c_2 \lambda_{D_m E}}} dy \right] \\
&= \Phi_1 + (1 - \Phi_1)(1 - \frac{M!}{(M-n)!(n-1)!} \sum_{k=0}^{n-1} C_k^{n-1}(-1)^k (\Phi_2 + \Phi_3)),
\end{aligned}
\tag{30}
$$

where γ_t is the threshold to detect s_n and Φ_1, Φ_2, Φ_3 are calculated as follows

$$
\begin{aligned}
\Phi_1 &= F_{X_1}\left(\frac{\gamma_t}{b_4 - b_3 \gamma_t}\right) \\
&= \begin{cases} \frac{M!}{(M-m)!(m-1)!} \sum_{k=0}^{m-1} (-1)^k C_k^{m-1} \frac{1}{M-m+k+1}\left[1 - e^{-\frac{\gamma_t(M-m+k+1)}{(b_4 - b_3 \gamma_t)\lambda_{SDm}}}\right], & \gamma_t < \frac{a_n}{a_m} \\ 1, & \gamma_t > \frac{a_n}{a_m} \end{cases}
\end{aligned}
$$

$$\Phi_2 = \int_0^\infty \frac{c_1\lambda_{mn}}{c_3\lambda_{SDn}(2^{R_S/(1-\alpha)}(1+y)-1)+c_1\lambda_{mn}(M-n+k+1)} e^{-\frac{(2^{R_S/(1-\alpha)}(1+y)-1)}{c_1\lambda_{mn}}}$$

$$\times \frac{c_2 c_4 \lambda_{DmE}\lambda_{SE}}{(c_4\lambda_{SE}y + c_2\lambda_{DmE})^2} e^{-\frac{y}{c_2\lambda_{DmE}}} dy$$

$$= c_1\lambda_{mn}c_2c_4\lambda_{DmE}\lambda_{SE} e^{-\frac{2^{R_S/(1-\alpha)}-1}{c_1\lambda_{mn}}} \int_0^\infty \frac{1}{(ay+b)^2}\frac{1}{cy+d} e^{-\left(\frac{1}{c_2\lambda_{DmE}}+\frac{2^{R_S/(1-\alpha)}}{c_1\lambda_{mn}}\right)y} dy$$

$$= c_1\lambda_{mn}c_2c_4\lambda_{DmE}\lambda_{SE} e^{-\frac{2^{R_s}-1}{c_1\lambda_{mn}}}\left[\int_0^\infty \frac{E_1}{cy+d}e^{-\mu y}dy + \int_0^\infty \frac{E_2}{ay+b}e^{-\mu y}dy\right.$$

$$\left.+ \int_0^\infty \frac{E_3}{(ay+b)^2}e^{-\mu y}dy\right]$$

$$= c_1\lambda_{mn}c_2c_4\lambda_{DmE}\lambda_{SE} e^{-\frac{2^{R_s}-1}{c_1\lambda_{mn}}}\left[\frac{E_1}{c}e^{\frac{d}{c}\mu}\Gamma(0,\frac{d}{c}\mu) + \frac{E_2}{a}e^{\frac{b}{a}\mu}\Gamma(0,\frac{b}{a}\mu)\right.$$

$$\left.+ \frac{E_3}{a^2}\mu e^{\frac{b}{a}\mu}\Gamma(-1,\frac{b}{a}\mu)\right]. \tag{31}$$

$$\Phi_3 = \int_0^\infty \frac{c_1\lambda_{mn}}{c_3\lambda_{SDn}(2^{R_s}(1+y)-1)+c_1\lambda_{mn}} e^{-\frac{(2^{R_s}(1+y)-1)}{c_1\lambda_{mn}}}$$

$$\times \frac{1}{(c_4\lambda_{SE}y + c_2\lambda_{DmE})} e^{-\frac{y}{c_2\lambda_{DmE}}} dy$$

$$= c_1\lambda_{mn} e^{-\frac{2^{R_s}-1}{c_1\lambda_{mn}}} \int_0^\infty \frac{1}{ay+b}\frac{1}{cy+d} e^{-\left(\frac{1}{c_2\lambda_{DmE}}+\frac{2^{R_s}}{c_1\lambda_{mn}}\right)y} dy$$

$$= c_1\lambda_{mn} e^{-\frac{2^{R_s}-1}{c_1\lambda_{mn}}}\left[\int_0^\infty \frac{F_1}{cy+d}e^{-\mu y}dy + \int_0^\infty \frac{F_2}{ay+b}e^{-\mu y}dy\right]$$

$$= c_1\lambda_{mn} e^{-\frac{2^{R_s}-1}{c_1\lambda_{mn}}}\left[\frac{F_1}{c}e^{\frac{d}{c}\mu}\Gamma(0,\frac{d}{c}\mu) + \frac{F_2}{a}e^{\frac{b}{a}\mu}\Gamma(0,\frac{b}{a}\mu)\right]$$

Denoted that $a = c_4\lambda_{SE}, b = c_2\lambda_{DmE}, c = c_3\lambda_{SDn}2^{R_S/(1-\alpha)}, d = c_3\lambda_{SDn}$ $(2^{R_S/(1-\alpha)}-1)+c_1\lambda_{mn}(M-n+k+1), \mu = \frac{2^{R_S/(1-\alpha)}}{c_1\lambda_{mn}} + \frac{1}{c_2\lambda_{DmE}}, E_1 = \frac{c^2}{(ad-bc)^2}, E_2 = -\frac{ac}{(ad-bc)^2}, E_3 = \frac{a}{(ad-bc)}, F_1 = -\frac{c}{ad-bc}, F_2 = \frac{a}{ad-bc}.$

Substituting Φ_1, Φ_2, Φ_3 into (30), we obtain the closed-form expression of SOP for the link $D_m - D_n$ in the case of using AN. This concludes the proof.

References

1. Ding, Z., Yang, Z., Fan, P., Poor, H.V.: On the performance of non-orthogonal multiple access in 5G systems with randomly deployed users. IEEE Signal Process. Lett. **21**(12), 1501–1505 (2014)
2. Dai, L., Wang, B., Yuan, Y., Han, S., Chih-Lin, I., Wang, Z.: Nonorthogonal multiple access for 5G: solutions, challenges, opportunities, and future research trends. IEEE Commun. Mag. **53**(9), 74–81 (2015)
3. Shimojo, T., Umesh, A., Fujishima, D., Minokuchi, A.: Special articles on 5G technologies toward 2020 deployment. NTT DOCOMO Tech. J. **17**(4), 50–59 (2016)

4. Islam, S.M.R., Avazov, N., Dobre, O.A., Kwak, K.S.: Power-domain non-orthogonal multiple access (NOMA) in 5G systems: potentials and challenges. IEEE Commun. Surv. Tutorials **19**(2), 721–742 (2017)
5. Lee, S., Duong, T.Q., da Costa, D.B., Ha, D.B., Nguyen, S.Q.: Underlay cognitive radio networks with cooperative non-orthogonal multiple access. IET Commun. **12**(3), 359–366 (2018)
6. Suraweera, H.A., Karagiannidis, G.K., Smith, P.J.: Performance analysis of the dual-hop asymmetric fading channel. IEEE Trans. Wirel. Commun. **8**(6), 2783–2788 (2009)
7. Kim, K.J., Duong, T.Q., Poor, H.V.: Performance analysis of cyclic prefixed single-carrier cognitive amplify-and-forward relay systems. IEEE Trans. Wirel. Commun. **12**(1), 195–205 (2013)
8. Ding, Z., Peng, M., Poor, H.V.: Cooperative non-orthogonal multiple access in 5G systems. IEEE Commun. Lett. **19**(8), 1462–1465 (2015)
9. Kim, J.B., Song, M.S., Lee, I.H.: Achievable rate of best relay selection for non-orthogonal multiple access-based cooperative relaying systems. In: International Conference on Information and Communication Technology Convergence (ICTC), Jeju, pp. 960–962. IEEE, South Korea (2016)
10. Liu, Y., Ding, Z., Elkashlan, M., Poor, H.V.: Cooperative nonorthogonal multiple access with simultaneous wireless information and power transfer. IEEE J. Sel. Areas Commun. **34**(4), 936–953 (2016)
11. Ha, D.B., Nguyen, Q.S.: Outage performance of energy harvesting DF relaying NOMA networks. Mob. Netw. Appl. **23**(6), 1572–1585 (2017)
12. Tran, D.D., Tran, H.V., Ha, D.B., Kaddoum, G.: Cooperation in NOMA networks under limited user-to-user communications: solution and analysis. In: IEEE Wireless Communications and Networking Conference (WCNC), 15–18 April 2018, Barcelona, Spain (2018)
13. Ha, D.B., Duong, T.Q., Tran, D.D., Zepernick, H.J., Vu, T.T.: Physical layer secrecy performance over Rayleigh/Rician fading channels. In: The 2014 International Conference on Advanced Technologies for Communications (ATC 2014), 15–17 October 2014, pp. 113–118, Hanoi, Vietnam (2013)
14. Wyner, A.: The wire-tap channel. Bell Syst. Tech. J. **54**(8), 1355–1387 (1975)
15. Bloch, M., Barros, J., Rodrigues, M.R., McLaughlin, S.W.: Wireless information-theoretic security. IEEE Trans. Inf. Tech. **54**(6), 2515–2534 (2008)
16. Ng, D.W.K., Schober, R.: Resource allocation for secure communication in systems with wireless information and power transfer. In: IEEE Globecom Workshops, Atlanta, USA, pp. 1251–1257 (2013)
17. Ha, D.B., Van, P.T., Vu, T.T.: Physical layer secrecy performance analysis over Rayleigh/Nakagami fading channels. In: The World Congress on Engineering and Computer Science 2014 (WCECS2014), 22–24 October 2014, San Francisco, USA (2014)
18. Ha, D.B., Vu, T.T., Duy, T.T., Bao, V.N.Q.: Secure cognitive reactive decode-and-forward relay networks: with and without eavesdropper. Wirel. Pers. Commun. (WPC) **85**(4), 2619–2641 (2015)
19. Tran, D.D., Ha, D.B.: Secrecy performance analysis of QoS-based non-orthogonal multiple access networks over Nakagami-m fading. In: The International Conference on Recent Advances in Signal Processing, Telecommunications and Computing (SigTelCom), HCMC, Vietnam (2018)
20. Lei, H., Zhang, J., Park, K.H., Xu, P., Ansari, I.S., Pan, G.: On secure noma systems with transmit antenna selection schemes. IEEE Access **5**, 17450–17464 (2017)

21. Liu, Y., Qin, Z., Elkashlan, M., Gao, Y., Hanzo, L.: Enhancing the physical layer security of nonorthogonal multiple access in large-scale networks. IEEE Trans. Wirel. Commun. **16**(3), 1656–1672 (2017)
22. Lv, L., Ding, Z., Ni, Q., Chen, J.: Secure MISO-NOMA transmission with artificial noise. IEEE Trans. Veh. Technol. **67**(7), 6700–6705 (2018)
23. Chen, J., Yang, L., Alouini, M.S.: Physical layer security for cooperative NOMA systems. IEEE Trans. Veh. Technol. **67**(5), 4645–4649 (2018)
24. Men, J., Ge, J.: Performance analysic of non-orthogonal multiple access in downlink cooperative network. IET Commun. **9**(18), 2267–2273 (2015)

Performance Analysis of Software Defined Network Concepts in Networked Embedded Systems

Bach Tran, Mohamed Elamin, Nghi Tran$^{(\boxtimes)}$, and Shivakumar Sastry

Department of Electrical and Computer Engineering, University of Akron,
Akron, OH 44325-3904, USA
{bxt1,mae72}@zips.uakron.edu, {nghi.tran,ssastry}@uakron.edu

Abstract. We present a novel approach to incorporating Software-Defined Networking (SDN) in networked embedded systems where the nodes have limited capabilities and the wireless links have limited bandwidth. The SDN controller was implemented on a Beagle Black board. The approach was validated through simulations and experiments on a physical testbed.

Keywords: Wireless-SDN · Networked embedded system · Control overhead · Performance analysis

1 Introduction

Networking and connectivity is an essential infrastructure for our economies, industries, societies and systems and have a central role in managing costs and productivity [1], and our health and wellness [2]. Modern networks confront a few challenges such as complexity and operational costs as the size of the applications increase [3]. The introduction of low-power, resource constrained nodes that are cost-effective for data gathering has pushed the networking issues to their limits because of the tight bandwidth available and the unreliability of the links. Such systems are asynchronous, decentralized and severely limited, and yet they must provide reliable and trustworthy service in a variety of application settings.

SDN helps to address some of these drawbacks by decoupling the control and data planes. Network intelligence and state are logically centralized and the underlying network infrastructure is abstracted from the applications [4]. Users are able to programmatically change the capabilities of the underlying network with minimal disruptions to the applications that rely on the networking infrastructure. A new abstraction layer is added between applications and router/switches. SDN was primarily designed for wired networks with point-to-point (P2P) connections between their nodes [3].

Many emerging networked systems such as Wireless Sensor Networks (WSN) [5,6], Internet of Things [7] and Advanced Manufacturing [8–10] rely on wireless

T. Q. Duong et al. (Eds.): INISCOM 2019, LNICST 293, pp. 99–108, 2019.
https://doi.org/10.1007/978-3-030-30149-1_8

links and have motivated the exploration of SDN services to programmatically alter the capabilities of networked embedded systems. Several challenges must be addressed to achieve this vision. The function of nodes in networked embedded systems is more than just sensing. Although the node has limited resources and computation power, it can still drive basic operations and control heavy machinery. Unlike the fully capable desktop systems used in common networking scenarios, the nodes have limited resources. The radio transceivers used in such systems have limited capabilities and, typically, infrastructure such as cell towers and WiFi are not available. The systems are expected to be deployed in ad-hoc manners and must support incremental growth and change. While the communications requirements in these systems usually follow a set of paths, there would be changes from time to time that require programmatic changes in the network capabilities.

In this paper, we present an approach for adapting key SDN concepts in wireless networked embedded systems. Starting with a well-known SDN controller called SDN-Wise [11] that was recently developed for the Raspberry PI, the controller was adapted to address the needs of networked embedded systems. A finite state machine was designed to execute on commercial mote platforms. The approach was validated both in simulation settings and through physical experiments.

The remainder of this paper is organized as follows: Sect. 2 reviews background in SDN and WSN. Section 3 describes the system's structure and role. Experimental results that demonstrate our system are discussed in Sect. 4, and conclusions are presented in Sect. 5.

2 Background

The section presents some of the approaches used to design and analyze SDN. Some of the key concepts of WSN and protocols proposed are reviewed.

Currently, networks support complex tasks using low-level network configuration commands. The complexity of configuring the systems and managing networking operations increase both as the number of nodes increase and as the communication demand increases. Such low-level commands are not suitable for continually changing environments. Moreover, system reconfiguration is done manually and on a per-case basis and this situation may lead to inconsistency in the network state [12]. Many of these issues can be addressed by using a new architecture that accounts for dynamic changes in data, computation and storage [13].

The key idea in SDN is to separate or decouple the control plane and the data plane. This idea was introduced in other systems and protocols. For example, in [14], the authors introduced a system that mimics forwarding decisions, made by regular hardware-based switches, in software and applied these rules for new packets. Their goal was to increase the system's functionality space without changing the underlying hardware. Thus, they made the network improvements cycle much faster. The 4D environment described in [15] divided the process into four planes: decision, dissemination, discovery and data. While all the decisions are made in the decisions plane using an autonomous system and decision elements, the data plane was responsible for only forwarding packets depending on the decision plane's output. That resulted in a system where controlling and forwarding are prevented from running on the same network element and gave an extra level of abstraction and simplicity. Several efforts to standardize the core networking ideas have helped to mature this idea of separating the control and data planes [16–18].

The controller is the central artifact of the control plane. This plane is logically centralized and is essential to differentiate between the operations needed to manage networking capabilities and the operations necessary to forward data packets between the nodes.

A single, centralized, controller is simple to design and implement. Such an architecture reduces the chances for logical inconsistencies. However, such a node has a high risk and is a single-point of failure. It suffers from scalability limitations because when the network starts growing, the computation power of the controller must grow at a higher rate to ensure the network's performance [19]. In an effort to target the drawbacks of centralized controllers the authors in [20] introduced the idea of decentralizing SDN's control plane both physically and logically by defining the concepts of a control hierarchy which consisted of main and secondary controllers. This helped the network to distribute the load among different physical controllers thus avoiding network's bottlenecks; and reduce the cost of needing high computation power. Many other architectures were proposed [21–23]. Nevertheless, these techniques increase the difficulty of maintaining a consistent network view and network decisions.

The main role of a flow-table is to represent the multi-hop connectivity structure between nodes. The flow-table has all the characteristics of a routing-table but there are two major differences. First, flow-tables are only updated with information coming from the controller. Second, while routing-tables are generally stateless, a flow-table can be designed to make decisions based on its state and change specific attributes in the packet.

The configuration commands that the controller sends are used to update flow-tables inside each SDN node. Each table contains *flow entries*, each with three sections. The *Matching* section has the rules and attributes that must be compared and checked for each incoming packet. The *Action* section specifies what must be done for each matched packet. The *Statistical Information* section gathers information about the flow-table entry [11,24]. Generally, a packet can match more than one rule and the relevant actions will determine its flow through the system. Unlike simulated wireless links, the physical wireless links present many challenges that affect the performance of the network. Although finding the shortest path is preferable in most cases, the throughput of the network is affected by traffic congestion in specific nodes, limited capabilities in the node, and the effects of hidden/exposed terminals. Thus, we may increase the network's throughput by choosing other paths in the network [25].

Many estimation techniques were proposed to target specific features such as rapid calculation, memory efficiency and performance by using information from the physical layer, link layer or the network layer. The physical layer can indicate the channel quality and the errors in the packets, but it will not take into consideration lost packets. The link layer can measure the delivery of the sent packets through acknowledgments. The network layer can indicate the importance of each link and how to share node's estimations with other nodes [26]. The authors in [25] introduced an estimator that depends on the link layer's information. They would calculate the number of transmissions required for a packet to successfully be sent plus the number acknowledgments received and then calculate the delivery ratio for each link. The HyperLQI [27] uses the success rate maintained by link layer acknowledgments in addition to the channel Link Quality Indicator (LQI), from the physical layer to give an estimation of each link. A link quality estimator that uses information from all the three networking layers is presented in [26].

3   System Structure

This section presents the versatile design of the physical Wireless-SDN node that can support the deployment of multiple network topologies, such as tree, ring and mesh. The CSMA/CA protocol B-MAC [28] was used in all the nodes. The IRIS from MEMSIC motes used as the platform for experiments consist of an ATmel Atmega1281, a low-power 8-bit microcontroller, with 128KB in-system programmable flash memory and 8KB of RAM. Each node includes an ATmel AT86RF230 wireless transceiver, which is a 2.4 GHz ZigBee IEEE 802.15.4 compliant module. The RF module has a maximum data rate of 250Kbps, an automatic reception acknowledgment feature and power detection [29].

SINK Node. The SINK node, which is connected to the controller through a serial (USB) connection, serves as the interface between the control and data planes. All the packets from every node to the controller, and vice versa, must go through the SINK. The SINK reorganizes those packets here so that the controller can handle. In addition, this node is responsible for initiating the topology formation. When the SINK powers-up, it send a PROXY packet to the controller to introduce its self as a valid and ready SINK node. After that, the SINK will start the topology discovery protocol by propagating BEACON packets periodically.

Beaconing and Topology Discovery. Every node must maintain a valid route to the SINK and, thus, the controller. The BEACON packets and the topology discovery protocol are designed to help in this task. Except for the SINK, every node will be powered up to idle state, where it listen to the channel waiting for a BEACON packet to be broadcast by any other neighbor.

Routes to the SINK. The BEACON packets contain the sender's distance from the SINK and its battery level. On the first receiving of this packet, a node will change its state to ACTIVE, register the sender as next hop to the SINK and start all the timers within it. When an ACTIVE node receives a BEACON packet, it will compare its distance to the SINK with the sender's distance and measures the quality of that link using the estimation. If both appear to be better, the node will update its route to the SINK based on the new one, otherwise it will just update its neighbor-table with this sender.

Neighbor-Table. Each node maintains a set of single hop's neighbors that can be directly reached, and keeps track of each neighbor in terms of TTLs and link quality. This process is memory intensive and must be managed carefully. A large number of neighbors will result in a strongly connected graph with short mean distance to each destination, but it will affect the performance of the nodes, increase search time, and cause a larger number of REPORT packets to be generated. On the other hand, a smaller neighborhood will result in longer paths and a higher chance of the node being unreachable.

Reports to Controller. The REPORT packets periodically inform the controller about the neighborhood of each node in the system.These packets contain a list of all the node's neighbors, available battery levels, distance from SINK and quality of each link.

Controller. The controller logic used in our experiments is studied in [11]. This controller uses the weighted version of Dijkstra algorithm to find the best route between two points using both the number of hops and link quality. It maintains a map of the network depending on both the REPORT packets received and an internal expiration timer. The system takes into account that not every link is a bidirectional one unless it was reported by both nodes that are at either end of the link. When a controller receives a REQUEST packet from the SINK, it will extract the source and destination information. Based on the latest map it executes the Dijkstra algorithm to find the shortest path between the nodes. The controller will configure all the nodes in the path by generating a single OpenPath packet that contains all their address to the final destination.

Time to Live and Flow-Table Entries. Because of the limited memory space and the goal of reducing searching time, each node will keep track of a certain number of the most active neighbors. The validation time associated with each entry in the Flow-table and neighbor-table is updated in every TTL's period. If the initial value is very large, the node will exhaust its resources on non-useful entries and will keep track of non-existent neighbors. On the other hand, if this value is too small, the network will suffer from a huge overhead of sent REQUEST packets and the node will not have a stable neighbor list. When the TTL timer triggers, the node will go through the Flow-table and the neighbor-table and remove all entries with TTL values less than the TTL period. A neighbor's TTL is reset when the node hears a packet generated from that specific neighbor, and a flow-table entry's TTL is reset only if that entry is used.

Wireless Link Estimator. In our topologies, we used a link quality estimator that collects information from physical and link layers, depending on both broadcast and unicast transmissions to give the overall evaluation. Each estimate is updated after a predefined number of packets. The node will evaluate the broadcast packets by detecting their energy on reception and comparing it to a threshold value. Unicast messages were measured by counting the number of sent and acknowledged packets on every link to calculating the success and failure rates. To avoid fast changing environment, we assigned weights to the new estimate New and the previous estimate $Estimation_{old}$ using the exponential moving average window as:

$$Estimation_{final} = \alpha New + (1 - \alpha)Estimation_{old}.$$

here α denotes the estimation factor. This average estimates are calculated separately for broadcast and unicast transmissions as follows:

- Broadcast
    1. After receiving a specified number of broadcasts from a source, the node calculates an immediate estimate.
    2. This value is combined with the previous broadcast-estimate values.
    3. At the end of broadcast estimation we obtain $Estimation_{final}$.
- Unicast
    1. After sending the specified number of unicast messages over a specific link, the node will keep track of the number of acknowledged and unacknowledged packets.
    2. The success rate is then calculated.
    3. The success rate will be used as the new estimate to find $Estimation_{final}$.

The estimator keeps track of a number of links that are not part of the neighbor-table. These are important because they will serve as a base to compare neighbor's links. At the end of each TTL cycle, the node will check all the available links in the table to figure out the best value and all the links of the neighbor will be compared to that value. In this way, the node only keeps the most reliable links in the neighbor-table.

4   Results

This section presents the simulation and experiment results obtained from the network with a collection of motes. The observed behavior of the network provides understanding on which the system should be designed in order to minimize the interaction between the control and data planes.

Simulation Approach. Two network topologies were designed to understand the baseline behavior of the Wireless-SDN network, one is a 16 node mesh and the other is a 14 node tree network, both with a single SINK labeled 1.0.1. We use the mesh topology to examine the effect of the time-to-live parameter and

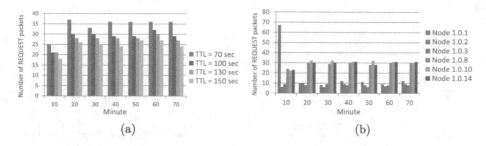

Fig. 1. Average number of REQUEST packets generated by an edge node in a mesh topology (a) and those generated by different nodes (intermediate and leaf nodes) in a tree topology (b).

Table 1. Major differences between simulated systems and physical systems.

	Simulation environment	Physical system
Links	Static	Dynamic
Link quality	Fixed	Changing
Transmission	P2P	Inherently broadcast
Memory	Virtually unlimited	Limited
Processing time	Negligible	Significant

the tree topology to examine the control overhead of a node with respect to its position. Figure 1a shows the average number of REQUEST packets generated from the edge node in a mesh topology for every 10 min. With lower value of TTL, each flow entry has a shorter time until it is removed from the flow table, therefore the node then need to send REQUEST packet to the SDN controller more frequently in order to update the expired flow entry. Figure 1b shows the difference between the number of REQUEST packets generated from different nodes in the tree topology for every 10 min. Node 1.0.2 and 1.0.3, which are the immediate child nodes of the root, have to generate less REQUEST packets than the leave node 1.0.8, 1.0.10, and 1.0.14. The main reason is that the intermediate nodes can update their flow-tables without generating additional REQUEST packets based on the OpenPath generated by the controller.

Experiments With IRIS Motes. The physical implementation of a Wireless-SDN using commercially available mote platforms presents a few additional challenges in addition to what can be realized in simulation environments, which are summarized in Table 1. Figure 2a clearly shows that the network goes through different phases. The highlighted section in this figure corresponds to the formation phase where the REQUEST packets are rapidly generated. A steady level of reporting will allow the SDN controller to beresponsive to new REQUEST packets. After the flow-tables are configured, the number of REQUEST and REPORT packets remains stable during the lifetime of the network. A compar-

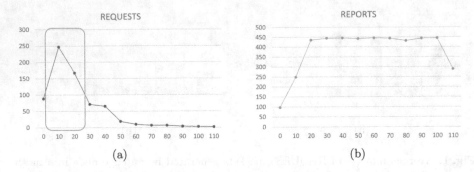

Fig. 2. The overhead in an SDN network comprises of REQUEST packets and REPORT packets.

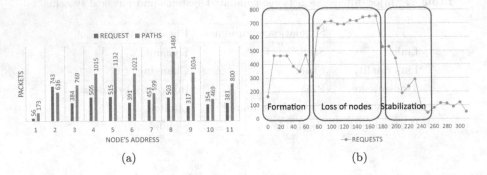

Fig. 3. A comparison between active and passive learning in topology discovery on the left (a) and three phases of a large network on the right (b).

ison between the number of REQUEST packets generated and the OpenPath packets received is presented Fig. 3a. Notice that the nodes that receive more OpenPath packets tend to generate fewer REQUEST packets. This result indicates the importance of node's position in the topology, i.e., the number of paths on which the node lies. It give the designers a suggestion of critical point of failure and load balancing algorithms. The system was tested in a larger network with 18-nodes and high packet generation rates, and its behavior is shown in Fig. 3b. The first section in this figure presents the number of REQUEST packets during the formation phase. In the second section, the network suffered from the loss of four nodes that led to the increase in Control overhead, and in the final one the network recovered as the depleted nodes re-connected to the topology.

5 Conclusion

This paper presented the design and implementation of a Wireless-Software Defined Network that is suitable for a variety of networked embedded systems. The design was validated in a simulation setting and through experiments using commercial motes. The system model can be adapted to different hardware platforms because the design relies on common features that are available in a variety of platforms. This paper presented the general behavior and key parameters to analyze the network performance in typical operational scenarios. Insights into these features allow us to have more control on the flow of packets through a network both at the system level and at the level of individual nodes. The results are encouraging and serve as a foundation for several related investigations in the future.

Acknowledgement. This material is based upon work supported by the US National Science Foundation (NSF) under Grant 1531070. Any opinions, findings, and conclusions or recommendations expressed in this material are those of the authors and do not necessarily reflect the views of the NSF.

References

1. Goel, R., Mehta, A., Dua, S.: Next generation networks: regulation and interconnection in modern communications era. In: Networks 2006, 12th International Telecommunications Network Strategy and Planning Symposium, pp. 1–5, November 2006
2. Lin, Y.-J., Lai, J.-S., Wu, Y.-J., Ou, M.-J.: A communication system for living alone elders. In: HEALTHCOM 2006 8th International Conference on e-Health Networking, Applications and Services, pp. 108–113, August 2006
3. Sezer, S., et al.: Are we ready for SDN? implementation challenges for software-defined networks. IEEE Commun. Mag. **51**, 36–43 (2013)
4. ONF, "The new norm for networks," white paper (2012). https://www.opennetworking.org
5. Dargie, W., Poellabauer, C.: Fundamentals of Wireless Sensor Networks: Theory and Practice. Wiley Publishing (2010)
6. Burri, N., von Rickenbach, P., Wattenhofer, R.: Dozer: ultra-low power data gathering in sensor networks. In: 2007 6th International Symposium on Information Processing in Sensor Networks, pp. 450–459, April 2007
7. Atzori, L., Iera, A., Morabito, G.: The Internet of Things: a survey. Comput. Netw. **54**(15), 2787–2805 (2010)
8. Archer, B., Sastry, S., Rowe, A., Rajkumar, R.: Profiling primitives of networked embedded automation. In: 2009 IEEE International Conference on Automation Science and Engineering, pp. 531–536, August 2009
9. Mahamadi, A., Chippa, M., Sastry, S.: Reservation based protocol for resolving priority inversions in composable conveyor systems. J. Syst. Architect. **74**, 14–29 (2017)
10. Hayslip, N., Sastry, S., Gerhardt, J.S,: Networked embedded automation. Assembly Autom. **26**(3), 235–241 (2006)
11. Galluccio, L., Milardo, S., Morabito, G., Palazzo, S.: SDN-wise: design, prototyping and experimentation of a stateful sdn solution for wireless sensor networks. In: 2015 IEEE Conference on Computer Communications (INFOCOM), pp. 513–521, April 2015
12. Kim, H., Feamster, N.: Improving network management with software defined networking. IEEE Commun. Mag. **51**, 114–119 (2013)

13. Metzger, A., Cassales Marquezan, C.: Future Internet apps: the next wave of adaptive service-oriented systems? In: Abramowicz, W., Llorente, I.M., Surridge, M., Zisman, A., Vayssière, J. (eds.) ServiceWave 2011. LNCS, vol. 6994, pp. 230–241. Springer, Heidelberg (2011). https://doi.org/10.1007/978-3-642-24755-2_22
14. Casado, M., Koponen, T., Moon, D., Shenker, S.: Rethinking packet forwarding hardware. In: HotNets, pp. 1–6 (2008)
15. Greenberg, A., et al.: A clean slate 4D approach to network control and management. ACM SIGCOMM Comput. Commun. Rev. **35**(5), 41–54 (2005)
16. Berman, M., et al.: GENI: a federated testbed for innovative network experiments. Comput. Netw. **61**, 5–23 (2014). Special issue on Future Internet Testbeds - Part I
17. ONF Overview. https://www.opennetworking.org/about/onf-overview. Accessed 30 Mar 2017
18. Linux Foundation: Opendaylight - an open source community and meritocracy for software defined networking, A Linux Foundation Collaborative Project
19. Hakiri, A., Gokhale, A., Berthou, P., Schmidt, D.C., Gayraud, T.: Software-defined networking: challenges and research opportunities for future Internet. Comput. Netw. **75**, 453–471 (2014). Part A
20. Santos, M.A.S., Nunes, B.A.A., Obraczka, K., Turletti, T., de Oliveira, B.T., Margi, C.B.: Decentralizing SDN's control plane. In: 39th Annual IEEE Conference on Local Computer Networks, pp. 402–405, September 2014
21. Ng, E., Cai, Z., Cox, A.: Maestro: a system for scalable openflow control, Rice University, Houston, TX, USA, TSEN Maestro-Technical report, TR10-08 (2010)
22. Hassas Yeganeh, S., Ganjali, Y.: Kandoo: a framework for efficient and scalable offloading of control applications. In: Proceedings of the First Workshop on Hot Topics in Software Defined Networks, pp. 19–24. ACM (2012)
23. Tootoonchian, A., Ganjali, Y.: Hyperflow: a distributed control plane for openflow. In: Proceedings of the 2010 Internet Network Management Conference on Research on Enterprise Networking, p. 3 (2010)
24. Openflow® switch specification ver 1.5.1. https://www.opennetworking.org/sdn-resources/technical-library April 2015. Accessed 11 Apr 2017
25. De Couto, D.S., Aguayo, D., Bicket, J., Morris, R.: A high-throughput path metric for multi-hop wireless routing. Wirel. Netw. **11**(4), 419–434 (2005)
26. Fonseca, R., Gnawali, O., Jamieson, K., Levis, P.: Four-bit wireless link estimation. In: HotNets (2007)
27. Gupta, A., Sharma, M., Marot, M., Becker, M.: HybridLQI: hybrid multihopLQI for improving asymmetric links in wireless sensor networks. In: 2010 Sixth Advanced International Conference on Telecommunications, pp. 298–305, May 2010
28. Polastre, J., Hill, J., Culler, D.: Versatile low power media access for wireless sensor networks. In: Proceedings of the 2nd International Conference on Embedded Networked Sensor Systems, pp. 95–107, ACM (2004)
29. Shukri, S., et al.: Analysis of RSSI-based DFL for human detection in indoor environment using IRIS mote. In: 2016 3rd International Conference on Electronic Design (ICED), pp. 216–221, August 2016

System Performance of Relay-Assisted Heterogeneous Vehicular Networks with Unreliable Backhaul over Double-Rayleigh Fading Channels

Cheng Yin[1(✉)], Ciaran Breslin[1], Emiliano Garcia-Palacios[1], and Hien M. Nguyen[2]

[1] Queen's University Belfast, Belfast, UK
{cyin01,cbreslin07}@qub.ac.uk, e.garcia@ee.qub.ac.uk
[2] Duy Tan University, Da Nang, Vietnam
nguyenminhhien2501@gmail.com

Abstract. This paper introduces for the first time a relay-assisted heterogeneous vehicular model including numerous stationary small cells, a mobile relay and a mobile receiver with unreliable backhaul. In this proposed system model, a macro-base station connected to the cloud communicates to numerous small cells through wireless backhaul links. A Bernoulli process is utilized to model the backhaul reliability. A relay using decode-and-forward protocol is considered to help the transmission from the stationary small cells to the mobile receiver. Moreover, at the mobile relay side, a selection combining protocol is applied to maximize the received signal-to-noise ratio. The links between stationary small cells and mobile relay are Rayleigh fading channels, and the link between mobile relay and mobile receiver is double-Rayleigh fading channel. A closed-form expression for outage probability is provided to evaluate the influence of the number of small cells and the backhaul reliability on the proposed system performance.

Keywords: Relay · Double-Rayleigh fading channels · Unreliable backhaul · Outage probability

1 Introduction

With the fast increasing amount of smart devices, future networks will be more dense and heterogeneous [5, 12]. In heterogeneous networks (HetNets), the adoption of backhaul to connect the macro-base station and small cells is gaining interest. The conventional wired backhaul can provide stable transmissions, but

This work was supported in part by the Newton Prize 2017 and by the Newton Fund Institutional Link through the Fly-by Flood Monitoring Project under Grant ID 428328486, which is delivered by the British Council.

T. Q. Duong et al. (Eds.): INISCOM 2019, LNICST 293, pp. 109–116, 2019.
https://doi.org/10.1007/978-3-030-30149-1_9

it costs a lot to deploy and maintain it. Alternatively, wireless backhaul is flexible and cost-effective, but because of non-line of sight (nLOS) and channel fading [6] wireless backhaul can not ensure a transmission as reliable as wired backhaul.

In recent years, the increasing number of vehicles has caused traffic congestion, traffic accidents and resource consumption. Therefore, future vehicular communication networks and vehicular ad-hoc networks (VANETs) which can provide information transmission through vehicle-to-vehicle (V2V) and vehicle-to-infrastructure (V2I) have been widely investigated [16]. Many applications are available to improve road safety and resource management in vehicular networks [1]. Hence, it is essential to study the performance of heterogeneous vehicular network. In vehicular networks, vehicles are mobile so channel models for static objects, such as, Rayleigh, Rician and Nakagami-m are not suitable for vehicular communications. Instead, double-Rayleigh fading channel was introduced and can be utilized for mobile transmission links [2].

In heterogeneous vehicular networks, relays also attract interest because they can help to improve the system coverage and overall system performance [3,11]. In vehicular networks, relays are likely to exist in abundance as vehicles are scattered around a geographical area. There are two well known relay protocols, amplify-and-forward (AF) and decode-and-forward (DF) [10]. Networks with DF relays can achieve better system performance than AF because of the lower interference [14]. The authors in [15] studied the impact of backhaul reliability on vehicular networks, but the benefit of using relays was not investigated. This motivates us to consider exploiting cooperative DF relays in heterogeneous vehicular networks.

In recent research, the influence of wireless backhaul on system performance over Rayleigh fading channels [14] and Nakagami-m fading channels [8] was investigated and the conclusion of this research shows that backhaul reliability is a crucial parameter which can influence the system model significantly [6–9,13,14]. Hence, in our proposed relay-assisted heterogeneous vehicular networks, the backhaul reliability parameter should be well studied.

Notation: For a random variable X, $F_X(\cdot)$ represents the cumulative distribution function (CDF) and $f_X(\cdot)$ represents the corresponding probability density function (PDF). $P[\cdot]$ denotes the probability of occurrence of an event. $\max(\cdot)$ represents the maximum of arguments, and $\min(\cdot)$ represents the minimum of arguments.

2   System Model

We propose a relay-assisted heterogeneous vehicular network including a Macro-Base station (BS) linked to the cloud, K stationary small cells $\{T_1, ... T_k, ... T_K\}$, a mobile DF relay R and a mobile receiver D as shown in Fig. 1. s_k represents the backhaul reliability for small cell T_k and it denotes the probability that the small cell T_k can decode the kth T_k's information successfully from BS through unreliable backhaul. The relay R selects a best small cell T_k with the highest SNR using selection combining protocol. All nodes are assumed to be equipped with single antenna. Because the small cells T_k are stationary and the relay is mobile,

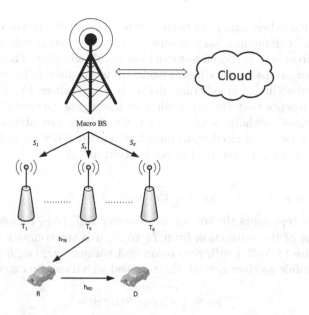

Fig. 1. Relay-assisted vehicular network system model with small cells and unreliable backhaul

we assume that the links from small cells T_k to the mobile relay R follow independent and identically distributed Rayleigh fading channels. Moreover, both the relay and receiver are mobile nodes, so we assume that the channel from the relay R to the destination D is independent and identically distributed double-Rayleigh fading channel. Assume that perfect CSI of the connections from T_k to R and R to D is available.

The channel from T_k to R is Rayleigh fading channel, and the CDF and PDF of Rayleigh fading channels are written as [13]

$$F_X(x) = 1 - \exp(-\lambda x), \tag{1}$$

$$f_X(x) = \lambda \exp(-\lambda x). \tag{2}$$

The channel from R to D is double-Rayleigh fading channel, and the CDF and PDF of double-Rayleigh fading channels are given as [1]

$$F_X(x) = 1 - 2\sqrt{x}\mathcal{K}_1\left(2\sqrt{x}\right), \tag{3}$$

$$f_X(x) = 2\mathcal{K}_0\left(2\sqrt{x}\right). \tag{4}$$

where $Kv(.)$ denotes the modified Bessel function of second kind with order v.

The unreliable backhaul links behave either a successful or failed transmission in a "one shot" communication fashion with no packet retransmission. Therefore, a Bernoulli process $\mathbb{I}_k$ is chosen to model backhaul reliability. The success probability of the transmission via wireless backhaul is s_k where $P(\mathbb{I}_{k^*} = 1) = s_k$, similarly, the probability of failed transmission is $1 - s_k$ where $P(\mathbb{I}_{k^*} = 0) = 1 - s_k$ [6]. This represents that the probability of the signal successfully transmitted through wireless backhaul is s_k, however, the failure probability is $1 - s_k$. x is assumed to be the desired transmitted signal from BS to mobile D. At the mobile relay R side, the received signal is given as

$$y_R = \sqrt{P_T} h_{\mathsf{TR}} x \mathbb{I}_k + n. \tag{5}$$

Where P_T represents the transmission power at T_k, h_{TR} represents the channel coefficient of the connection from T_k to R, n is the complex additive white Gaussian noise (AWGN) with zero mean and variance σ^2, i.e., $n \sim CN(0, \sigma^2)$.

At the mobile receiver side D, the received signal can be expressed as

$$y_D = \sqrt{P_R d_{RD}^{-\beta}} h_{\mathsf{RD}} x + n. \tag{6}$$

where P_R represents the transmit power at R, h_{RD} is the channel coefficient of the link from R to D, d_{RD} represents the distance from R to D and β is the path loss exponent.

Hence, the received SNR at the mobile relay R is given as

$$\gamma_R = \gamma_I |h_{\mathsf{TR}}|^2 \mathbb{I}_k, \tag{7}$$

where $\gamma_I = \frac{P_T}{n}$.

The selection combining protocol [4] is applied at R to select the best T_{k^*} with highest SNR to transmit the signal,

$$k^* = \max_{k=1,\ldots,K} \arg(\gamma_R). \tag{8}$$

Therefore, the received SNR at R with the selected T_{k^*} using selection combining protocol can be rewritten as

$$\gamma_{T_{K^*}R} = \gamma_I |h_{\mathsf{T}_{k^*}\mathsf{R}}|^2 \mathbb{I}_{k^*}, \tag{9}$$

where $|h_{\mathsf{T}_{k^*}\mathsf{R}}|^2$ represents the channel coefficient from the chosen T_{k^*} to R. Similarly, the received SNR at the mobile receiver D can be given as

$$\gamma_D = \frac{\gamma_S |h_{\mathsf{RD}}|^2}{d_{RD}^{\beta}}, \tag{10}$$

where $\gamma_S = \frac{P_R}{n}$.

We assume that the relay R utilizes the DF protocol, so the overall SNR can be derived as

$$\gamma_{T_{K^*}D} = \min(\gamma_{T_{K^*}R}, \gamma_D). \tag{11}$$

3   Outage Probability Analysis

Outage probability is a key performance metric to figure out the system performance and is described as the probability that the instantaneous mutual information rate is below a certain threshold θ. In this section, outage probability is utilized to investigate the proposed relay-assisted vehicular system performance.

Firstly, without considering the backhaul reliability, the CDF of SNR from T_k to R can be derived as,

$$F_{\gamma_T}(x) = P[\gamma_I |h_{TR}|^2 < x]$$
$$= 1 - \exp(-\frac{\lambda x}{\gamma_I}). \tag{12}$$

We now take into account the unreliable backhaul. We assume that success probability s for each link from BS to T_k i.e., $s_k = s$, $\forall k$. The PDF of γ_R is modeled by the mixed distribution [6],

$$f_{\gamma_R}(x) = (1 - s)\delta(x) + s\frac{\partial F_{\gamma_T}(x)}{\partial x}, \tag{13}$$

where $\delta(x)$ denotes the Dirac delta function. As stated in (15), the CDF of γ_R is given as

$$F_{\gamma_R}(x) = \int_0^x f_{\gamma_R}(t)dt. \tag{14}$$

According to selection combining protocol, T_{k^*} with the highest SNR γ_R will be chosen, because all random variables γ_R are independent and identically distributed. Hence, the CDF of SNR γ_{TR} can be given as

$$F_{\gamma_{T_{K^*}R}}(x) = F_{\gamma_R}(x)^k$$
$$= 1 + \sum_{k=1}^{K} \binom{K}{k}(-1)^k s^k \exp(-\frac{\lambda k x}{\gamma_I}). \tag{15}$$

Moreover, the CDF of SNR from R to D is written as,

$$F_{\gamma_D}(x) = 1 - 2\sqrt{\frac{x d_{RD}^\beta}{\gamma_S}} \mathcal{K}_1\left(2\sqrt{\frac{x d_{RD}^\beta}{\gamma_S}}\right). \tag{16}$$

With the help of (11) (15) and (16), the CDF of our proposed model from BS to D is given as,

$$F_{\gamma_{T_{K*D}}}(x) = 1 - (1 - F_{\gamma_{T_{K*R}}}(x))(1 - F_{\gamma_D}(x))$$

$$= 1 + 2\sum_{k=1}^{K}\binom{K}{k}(-1)^k s^k \exp(-\frac{\lambda k x}{\gamma_I})\sqrt{\frac{x d_{RD}^{\beta}}{\gamma_S}}$$

$$\mathcal{K}_1\left(2\sqrt{\frac{x d_{RD}^{\beta}}{\gamma_S}}\right).$$

(17)

4 Numerical Results

Numerical results of outage probability are investigated to study both the effect of backhaul reliability and number of stationary small cells on our proposed system performance. In this section, we have the following assumptions. The outage probability threshold is $\theta = 1$ bits/s/Hz. In a Cartesian coordinate system, the location of small cells, mobile relay and mobile receiver are: $T_k = (0, 0)$, $R = (0.8, 0)$ and $D = (1, 0)$. The path loss exponent is $\beta = 4$ and $\gamma_I = \gamma_S$. In the figures, 'Sim' is the results for simulation and 'Ana' is the results for analytical. The 'Sim' and 'Ana' match very well.

In Fig. 2, the influence of the number of small cells on system performance is presented. The backhaul reliability s is fixed at 0.90. We assume the number of small cells is $K = 1$, $K = 2$ and $K = 3$. In the figure, we can observe that when K increases, the outage probability drops. This is because the system can gain a superior performance because of the correlation of numerous signals at the mobile receiver side.

In Fig. 3, the effect of backhaul reliability s on system performance is studied. We assume that $K = 3$. The level of backhaul reliability is various: $s = 0.90$, $s = 0.80$ and $s = 0.70$. We can observe in the figure that when the value of s increases, the outage probability decreases, the system can achieve a significant superior performance. This is because when the signal has a higher probability to transmit successfully through the wireless backhaul the system can perform better.

As shown in Figs. 2 and 3, both the number of stationary small cells and the level of backhaul reliability have a significant influence on our proposed system performance. More specifically, in this relay-assisted heterogeneous vehicular network, adding more stationary small cells and increasing the backhaul reliability can result in a significant better system performance.

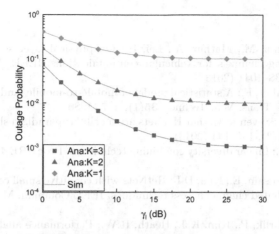

Fig. 2. The influence of the number of small cells on system performance (s = 0.90)

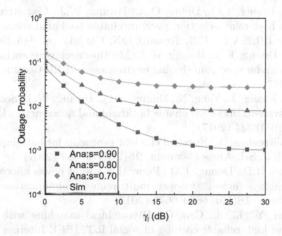

Fig. 3. The influence of backhaul reliability on system performance (K = 3)

5   Conclusion

We propose a relay-assisted heterogeneous vehicular network with a macro-base station, numerous stationary small cells, a mobile relay and a mobile receiver under unreliable backhaul over both Rayleigh fading channels and double-Rayleigh fading channels. At the mobile relay, selection combining protocol is utilized to select a best stationary small cell with the highest SNR. Outage probability is the metric to evaluate the system performance and the closed-form expression is provided. Our results prove that wireless backhaul reliability can affect the system performance significantly and this parameter should be considered when studying heterogeneous vehicular networks. Moreover, increasing the number of stationary small cells in our proposed system can help the system achieve a better performance.

References

1. Ai, Y., Cheffena, M., Mathur, A., Lei, H.: On physical layer security of double Rayleigh fading channels for vehicular communications. IEEE Wirel. Commun. Lett. **7**(6), 1038–1041 (2018)
2. Akki, A.S., Haber, F.: A statistical model of mobile-to-mobile land communication channel. IEEE Trans. Veh. Technol. **35**(1), 2–7 (1986)
3. Andrews, J.G.: Seven ways that HetNets are a cellular paradigm shift. IEEE Commun. Mag. **51**(3), 136–144 (2013)
4. Brennan, D.G.: Linear diversity combining techniques. Proc. IRE **47**(6), 1075–1102 (1959)
5. ElSawy, H., Hossain, E., Kim, D.I.: HetNets with cognitive small cells: user offloading and distributed channel access techniques. IEEE Commun. Mag. **51**(6), 28–36 (2013)
6. Khan, T.A., Orlik, P., Kim, K.J., Heath, R.W.: Performance analysis of cooperative wireless networks with unreliable backhaul links. IEEE Commun. Lett. **19**(8), 1386–1389 (2015)
7. Nguyen, H.T., Duong, T.Q., Dobre, O.A., Hwang, W.J.: Cognitive heterogeneous networks with best relay selection over unreliable backhaul connections. In: Proceedings of the IEEE VTC-Fall, Toronto, ON, Canada, pp. 1–5, September 2017
8. Nguyen, H.T., Duong, T.Q., Hwang, W.J.: Multiuser relay networks over unreliable backhaul links under spectrum sharing environment. IEEE Commun. Lett. **21**(10), 2314–2317 (2017)
9. Nguyen, H.T., Zhang, J., Yang, N., Duong, T.Q., Hwang, W.J.: Secure cooperative single carrier systems under unreliable backhaul and dense networks impact. IEEE Access **5**, 18310–18324 (2017)
10. Rankov, B., Wittneben, A.: Spectral efficient protocols for half-duplex fading relay channels. IEEE J. Sel. Areas Commun. **25**(2), 379–389 (2007)
11. Sheng, Z., Tuan, H.D., Duong, T.Q., Poor, H.V.: Joint power allocation and beamforming for energy-efficient two-way multi-relay communications. IEEE Trans. Wireless Commun. **16**(10), 6660–6671 (2017)
12. Wang, B., Sun, Y., Li, S., Cao, Q.: Hierarchical matching with peer effect for low-latency and high-reliable caching in social IoT. IEEE Internet Things J. **6**(1), 1193–1209 (2019)
13. Yin, C., Garcia-Palacios, E., Vo, N.S., Duong, T.Q.: Cognitive heterogeneous networks with multiple primary users and unreliable backhaul connections. IEEE Access **7**, 3644–3655 (2019)
14. Yin, C., Nguyen, H.T., Kundu, C., Kaleem, Z., Garcia-Palacios, E., Duong, T.Q.: Secure energy harvesting relay networks with unreliable backhaul connections. IEEE Access **6**, 12074–12084 (2018)
15. Yin, C., Yang, L., Garcia-Palacios, E.: Outage probability of vehicular networks under unreliable backhaul. EAI Endorsed Trans. Ind. Netw. Intell. Syst. **5**(17) (2018). https://doi.org/10.4108/eai.19-12-2018.156077
16. Zheng, K., Zheng, Q., Chatzimisios, P., Xiang, W., Zhou, Y.: Heterogeneous vehicular networking: a survey on architecture, challenges, and solutions. IEEE Commun. Surv. Tutorials **17**(4), 2377–2396 (2015)

On Secure Cooperative Non-orthogonal Multiple Access Network with RF Power Transfer

Duy-Hung Ha[1], Dac-Binh Ha[2(✉)], and Miroslav Voznak[1]

[1] Faculty of Electrical Engineering and Computer Science,
VSB - Technical University of Ostrava,
17. lisltopadu 2172/15, 708 00 Ostrava, Czechia
haduyhung@tdtu.edu.vn, miroslav.voznak@vsb.cz
[2] Faculty of Electrical and Electronics Engineering, Duy Tan University,
Da Nang, Vietnam
hadacbinh@duytan.edu.vn

Abstract. In this paper, we investigate the secrecy performance of the cooperative non-orthogonal multiple access (NOMA) network with radio frequency (RF) power transfer. Specifically, this considered network consists of one RF power supply station, one source and multiple energy constrained NOMA users in the presence of a passive eavesdropper. The better user helps the source to forward the message to worse user by using the energy harvested from the power station. The expression of secrecy outage probability for the scenario of wiretaping from user-to-user link is derived by using the statistical characteristics of signal-to-noise ratio (SNR) and signal-to-interference-plus-noise ratio (SINR) of transmission links. In order to understand more detail about the behaviour of this considered system, the numerical results are provided according to the system key parameters, such as the transmit power, number of users, time switching ratio and power allocation coefficients. The simulation results are also provided to confirm the correctness of our analysis.

Keywords: Non-orthogonal multiple access · Relaying network ·
Physical layer secrecy · RF power transfer · Secrecy outage probability

1 Introduction

The next generation wireless networks have been developing to satisfy human being non-stop growing need, i.e., big data rate (5G is 100 times compare to 4G), large number of users and secure transmission. Some emerging techniques, such as relaying, NOMA, MIMO techniques etc., are deployed to meet these requirements. Relaying technique with amplify-and-forward (AF) or decode-and-forward (DF) scheme allows to extend the coverage and improve the performance of wireless networks [1–3]. On the other hand, NOMA technique is employed in

© ICST Institute for Computer Sciences, Social Informatics and Telecommunications Engineering 2019
Published by Springer Nature Switzerland AG 2019. All Rights Reserved
T. Q. Duong et al. (Eds.): INISCOM 2019, LNICST 293, pp. 117–129, 2019.
https://doi.org/10.1007/978-3-030-30149-1_10

power domain to achieve multiple access strategies. It has the potential to be integrated with conventional orthogonal multiple access, i.e., frequency division multiple access (FDMA), time division multiple access (TDMA), and code division multiple access (CDMA). The combination between relaying and NOMA applied in wireless network is studied in a number of works [4–12].

Moreover, due to diverse functions of wireless devices, the energy is the most concerned issue to last long the lifetime of wireless devices and extend the coverage of network. Wireless energy harvesting is approach that the wireless devices can harvest the energy from the environment (e.g., via solar, wind, thermal, and RF power sources) and convert it into electrical energy for the energy constrained devices. Among them, RF power transfer is the emerging technique that allows the energy constrained wireless devices to harvest the energy from RF sources (e.g., base station, TV/radio broadcast station, microwave station, satellite earth station etc.). A significant works [13–18] have studied the impact of RF energy harvesting (EH) on the performance of wireless networks when they took into account the RF energy harvesting in their system. This service of RF energy harvesting is predicted to be available on the future mobile networks [19].

In wireless networks, due to the broadcast nature of wireless links, the data transmission is vulnerable to be attacked and wiretapped. Although we have a number of solutions (e.g., the public-key system developed by Rivest, Shamir, and Adleman (RSA) and the data encryption standard (DES), etc.) to protect the transmit data, these solutions have not covered all of scenarios of wireless networks (e.g., error physical layer link between transmitter and receiver, powerfull computational power of eavesdropper with efficient algorithms). Physical layer security (PLS) is a novel approach that can employs the random variation characteristics of wireless links to enhance the secure transmission of wireless communication [20–22]. There are a number of works in recent to investigate PLS in NOMA relaying network without RF energy harvesting [23–29]. The work of [23] evaluated the PLS of simple NOMA model of large-scale networks through the secrecy outage probability (SOP). The better PLS performance of overall communication process has been proven in case that there is not much difference in the level of priority between legitimate users of downlink NOMA system in [24], in which users' QoS requirements to perform NOMA. The work in [25] provided the secrecy performance analysis of a two-user downlink NOMA systems with two considered SISO and MISO schemes. The authors concluded that the secrecy performance for the far user with fixed power allocation scheme degraded as the transmit power beyond the threshold and then reaches a floor as the interference from the near user increases. In [26], the artificial noise is deployed at the base station to enhance the security ability of a beamforming-aided multiple-antenna system. The PLS for cooperative NOMA systems is studied in [27], in which the AF and DF schemes are considered. They found that AF and DF schemes nearly achieve the same secrecy performance and it is independent of the channel conditions between the relay and the worse user.

Unlike above works, in this work we study the PLS performance for the cooperative NOMA network with RF energy harvesting. This considered network consists of one RF power supply station, one source and multiple energy constrained NOMA users in the presence of a passive eavesdropper. The better user harvests the RF energy from the power station to help the source to forward the message to worse user. The expression of secrecy outage probability for the scenario of wiretaping from user-to-user link is derived by using the statistical characteristics of signal-to-noise ratio (SNR) and signal-to-interference-plus-noise ratio (SINR) of transmission links. In order to understand more detail about the behaviour of this considered system, the numerical results are provided according to the system key parameters, such as the transmit power, number of users, time switching ratio and power allocation coefficients. The correctness of our analysis is confirmed by simulation results.

The remain of this paper is organized as follows. The system model is presented in Sect. 2. Secrecy performance analysis is provided in Sect. 3. The numerical results are shown in Sect. 4. Finally, we conclude our work in Sect. 5.

2   System Model Description

Figure 1 depicts a downlink RF EH cooperative NOMA system. A power supply station P intends to transfer RF power to energy-constrained users. A source S, i.e., base station, intends to transmit information to M energy constrained mobile users denoted as D_i, $(1 \leq i \leq M)$ in the presence of a passive eavesdropper E. In this considered system, we can divided M users into multiple pairs, such as $\{D_m, D_n\}$ $(m < n)$, to perform NOMA [4]. We design that the better user D_m forward the information of the poor user D_n after use applying successive interference cancellation (SIC) to detect the D_m's signal. We investigate the scenario that D_n and E can not hear from S due to the severe shadowing environment. In other words, E only tries to extract the message s_n of D_n from D_m.

Without loss of generality, assuming that all the channel gains between S and D_i follow the order of $|h_{SD_1}|^2 \geq ... \geq |h_{SD_m}|^2 \geq |h_{SD_n}|^2 \geq ... \geq |h_{SD_M}|^2$, where $|h_{SD_m}|^2$ and $|h_{SD_n}|^2$ are denoted as the ordered channel gains of the mth user and the nth user, respectively. And $|h_{PD_m}|^2$, $|h_{mn}|^2$, $|h_{SE}|^2$, $|h_{D_mE}|^2$ are the channel gains of the links $P - D_m$, $D_m - D_n$, $S - E$ and $D_m - E$, respectively. All the user nodes are single-antenna devices and operate in half-duplex mode, i.e., sensor nodes. All wireless links are assumed to undergo independent frequency non-selective Rayleigh block fading and additive white Gaussian noise (AWGN) with zero mean and the same variance σ^2, i.e., $\sim \mathcal{CN}(0, \sigma^2)$. We also denote $d_{PD_m}, d_{SD_m}, d_{mn}, d_{D_mE}$ as the Euclidean distances of $P - D_m, S - D_m, D_m - D_n, D_m - E$, respectively and θ denote the path-loss exponent. Let $X_1 \triangleq |h_{PD_m}|^2$, $X_2 \triangleq |h_{SD_m}|^2$, $X_3 \triangleq |h_{mn}|^2$, and $Z_3 \triangleq |h_{D_mE}|^2$.

The triple-phase protocol for this RF EH cooperative NOMA system is proposed as follows

(1) In the first phase: P transfers RF energy to the users with power P_0 in the time αT ($\alpha \in (0, 1)$: time switching ratio; T: block time).

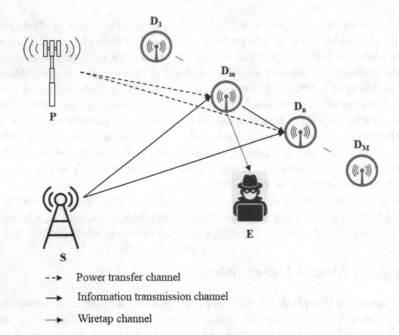

D_1

D_m

D_n

D_M

P

S

E

- -▶   Power transfer channel
——▶   Information transmission channel
·······▶   Wiretap channel

Fig. 1. System model for secured RF EH cooperative NOMA network

(2) In the second phase: S transmits information signal $x = \sqrt{a_m}s_m + \sqrt{a_n}s_n$ with power P_S to user pair $\{D_m, D_n\}$ in the time of $(1 - \alpha)T/2$, where s_m and s_n are the message for the mth user D_m and the nth user D_n, respectively; a_m and a_n are the power allocation coefficients satisfied the conditions: $0 < a_m < a_n$ and $a_m + a_n = 1$ by following the NOMA scheme.

(3) In the third phase: Applying NOMA, D_m uses SIC to detect message s_n and subtracts this component from the received signal to obtain its own message s_m, then re-encodes and forwards s_n to D_n in the remain time of $(1-\alpha)T/2$ with the energy harvested from P.

Next, we present the RF EH NOMA relaying-based transmission in mathematical manner.

2.1   The First Phase

In this phase, the energy harvested by D_m in the time of αT can be expressed as

$$E_{D_m} = \frac{\eta P_0 |h_{PD_m}|^2 \alpha T}{d_{PD_m}^\theta}, \tag{1}$$

where η denotes as the energy conversion efficiency $(0 < \eta < 1)$.

2.2 The Second Phase

In the second phase, the source S broadcasts information to the user pair in duration of $(1 - \alpha)T/2$. The received signal at D_m is given by

$$y_{SD_m} = \sqrt{\frac{P_S}{d^\theta_{SD_m}}}(\sqrt{a_m}s_m + \sqrt{a_n}s_n)h_{SD_m} + n_{SD_m}, \qquad (2)$$

where $n_{SD_m} \sim \mathcal{CN}(0, \sigma^2)$.

The instantaneous SINR at D_m to detect s_n transmitted from S can be written as

$$\gamma^{s_n}_{SD_m} = \frac{a_n\bar{\gamma}_S|h_{SD_m}|^2}{a_m\bar{\gamma}_S|h_{SD_m}|^2 + d^\theta_{SD_m}} = \frac{b_2 X_2}{b_1 X_2 + 1}, \qquad (3)$$

where $\bar{\gamma}_S = \frac{P_S}{\sigma^2}$, $b_1 = \frac{a_m\bar{\gamma}_S}{d^\theta_{SD_m}}$, $b_2 = \frac{a_n\bar{\gamma}_S}{d^\theta_{SD_m}}$.

2.3 The Third Phase

In this phase, D_m uses the harvested energy E_{Dm} as (1) to forward s_n to D_n in duration of $(1 - \alpha)T/2$. Here, we ignore the processing power required by the transmit/receive circuitry of D_m. The transmit power of D_m is given by

$$P_{D_m} = \frac{\eta\alpha P_0|h_{PD_m}|^2}{(1 - \alpha)d^\theta_{PD_m}}. \qquad (4)$$

The received signal at D_n is expressed as

$$y_{mn} = \sqrt{\frac{P_{D_m}}{d^\theta_{mn}}}h_{mn}s_n + n_{mn}, \qquad (5)$$

where $n_{mn} \sim \mathcal{CN}(0, \sigma^2)$. From (4) and (5), the instantaneous SNR at D_n in the last phase is as follows

$$\gamma_{mn} = \frac{P_{D_m}|h_{mn}|^2}{\sigma^2 d^\theta_{mn}} - c_1 X_1 X_3, \qquad (6)$$

where $c_1 = \frac{\eta\alpha\bar{\gamma}_0}{(1-\alpha)d^\theta_{PD_m}d^\theta_{mn}}$, $\bar{\gamma}_0 = \frac{P_0}{\sigma^2}$ is denoted as the average transmit SNR of $D_m - D_n$ link.

Due to we consider the scenario that the passive eavesdropper E tries to extract the poor user's message s_n from the links $D_m - D_n$ without any attacks, the received signal at E is written as

$$y_{D_m E} = \sqrt{\frac{P_{D_m}}{d^\theta_{D_m E}}}h_{D_m E}s_n + n_{D_m E}, \qquad (7)$$

where $n_{D_m E} \sim \mathcal{CN}(0, \sigma_E^2)$. Similarly, the instantaneous SNR at E in this phase is given by

$$\gamma_{D_m E} = c_2 X_1 Z_3, \tag{8}$$

where $c_2 = \frac{\eta \alpha \bar{\gamma}_{0E}}{(1-\alpha) d_{PD_m}^\theta d_{D_m E}^\theta}$, $\bar{\gamma}_{0E} = \frac{P_0}{\sigma_E^2}$ is denoted as the average transmit SNR of $D_m - E$ link.

The i.i.d. Rayleigh channel gains ($|h_{PDm}|^2$, $|h_{SDm}|^2$, $|h_{mn}|^2$, and $|h_{DmE}|^2$) follow exponential distributions with parameters λ_{PDm}, λ_{SDm}, λ_{mn}, and λ_{DmE}, respectively. According to [30], the probability density function (PDF) and the cumulative distribution function (CDF) of ordered random variable X_2 are respectively written as follows

$$f_{X_2}(x) = \frac{M!}{(M-m)!(m-1)!} \frac{1}{\lambda_{SDm}} \sum_{k=0}^{m-1} C_k^{m-1}(-1)^k e^{\frac{-x(M-m+k+1)}{\lambda_{SDm}}}, \tag{9}$$

$$F_{X_2}(x) = \frac{M!}{(M-m)!(m-1)!} \sum_{k=0}^{m-1} \frac{C_k^{m-1}(-1)^k}{M-m+k+1} \left[1 - e^{\frac{-x(M-m+k+1)}{\lambda_{SDm}}} \right]. \tag{10}$$

Under Rayleigh fading, the PDF and CDF of random variable V are respectively expressed as

$$f_V(x) = \frac{1}{\lambda} e^{-\frac{x}{\lambda}}, \tag{11}$$

$$F_V(x) = 1 - e^{-\frac{x}{\lambda}}, \tag{12}$$

where $V \in (X_1, X_3, Z_3)$, $\lambda \in \{\lambda_{PD_m}, \lambda_{mn}, \lambda_{DmE}\}$.

3   Secrecy Performance Analysis

In this section, we analyze the secrecy performance by derivation the expression of secrecy outage probability. Notice that, in this considered system we only consider the case of the eavesdropper tries to hear the message of D_n at D_m without any attacks. Therefore, the instantaneous secrecy capacity for $D_m - D_n$ is given by

$$C_S = \begin{cases} \frac{(1-\alpha)}{2} \log_2 \left(\frac{1+\gamma_{mn}}{1+\gamma_{D_m E}} \right), & \gamma_{mn} > \gamma_{D_m E} \\ 0, & \gamma_{mn} \leq \gamma_{D_m E} \end{cases}, \tag{13}$$

Here, for simplicity we assume $B = 1\,\mathrm{Hz}$. SOP is defined as the probability that the instantaneous secrecy capacity falls below a predetermined secrecy rate threshold $R_S > 0$ or $SOP = Pr(C_S < R_S)$.

For further calculation, we derive the following proposition.

Proposition 1. *Under Rayleigh fading, the joint CDF of γ_{mn} and $\gamma_{D_m E}$ is given by*

$$F_{\gamma_{mn}, \gamma_{D_m E}}(x, y) = 1 - u\mathcal{K}_1(u) - v\mathcal{K}_1(v) + t\mathcal{K}_1(t), \tag{14}$$

where $u = 2\sqrt{\frac{x}{c_1 \lambda_{PD_m} \lambda_{mn}}}$, $v = 2\sqrt{\frac{y}{c_2 \lambda_{PD_m} \lambda_{D_m E}}}$, $t = 2\sqrt{\frac{c_2 \lambda_{D_m E} x + c_1 \lambda_{mn} y}{c_1 c_2 \lambda_{PD_m} \lambda_{mn} \lambda_{D_m E}}}$, $\mathcal{K}_v(.)$ is the modified Bessel function of the second kind and v^{th} order [31].

Proof. See Appendix A.

In this considered scenario, the secrecy outage event occurs when D_m cannot detect s_n or D_m can detect s_n but the secrecy capacity is below the secrecy threshold. Therefore, the secrecy outage probability at the D_m is calculated as follows

$$
\begin{aligned}
SOP &= \Pr(\gamma_{SD_m}^{s_n} < \gamma_t) + \Pr(\gamma_{SD_m}^{s_n} > \gamma_t, C_{S_3} < R_S) \\
&= \Pr(\gamma_{SD_m}^{s_n} < \gamma_t) + \Pr(\gamma_{SD_m}^{s_n} > \gamma_t)\Pr(C_{S_3} < R_S),
\end{aligned}
\tag{15}
$$

where γ_t is denoted as the SNR threshold to ensure successful detection at the D_m for a given target data rate R and $\gamma_t = 2^{\frac{2R}{1-\alpha}} - 1$.

Theorem 1. *Under Rayleigh fading, the SOP of the link $D_m - D_n$ is given by*

$$
SOP = \Phi_1 + (1 - \Phi_1)\Phi_2,
\tag{16}
$$

where

$$
\Phi_1 = \begin{cases} \frac{M!}{(M-m)!(m-1)!} \sum\limits_{k=0}^{m-1} \frac{C_k^{m-1}(-1)^k}{M-m+k+1}\left[1 - e^{-\frac{\gamma_t(M-m+k+1)}{(b_2 - b_1\gamma_t)\lambda_{SD_m}}}\right], & \gamma_t < \frac{a_n}{a_m} \\ 1, & \gamma_t > \frac{a_n}{a_m} \end{cases}
$$

$$
\Phi_2 = 1 - \frac{2c_1 \lambda_{mn}}{c_1 \lambda_{mn} + c_2 \lambda_{D_m E} \Omega_S}\sqrt{\frac{\Omega_S - 1}{c_1 \lambda_{PD_m} \lambda_{mn}}}\mathcal{K}_1\left(2\sqrt{\frac{\Omega_S - 1}{c_1 \lambda_{PD_m} \lambda_{mn}}}\right).
$$

where $\Omega_S = 2^{\frac{2R_S}{1-\alpha}}$.

Proof. See Appendix B.

4 Numerical Results and Discussion

In this section, we provide the numerical results to clarify the physical layer secrecy performance of proposed protocol for this considered RF EH NOMA relaying system. Further more, Monte-Carlo simulation results are also provided to verify our analytical results.

Figure 2 plots the curves of SOP of this system at D_m versus the transmit power of P and different average transmit SNR of link $D_m - D_n$. This result shows that when we increase the transmit power from P to the better user D_m, SOP of the system decreases. This means that we can improve the secrecy performance by increasing the transmit power to provide more energy to legitimate users.

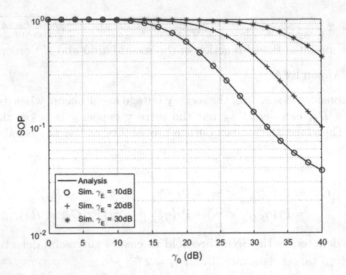

Fig. 2. SOP vs. the transmit power of P and different average transmit SNR of link $D_m - D_n$ with $\gamma_S = 20\,\text{dB}$, $a_n = 0.9$, $M = 4$, $m = 2$, $n = 3$, $R = 1\,\text{bps/Hz}$, $R_S = 1\,\text{bps/Hz}$, $\alpha = 0.3$, $\eta = 0.9$, $d_{PD_m} = d_{SD_m} = d_{mn} = d_{D_m E} = 1$, $\theta = 2$.

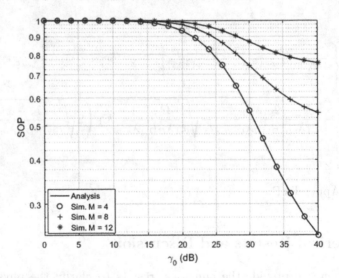

Fig. 3. SOP vs. the transmit power of P with different number of users M with $\gamma_S = 20\,\text{dB}$, $\gamma_E = 20\,\text{dB}$, $a_n = 0.9$, $m = 2$, $n = 3$, $R = 1\,\text{bps/Hz}$, $R_S = 1\,\text{bps/Hz}$, $\alpha = 0.3$, $\eta = 0.9$, $d_{PD_m} = d_{SD_m} = d_{mn} = d_{D_m E} = 1$, $\theta = 2$.

The curves of SOP of this system at D_m versus the transmit power of P with different number of users M are plotted in Fig. 3. From this figure, we can see that the secrecy performance degrades when increasing the number of users.

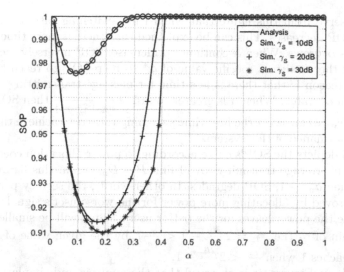

Fig. 4. SOP vs. time switching ratio α with different average transmit SNR of link $S - D_m$ with $\gamma_0 = 20\,\text{dB}$, $\gamma_E = 20\,\text{dB}$, $a_n = 0.9$, $M = 4$, $m = 2$, $n = 3$, $R = 1\,\text{bps/Hz}$, $R_S = 1\,\text{bps/Hz}$, $\eta = 0.9$, $d_{PD_m} = d_{SD_m} = d_{mn} = d_{D_m E} = 1$, $\theta = 2$.

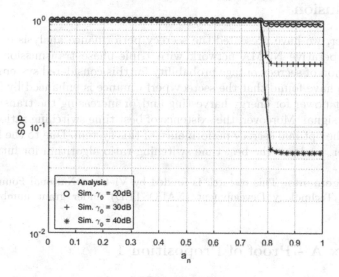

Fig. 5. SOP vs. power allocation coefficient a_n with different average transmit SNR of link $P - D_m$ with $\gamma_S = 20\,\text{dB}$, $\gamma_E = 20\,\text{dB}$, $M = 4$, $m = 2$, $n = 3$, $R = 1\,\text{bps/Hz}$, $R_S - 1\,\text{bps/Hz}$, $\eta = 0.9$, $d_{PD_m} = d_{SD_m} = d_{mn} = d_{D_m E} = 1$, $\theta = 2$.

This can be explained that the more number of users the more opportunities to wiretape the message of s_n.

The numerical results for SOP of this system at D_m versus time switching ratio α with different average transmit SNR of link $S - D_m$ are provided in Fig. 4.

We can observe from this figure that when time switching ratio α is small, α increases SOP decreases. This can be explained that there is more time to power the users as α grows. When α continue to increase, SOP inversely increases. It means that there exists a specific value of α^* to help SOP to reach the lowest value. The reason is that there is less time for message transmitting when α is greater than α^* value. When α is greater than $1 - \dfrac{2R}{\log_2\left(\frac{a_n}{a_m}+1\right)}$ then SOP reaches 1. Obviously, we can select the best time switching ratio α to achieve the optimal secrecy performance of this system.

Figure 5 depicts the SOP's curve according to power allocation coefficient a_n with different average transmit SNR of link $P - D_m$. From this figure, we can see that when $a_n \to 1$, SOP degrades. In other words, the secrecy performance can be improved by allocating more power for the worse user's signal. However, at that time the power leaves for the better user's signal will be smaller. Due to the constrain of γ_t (i.e., $\gamma_t = 2^{\frac{2R}{1-\alpha}} - 1 < \frac{a_n}{a_m}$), by the given value of R and α, the SOP reaches 1 when $\frac{a_n}{a_m} < 2^{\frac{2R}{1-\alpha}} - 1$.

From above Figures, it is observed that the analysis and simulation results are matching very well. It means that the correctness of our analysis is verified.

5   Conclusion

In this paper, we have presented the secrecy performance analysis of downlink RF EH cooperative NOMA network with triple-phase transmission protocol. The expression of secrecy outage probability for this considered system has been derived. We have found that the secrecy performance is enhanced by increasing the transmit power for energy harvesting and/or increasing the transmit power for message signal. Moreover, the existence of best time switching ratio is proven to achieve the optimal secrecy performance of this system. Due to the limitation of this paper, we leave the best time switching ratio algorithm for future work.

Acknowledgements. This research is funded by Vietnam National Foundation for Science and Technology Development (NAFOSTED) under grant number 102.04-2017.301.

Appendix A - Proof of Proposition 1

Here, we derive the expression of the joint CDF of γ_{mn} and $\gamma_{D_m E}$ as follows

$$
\begin{aligned}
F_{\gamma_{mn},\gamma_{D_m E}}(x,y) &= \int_0^\infty F_{\gamma_{mn},\gamma_{D_m E}|X_1}(x,y|z) f_{X_1}(z) dz \\
&= \int_0^\infty F_{\gamma_{mn}|X_1}(x|z) F_{\gamma_{D_m E}|X_1}(y|z) f_{X_1}(z) dz \\
&= \int_0^\infty \left(1 - e^{-\frac{x}{c_1 \lambda_{mn} z}}\right)\left(1 - e^{-\frac{y}{c_2 \lambda_{D_m E} z}}\right) \frac{1}{\lambda_{PD_m}} e^{-\frac{z}{\lambda_{PD_m}}} dz \\
&= 1 - u\mathcal{K}_1(u) - v\mathcal{K}_1(v) + t\mathcal{K}_1(t),
\end{aligned}
\tag{17}
$$

where $u = 2\sqrt{\frac{x}{c_1 \lambda_{PD_m} \lambda_{mn}}}$, $v = 2\sqrt{\frac{y}{c_2 \lambda_{PD_m} \lambda_{D_m E}}}$, $t = 2\sqrt{\frac{c_2 \lambda_{D_m E} x + c_1 \lambda_{mn} y}{c_1 c_2 \lambda_{PD_m} \lambda_{mn} \lambda_{D_m E}}}$. This concludes the proof.

Appendix B - Proof of Theorem 1

By means of (15), Φ_1 and Φ_2 are respectively calculated as follows

$$\Phi_1 = \Pr\left(\frac{b_2 X_2}{b_1 X_2 + 1} < \gamma_t\right) = F_{X_2}\left(\frac{\gamma_t}{b_2 - b_1 \gamma_t}\right)$$

$$= \begin{cases} \frac{M!}{(M-m)!(m-1)!} \sum_{k=0}^{m-1} (-1)^k C_k^{m-1} \frac{1}{M-m+k+1} \left[1 - e^{-\frac{\gamma_t(M-m+k+1)}{(b_2 - b_1 \gamma_t)\lambda_{SDm}}}\right], & \gamma_t < \frac{a_n}{a_m} \\ 1, & \gamma_t > \frac{a_n}{a_m} \end{cases}$$

$$\Phi_2 = \Pr\left(\frac{1+\gamma_{mn}}{1+\gamma_{D_m E}} < \Omega_S\right) = \int_0^\infty \left[\frac{\partial F_{\gamma_{mn}, \gamma_{D_m E}}(x,y)}{\partial y}\right]_{x=\Omega_S(1+y)-1} dy$$

$$= -\frac{2}{c_2 \lambda_{PD_m} \lambda_{D_m E}} \int_0^\infty \mathcal{K}_0\left(\sqrt{\frac{y}{c_2 \lambda_{PD_m} \lambda_{D_m E}}}\right) dy$$

$$+ \frac{2}{c_2 \lambda_{PD_m} \lambda_{D_m E}} \int_0^\infty \mathcal{K}_0\left(\sqrt{\frac{(c_1 \lambda_{mn} + c_2 \lambda_{D_m E} \Omega_S)y + c_2 \lambda_{D_m E}(\Omega_S - 1)}{c_1 c_2 \lambda_{PD_m} \lambda_{mn} \lambda_{D_m E}}}\right) dy$$

$$= \int_0^\infty v \mathcal{K}_0(v) dv - \frac{2c_1 \lambda_{mn}}{c_1 \lambda_{mn} + c_2 \lambda_{D_m E} \Omega_S} \int_{2\sqrt{\frac{\Omega_S - 1}{c_1 \lambda_{PD_m} \lambda_{mn}}}}^\infty s \mathcal{K}_0(s) ds$$

$$= 1 - \frac{2c_1 \lambda_{mn}}{c_1 \lambda_{mn} + c_2 \lambda_{D_m E} \Omega_S} \sqrt{\frac{\Omega_S - 1}{c_1 \lambda_{PD_m} \lambda_{mn}}} \mathcal{K}_1\left(2\sqrt{\frac{\Omega_S - 1}{c_1 \lambda_{PD_m} \lambda_{mn}}}\right), \tag{18}$$

where $\Omega_S = 2^{\frac{2R_S}{1-\alpha}}$, $s = \sqrt{\frac{(c_1 \lambda_{mn} + c_2 \lambda_{D_m E} \Omega_S)y + c_2 \lambda_{D_m E}(\Omega_S - 1)}{c_1 c_2 \lambda_{PD_m} \lambda_{mn} \lambda_{D_m E}}}$.
This is the end of our proof.

References

1. Wu, H., Wang, C., Tzeng, N.F.: Novel self-configurable positioning technique for multihop wireless networks. IEEE/ACM Trans. Netw. **13**(3), 609–621 (2005)
2. Kim, K.J., Duong, T.Q., Poor, H.V.: Performance analysis of cyclic prefixed single-carrier cognitive amplify-and-forward relay systems. IEEE Trans. Wirel. Commun. **12**(1), 195–205 (2013)
3. Thanh, T.L., Hoang, T.M.: Cooperative spectrum-sharing with two-way AF relaying in the presence of direct communications. EAI Endorsed Trans. Ind. Netw. Intell. Syst. **5**(14), 1–9 (2018)
4. Ding, Z., Peng, M., Poor, H.V.: Cooperative non-orthogonal multiple access in 5G systems. IEEE Commun. Lett. **19**(8), 1462–1465 (2015)
5. Do, N.T., Costa, D.B.D., Duong, T.Q., An, B.: A BNBF user selection scheme for NOMA-based cooperative relaying systems with SWIPT. IEEE Commun. Lett. **21**(3), 664–667 (2017)

6. Islam, S.M.R., Avazov, N., Dobre, O.A., Kwak, K.S.: Power-domain non-orthogonal multiple access (NOMA) in 5G systems: potentials and challenges. IEEE Commun. Surv. Tutor. **19**(2), 721–742 (2017)
7. Nguyen, V.D., Tuan, H.D., Duong, T.Q., Poor, H.V., Shin, O.S.: Precoder design for signal superposition in MIMO-NOMA multicell networks. IEEE J. Sel. Areas Commun. **35**(12), 2681–2695 (2017)
8. Tran, D.D., Tran, H.V., Ha, D.B., Kaddoum, G.: Cooperation in NOMA networks under limited user-to-user communications: solution and analysis. In: IEEE Wireless Communications and Networking Conference (WCNC), Barcelona, Spain, 15–18 April 2018 (2018)
9. Lee, S., Duong, T.Q., da Costa, D.B., Ha, D.B., Nguyen, S.Q.: Underlay cognitive radio networks with cooperative non-orthogonal multiple access. IET Commun. **12**(3), 359–366 (2018)
10. Do, T.N., da Costa, D.B., Duong, T.Q., An, B.: Improving the performance of cell-edge users in NOMA systems using cooperative relaying. IEEE Trans. Commun. **66**(5), 1883–1901 (2018)
11. Do, T.N., da Costa, D.B., Duong, T.Q., An, B.: Improving the performance of cell-edge users in MISO-NOMA systems using TAS and SWIPT-based cooperative transmissions. IEEE Trans. Green Commun. Netw. **2**(1), 49–62 (2018)
12. Tuan, H.D., Nasir, A.A., Nguyen, H.H., Duong, T.Q., Poor, H.V.: Non-orthogonal multiple access with improper gaussian signaling. IEEE J. Sel. Topics Signal Process. **13**, 496–507 (2019)
13. Nasir, A.A., Zhou, X., Durrani, S., Kennedy, R.A.: Relaying protocols for wireless energy harvesting and information processing. IEEE Trans. Wireless Commun. **12**(7), 3622–3636 (2013)
14. Ha, D.B., Tran, D.D., Tran-Ha, V., Hong, E.K.: Performance of amplify-and-forward relaying with wireless power transfer over dissimilar channels. Elektronika ir Elektrotechnika J. **21**(5), 90–95 (2015)
15. Du, G., Xiong, K., Qiu, Z.: Outage analysis of cooperative transmission with energy harvesting relay: time switching versus power splitting. Math. Probl. Eng. **2015**, 1–9 (2015)
16. Chen, X., Zhang, Z., Chen, H.H., Zhang, H.: Enhancing wireless information and power transfer by exploiting multi- antenna techniques. IEEE Commun. Mag. **53**(4), 133–141 (2015)
17. Cvetkovic, A., Blagojevic, V., Ivanis, P.: Performance analysis of nonlinear energy-harvesting DF relay system in interference-limited Nakagami-m fading environment. ETRI J. **39**(6), 803–812 (2017)
18. Xu, K., Shen, Z., Wang, Y., Xia, X.: Beam-domain hybrid time-switching and power splitting SWIPT in full-duplex massive MIMO system. EURASIP J. Wirel. Commun. Netw., 1–21 (2018)
19. Lu, X., Wang, P., Niyato, D., Kim, D.I., Han, Z.: Wireless networks with RF energy harvesting: a contemporary survey. IEEE Commun. Surv. Tutor. **17**(2), 757–789 (2014)
20. Shannon, C.E.: Communication theory of secrecy systems. Bell Syst. Tech. J. **28**, 656–715 (1949)
21. Wyner, A.: The wire-tap channel. Bell Syst. Tech. J. **54**(8), 1355–1387 (1975)
22. Bloch, M., Barros, J., Rodrigues, M.R., McLaughlin, S.W.: Wireless information-theoretic security. IEEE Trans. Inf. Tech. **54**(6), 2515–2534 (2008)
23. Liu, Y., Qin, Z., Elkashlan, M., Gao, Y., Hanzo, L.: Enhancing the physical layer security of nonorthogonal multiple access in large-scale networks. IEEE Trans. Wirel. Commun. **16**(3), 1656–1672 (2017)

24. Tran, D.D., Ha, D.B.: Secrecy performance analysis of QoS-based non-orthogonal multiple access networks over nakagami-m fading. In: The International Conference on Recent Advances in Signal Processing, Telecommunications and Computing (SigTelCom), HCMC, Vietnam (2018)
25. Lei, H., Zhang, J., Park, K.H., Xu, P., Ansari, I.S., Pan, G.: On secure NOMA systems with transmit antenna selection schemes. IEEE Access 5, 17450–17464 (2017)
26. Lv, L., Ding, Z., Ni, Q., Chen, J.: Secure MISO-NOMA transmission with artificial noise. IEEE Trans. Veh. Technol. 67(7), 6700–6705 (2018)
27. Chen, J., Yang, L., Alouini, M.S.: Physical layer security for cooperative NOMA systems. IEEE Trans. Veh. Technol. 67(5), 4645–4649 (2018)
28. Kieu, T.N., Tran, D.D., Ha, D.B., Voznak, M.: On secure QoS-based NOMA networks with multiple antennas and eavesdroppers over Nakagami-m fading. IETE J. Res., 1–13 (2019)
29. Tran, D.D., Tran, H.V., Ha, D.B., Kaddoum, G.: Secure transmit antenna selection protocol for MIMO NOMA networks over Nakagami-m channels. IEEE Syst. J., 1–12 (2019)
30. Men, J., Ge, J.: Performance analysic of non-orthogonal multiple access in downlink cooperative network. IET Commun. 9(18), 2267–2273 (2015)
31. Gradshteyn, I., Ryzhik, I.: Table of Integrals, Series, and Products. Elsevier Academic Press, Cambridge (2007)

A Data-Driven Approach for Network Intrusion Detection and Monitoring Based on Kernel Null Space

Thu Huong Truong[1(✉)], Phuong Bac Ta[1], Quoc Thong Nguyen[2],
Huu Du Nguyen[3], and Kim Phuc Tran[4]

[1] School of Electronics and Telecommunications,
Hanoi University of Science and Technology, Hanoi, Vietnam
huong.truongthu@hust.edu.vn
[2] Division of Artificial Intelligence, Dong A University, Da Nang, Vietnam
[3] Faculty of Information Technology,
Vietnam National University of Agriculture, Hanoi, Vietnam
[4] GEMTEX Laboratory, Ecole Nationale Sup des Arts et Industries Textiles,
BP 30329 59056, Roubaix Cedex 1, France

Abstract. In this study, we propose a new approach to determine intrusions of network in real-time based on statistical process control technique and kernel null space method. The training samples in a class are mapped to a single point using the Kernel Null Foley-Sammon Transform. The Novelty Score are computed from testing samples in order to determine the threshold for the real-time detection of anomaly. The efficiency of the proposed method is illustrated over the KDD99 data set. The experimental results show that our new method outperforms the OCSVM and the original Kernel Null Space method by 1.53% and 3.86% respectively in terms of accuracy.

Keywords: Network security · Kernel Quantile Estimator ·
One-class classification · Kernel Null Space · Support vector machine

1 Introduction

Nowadays, every computer system has the security policies but these policies have not been strong enough to prevent or detect all new types of attacks. Therefore, building one monitoring system is essential to alarm novelties early. Detecting incoming intrusion early helps systems reduce the damage and protect the crucial information. Intrusion detection system (IDS) is the key to resolve these problems and attract a lot of researchers to work on the issue [3]. IDS has been used in a great number of applications such as network intrusion, fraud detection and security systems.

Currently, there are two families of mechanisms in IDS: signature-based IDS and anomaly-based IDS. In this paper, we focus on developing an anomaly-based

© ICST Institute for Computer Sciences, Social Informatics and Telecommunications Engineering 2019
Published by Springer Nature Switzerland AG 2019. All Rights Reserved
T. Q. Duong et al. (Eds.): INISCOM 2019, LNICST 293, pp. 130–140, 2019.
https://doi.org/10.1007/978-3-030-30149-1_11

IDS solution, in which the designed IDS system is trained based on knowledge of normal traffic only; the system does not need to be trained with attack data traces in advance to know if incoming traffic is anomaly or normal. This characteristic is good for the attack detection aspect because attack manners may vary over time; so, we may be in the situation that the system was not trained with an attack pattern before. In case of never-seen-before attacks, an IDS system based on training of attack and normal data traces may not be effective any more.

Among the anomaly-based IDS solution family, Novelty Detection is a research direction attracting a great number of researchers. A model is built from normal data to detect unknown abnormality by novelty detection algorithms such as OCSVM [6,11] and Kernel Null Space [1,2,4]. There is also an approach in intrusion detection using Statistical Process Control [7].

Our proposed solution aims at improving the performance of the Kernel Null Space method [2] in terms of accuracy. To be more specific, we propose using a Control-Chart based method called Kernel Quantile Estimator to determine the detection threshold dynamically driven by each specific training data set instead of using a fixed threshold as described in the existing Kernel Null Space solutions [1,2,4]. The Control Chart Based on a Kernel Estimator of the Quantile Function was also developed in [5]. In addition, we also optimize the kernel parameter of the kernel function to improve the performance of novelty detection.

The rest of the paper is organized as follows: Sect. 2 elaborates the related work. Our proposed Kernel Null Space solution for Novelty Detection is provided in Sect. 3, followed by the performance evaluation in Sect. 4. Finally, conclusion is given in Sect. 5.

2   Related Work

Generally, the novelty detection issues can be divided into two types based on the number of known classes during the training phase: one-class and multi-classes. Since our work focuses on one-class classification, we will review the state of the art for the family of one-class novelty detection. To the best of our knowledge, Kernel Null Space has the highest performance in novelty detection and there are only three studies dealing with one-class classification in novelty detection using this method [1,2,4]. In [2], the authors proposed Kernel Null Space for novelty detection but they made the experiment with a fixed threshold and a fixed kernel parameter of the kernel function. In paper [1], the authors also improved the performance of the original method. However, they only concentrated on decreasing the timing operating of the algorithm, the accuracy remains unchanged. Following this trend, paper [4] improved the solution proposed in [2] by decreasing the complexity of the kernel null space method without taking the accuracy into account.

From another approach, the OCSVM method, which detects novelty by finding the boundary of training data with maximum margin, is often used to solve the one-class novelty detection problem, for example, in [11,12]. The OCSVM method has received more extensive attention since it can easily handle nonlinear data with kernel trick and also achieve a high level of detection accuracy [11].

As mentioned in Sect. 1, the solution we will explain throughout the paper is to improve the accuracy of the Kernel Null Space method [2] in the favor of anomaly detection. We propose a solution combining Kernel null space and Control chart to automatically define an efficient detection threshold stemming from each training data trace.

Simultaneously, we also use the optimizing parameter method proposed in [11] to increase the accuracy for the algorithm. Our proposed solution is proved to outperform the Kernel Null Space methods in [1,2,4] and OCSVM in [11,12] in terms of Accuracy.

3   Intrusion Detection Scheme Using the Enhanced Kernel Null Space Method

For an intrusion detection system to work accurately, we propose a so-called an enhanced Kernel Null Space method to improve the accuracy of detecting novelty samples. The scheme is elaborated as follows:

- Pre-process and normalize the attributes of the data set.
- Design an enhanced Kernel Null Space method to analyze data inputs.

In this method, the threshold is computed by Kernel Quantile Estimator [9] for a given probability q. Figure 1 shows the process of the detection scheme, in which Internet raw data coming to the detection system will be pre-processed, and analyzed to test if it is a novelty (i.e. anomalies).

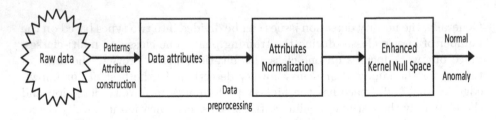

Fig. 1. Intrusion detection process

3.1   Pre-processing and Normalizing Data Attributes

In order to do the comparison with different intrusion detection methods, in the experiment, we use the NSL-KDD data set [10] which is commonly used. Each sample in this NSL-KDD corresponds to a real connection in the simulated military network, containing 41 attributes with Normal and Attack-type labels. In the data set, there are 39 types of attacks divided in 4 groups:

- DoS - Denial of services, e.g. syn flood.
- R2L: Unauthorized access from a remote machine, e.g. guessing password.
- Probing: surveillance and other probing, e.g. port scanning.
- U2R: unauthorized access to local super user (root) privileges, e.g. buffer overflow.

To make the data set simpler, reducing the redundancy without losing the information, we pre-process the data set as follows:

- Conversion from the Symbolic type to the Numeric type: there are 3 attributes in the Symbolic manner such as: Protocol, Service, Flag which are needed to be converted to the Numeric type to be compatible with the inputs of the algorithm. The symbolic values are labeled as in Table 1.

Table 1. Symbolic-typed attributes

Attribute	Symbolic	Corresponding numeric value
Protocol_type	UDP	1
	TCP	2
	ICMP	3
Flag	OTH	1
	REJ	2
	RTSO	3
	RTSOSO	4
	RSTR	5
	S0	6
	S1	7
	S2	8
	S3	9
	SF	10
	SH	11
Service	65 values	From 1 to 65

- Normalization: Normalization of data in the NSL-KDD data set is necessary since there are many big values in comparison with much smaller values in the set. We apply the Min-max normalization method to turn all values to the range [0, 1] as follows:

$$\hat{v}_i = \frac{v_i - min(v_i)}{max(v_i) - min(v_i)}, \text{ for } i = 1, 2, \ldots, 41 \tag{1}$$

where:

v_i: value of one attribute before normalization.
$\hat{v}_i$: value of one attribute after normalization.
$i = 1, \ldots, 41$: 41 attributes.

3.2 Enhanced Kernel Null Space

Before describing the enhanced Kernel Null Space, we briefly re-call the One-Class Classification using Kernel Null Space proposed in [2]. Let us consider a dataset of N training samples $\{x_1, x_2, \ldots, x_N\}$, with each $x_i \in R^D$, and D is the number of observed features. In the one-class setting, all the training samples belong to a single target class. The input features $X = [x_1, x_2, \ldots, x_N]$ are separated from the origin in the high-dimensional kernel feature space similar to one-class SVM [8]. As described in [2], a single null projection direction is computed to map all samples on a single target value s. A test sample x^* is projected on the null projection direction to obtain the value s^*. Figure 2 illustrates the one-class approach with kernel null space. The novelty score of x^* is the distance between s and s^*:

$$NoveltyScore(x^*) = \mid s - s^* \mid . \tag{2}$$

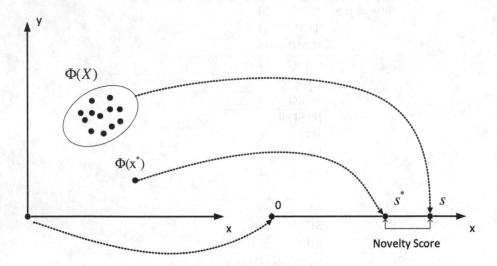

Fig. 2. The samples are separated from the origin in the kernel feature space with a mapping Φ, then mapped on a point s, and the novelty score of a testing sample x^* is the distance of its projection s^* to s.

A large novelty score indicates more likely novelty. In [2] and [1], a hard decision threshold $\theta_{threshold}$ is used to determine whether the test sample x^* belongs to the target class or not. Determining the threshold plays a very important role to the performance of the novelty detection process. To the best of our knowledge, this threshold has been selected heuristically up till now. Therefore, in this study, we propose an intrusion detection scheme based on an enhanced version of this Kernel Null Space method.

The procedure of the enhanced Kernel Null Space method is illustrated in Fig. 3 with two phases: the training phase and the detection phase.

In the training phase: training data samples $\{x_1, x_2, \ldots, x_N\}$, which have been already pre-processed, will be mapped on a point s in the Null Space F. The intrusion detection system uses another data set called the validation set that comprises other normal data samples $\{y_1, y_2 \ldots, y_M\}$. Each sample y_i of the validation set is mapped on a point $\hat{s}_i$ in the feature null space, for which $NoveltyScore(y_i)$ is calculated. After mapping all samples of the validation set and calculating Novelty scores for all of them, a set $\{NoveltyScore(y_i)\}$ is formed. Based on this set of novelty scores, we use the Kernel Quantile Estimator to derive the threshold $\theta_{threshold}$, which will be described in Sect. 3.2.

During the detection phase in real time, when a test data sample x^* comes, the system maps it on a point s^* and then calculate its $NoveltyScore(x^*)$. Then by comparing the $NoveltyScore(x^*)$ with $\theta_{threshold}$ found in the training phase, x^* can be classified as Normal or Anomaly.

In the following subsections, we will elaborate how we achieve an optimal kernel parameter on the given training data set and how to calculate threshold $\theta_{threshold}$ by Kernel Quantile Estimator.

Determination of Kernel and Kernel Parameter. In this paper, we select the Gaussian kernel (or Radial Basic Function (RBF)) for Kernel Null Space which is commonly used.

$$k(x, y) = \exp\left(\frac{-\| x - y \|^2}{2\sigma^2}\right) \tag{3}$$

where: σ stands for the kernel parameter in $[0, 1]$.

Using the method proposed in [11], the optimal sigma σ^* is estimated from the data set $\{x_1, x_2, \ldots, x_N\}$. The optimal σ^* is the one that maximizes the objective function $J(\sigma)$

$$J(\sigma) = \frac{2}{N} \sum_{i=1}^{n} \exp\left(-\frac{Near(x_i)}{2\sigma^2}\right) - \frac{2}{N} \sum_{i=1}^{n} \exp\left(-\frac{Far(x_i)}{2\sigma^2}\right) \tag{4}$$

Denote the nearest and farthest neighbors distances as:

$$Near(x_i) = \min_{j \neq i} \| x_i - x_j \|^2$$

$$Far(x_i) = \max_{i} \| x_i - x_j \|^2$$

Threshold Calculation Based on Kernel Quantile Estimator. As mentioned, the threshold for the Novelty Score is the crucial key for the accuracy in anomaly detection. A common method to choose a good threshold that we have observed up till now is checking various discrete threshold values in the increasing order until the test system outputs highest accuracy. But when we have to cope with continuous values, that heuristic check-up hardly finds a good threshold we can not check all continuous values.

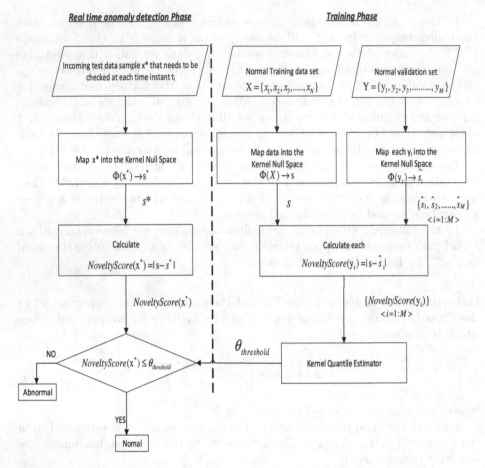

Fig. 3. Detection procedure using Kernel Null Space and Kernel Quantile Estimator

The set of the novelty scores is denoted by $\{NS_1, NS_2, \ldots, NS_M\}$ and investigated for the probability density distribution. As observed in Fig. 4, the Novelty Score values $\{NS_1, NS_2, \ldots, NS_M\}$ can not be approximated by a normal distribution, i.e. the underlying distribution of the sample is unknown. In this case, nonparametric methods could be used to explore this unknown underlying.

In this paper, we use the **Kernel Quantile Estimator** [9] to estimate $\theta_{threshold}$ over the set of Novelty Score values.

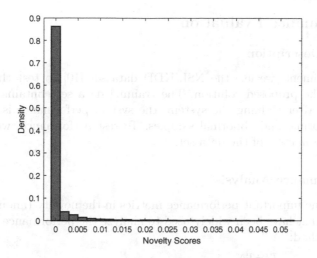

Fig. 4. Probability density distribution of novelty scores

Let $NS_{(1)} \leq NS_{(2)} \leq \ldots \leq NS_{(M)}$ denote the corresponding order statistics of the novelty scores. Suppose that $K(.)$ is a density function symmetric about Zero and that $h \to 0$ as $n \to \infty$, the Kernel Quantile Estimator can be calculated as follows [9]:

$$KQ_p = \sum_{i=1}^{N} \left[\int_{\frac{i-1}{n}}^{\frac{i}{n}} K_h(t - p)dt \right] NS_{(i)} \qquad (5)$$

where $h > 0$ is the bandwidth. The bandwidth h controls the smoothness of the estimator for a given sample of size n. $K_h(.) = \frac{1}{h}K(\frac{.}{h})$. And p is the proportion of the quantile.

Here we use the standard Gaussian kernel for the resulting estimate KQ_p which is a smooth unimodal,

$$K(u) = \frac{1}{\sqrt{2\pi}} \exp(-\frac{u^2}{2}) \qquad (6)$$

The selection of h is important in kernel density estimation: a large h will lead to an over-smoothed density estimate, while a small h will produce a ragged density with many spikes at the observations. As described in [9], the bandwidth computed as

$$h_{opt} = \left(\frac{pq}{n+1} \right)^{\frac{1}{2}} \qquad (7)$$

where: $q = 1 - p$

For a lot of continuous distributions used in statistics, specific quantiles such as the $p = 0.95$, 0.975, and 0.99 quantiles are tabulated. Therefore, in our experiment, we have investigated 3 cases of q: 0.05, 0.025 and 0.01 respectively. These 3 q values corresponds to 3 threshold value $KQ(p = 1 - q)$ (i.e. $\theta_{threshold}$).

4   Performance Evaluation

4.1   Data Description

In this experiment, we use the NSL-KDD data set [10] to test the detection accuracy of the proposed solution. The training data set contains 13449 normal samples After training the system, the system performance is checked by using 6000 normal and abnormal samples. To test performance, we use all 41 attributes/parameters of the data set.

4.2   Performance Analysis

There are some important performance metrics in the novelty (anomaly) detection domain that have been widely used to analyze the performance of a certain detection method:

- Accuracy $= \frac{TP+TN}{TP+FP+TN+FN}$
- ReCall - True Positive Rate or Sensitivity $= \frac{TP}{TP+FN}$
- FPR - False Positive Rate: $FPR = \frac{FP}{FP+TN}$

Where TP (True Positive) is the number of anomalies correctly diagnosed as anomalies; TN (True negative) is the number of normal events correctly diagnosed as normal; FP (False Positive) is the number of normal events incorrectly diagnosed as anomalies; and FN (False Negative) is the number of anomalies incorrectly diagnosed as normal events.

In our test, we compare the performance of the enhanced Kernel Null Space with the original Kernel Null Space in which the threshold is heuristically selected and fixed at 0.05 [2] and with the One Class Support Vector Machine method (OCSVM) [11].

As mentioned in Sect. 3.2, we have tested with 3 different q values: 0.01, 0.025 and 0.05 and found out that $q = 0.025$ brings best performance. The results are shown in Table 2:

As can be seen in Table 2, with the normalized and pre-processed 41-attribute data set $\{X_1, X_2, \ldots, X_N\}$, the optimal kernel parameter estimated is $\sigma^* = 0.5957$. Subsequently, from the given data set of Novelty scores $\{NS_1, NS_2, \ldots, NS_M\}$, supposed that $q = 0.025$, the threshold is $\theta_{threshold} = 0.0233$.

Table 2. Performance comparison

$\sigma = 0.5957$	Kernel Null Space			OCSVM	Origin
	q = 0.05	q = 0.025	q = 0.01		Kernel Null Space
	$\theta_{threshold} =$	$\theta_{threshold} =$	$\theta_{threshold} =$		with fixed
	0.0097	0.0233	0.0514		threshold = 0.05
Accuracy	0.9548	0.9598	0.92	0.9445	0.9212
FPR	0.0443	0.018	0.006	0.0433	0.006
Recall	0.954	0.9377	0.846	0.9323	0.8483

The obtained results show that; the enhanced Kernel Null Space slightly outperforms the OCSVM and the original Kernel Null Space methods in both terms of Accuracy and False Positive rate while a bit inferior to the Original Kernel Null Space method in terms of Recall.

5 Conclusion and Future Work

In this research, we have proposed and elaborated an Intrusion Detection System using the so-called enhanced Kernel Null Space method with data-driven threshold retrieval. The proposed solution with data-driven findings such as $q = 0.025$ and $\sigma = 0.5957$ is proved to outperform the current OCSVM and Original Kernel Null Space methods in terms of Detection Accuracy and False Positive Rate.

In the future, we would like to address the intrusion detection and the monitoring problem using deep learning, targeting on time series data with uncertainties. We also focus on the detection ability of our proposed approach for large stream data.

References

1. Bodesheim, P., Freytag, A., Rodner, E., Denzler, J.: Local novelty detection in multi-class recognition problems. In: 2015 IEEE Winter Conference on Applications of Computer Vision (WACV), pp. 813–820. IEEE (2015)
2. Bodesheim, P., Freytag, A., Rodner, E., Kemmler, M., Denzler, J.: Kernel null space methods for novelty detection. In: Proceedings of the IEEE Conference on Computer Vision and Pattern Recognition, pp. 3374–3381 (2013)
3. Borkar, A., Donode, A., Kumari, A.: A survey on intrusion detection system (IDS) and internal intrusion detection and protection system (IIDPS). In: International Conference on Inventive Computing and Informatics (ICICI), pp. 949–953. IEEE (2017)
4. Liu, J., Lian, Z., Wang, Y., Xiao, J.: Incremental kernel null space discriminant analysis for novelty detection. In: Proceedings of the IEEE Conference on Computer Vision and Pattern Recognition, pp. 792–800 (2017)
5. Mercado, G.R., Conerly, M.D., Perry, M.B.: Phase i control chart based on a kernel estimator of the quantile function. Qual. Reliab. Eng. Int. **27**(8), 1131–1144 (2011)
6. Nguyen, Q.T., Tran, K.P., Castagliola, P., Truong, T.H., Nguyen, M.K., Lardjane, S.: Nested one-class support vector machines for network intrusion detection. In: 2018 IEEE Seventh International Conference on Communications and Electronics (ICCE), pp. 7–12. IEEE (2018)
7. Park, Y., Baek, S.H., Kim, S.H., Tsui, K.L.: Statistical process control-based intrusion detection and monitoring. Qual. Reliab. Eng. Int. **30**(2), 257–273 (2014)
8. Schölkopf, B., Platt, J.C., Shawe-Taylor, J., Smola, A.J., Williamson, R.C.: Estimating the support of a high-dimensional distribution. Neural Comput. **13**(7), 1443–1471 (2001)
9. Sheather, S.J., Marron, J.S.: Kernel quantile estimators. J. Am. Stat. Assoc. **85**(410), 410–416 (1990)
10. Tavallaee, M., Bagheri, E., Lu, W., Ghorbani, A.A.: A detailed analysis of the KDD cup 99 data set. In: 2009 IEEE Symposium on Computational Intelligence for Security and Defense Applications, CISDA 2009, pp. 1–6. IEEE (2009)

11. Trinh, V.V., Tran, K.P., Huong, T.T.: Data driven hyperparameter optimization of one-class support vector machines for anomaly detection in wireless sensor networks. In: 2017 International Conference on Advanced Technologies for Communications (ATC), pp. 6–10, October 2017. https://doi.org/10.1109/ATC.2017.8167642
12. Wang, Y., Wong, J., Miner, A.: Anomaly intrusion detection using one class SVM. In: Information Assurance Workshop, pp. 358–364 (2004)

Social Coalition-Aware Task Assignment in Flying Internet of Things

Bowen Wang⬤, Yanjing Sun(✉)⬤, and Song Li⬤

China University of Mining and Technology, Xuzhou 221116, China
{bowenwang,yjsun,lisong}@cumt.edu.cn

Abstract. In this paper, we propose a social coalition-aware framework for task assignment in Flying Internet of Things (Flying IoT), where the intelligent unmanned aerial vehicles (UAVs), as a rising form of new IoT devices, can execute diverse tasks in a self-organized way. In this respect, the Social IoT (SIoT) paradigm promises to enable smart objects to autonomously work socially with surround objects, and thus build the social networks of objects without human intervention. The proposed framework aims at exploiting the extracted social attributes of drones for improving the efficiency of task coordination in a multi-hop delivery way. To measure the collaborative effect, a social-aware coalition game is formulated, and then we adopt the Shapley value to capture the relative importance of coalition members. The member with highest value will be elected as a leader for each coalition. Simulation results demonstrate that the proposed scheme can notably enhance the efficiency of task diffusion process compared to other benchmarks.

Keywords: Flying IoT · Social IoT · Task assignment · Game theory · Shapley value

1 Introduction

Internet of Things (IoT) promises to revolutionize humans' daily life by interconnect any possible smart objects, which can embrace some form of intelligence by providing a rich set of useful information to be gathered by heterogeneous objects [1]. However, the densification and heterogeneity of IoT devices have brought challenges to the scalability of IoT network architecture [2]. The unmanned aerial vehicle (UAV), as an emerging form of new smart devices, can collaboratively execute some dangerous tasks, such as environment sensing and emergency communication in disaster rescue [3]. However, it is impractical that

Supported by Key Research & Development Project for Science and Technology of Xuzhou, China (KC18105), the National Natural Science Foundation of China [51734009, 61771417, 51804304], Outstanding Innovation Scholarship for Doctoral Candidate of "Double First Rate" Construction Disciplines of CUMT, and joint Ph.D. program of "double first rate" construction disciplines of CUMT.

T. Q. Duong et al. (Eds.): INISCOM 2019, LNICST 293, pp. 141–151, 2019.
https://doi.org/10.1007/978-3-030-30149-1_12

all UAVs can directly communicate with ground controllers due to the limited power, which poses a challenge to the task implement [4]. Assume that intelligent UAVs can self-organize their decision making to free ground controllers from the tedious task of centralized control, the so-called Flying IoT are potential to play a crucial part in both civilian and military fields [5]. For better co-ordination and smoother task execution, drones need to collaborate on one mission. In this scenario, UAVs can be divided into several coalitions based on diverse mission types. In general, one UAV who communicates well with the ground controller can be elected as the coalition leader to guide the task implementation by broadcast task to its coalition members [4]. However, except for physical condition, some valuable social attributes should also be considered for UAVs to determine the importance in a coalition. In this respect, drones can establish social ties and collaborate socially with their surrounding drones, even though their owners have not established any relation on a social network, which conforms to the concept of Social Internet of Things (SIoT) [6]. According to the paradigm of SIoT, the UAVs can build the social networks among objects rather than participate in their owners' social networks, which inspires us to leverage the social context to improve the efficiency of task coordination.

Due to the dynamics of large-scale UAV networks, it is challenging to dynamically select a suitable sets of coalition leaders by accurately measuring their importance [4]. Game theory as a powerful mathematical tool, can analyze the users' rational behavior in wireless communications [7]. As one the most common characteristic function in cooperative game, the Shapley value plays an important role in value or cost-sharing game theoretic applications [8]. Due to its inherent ability to measure the contribution of the single players, the Shapley-value has been well applied into network and social-aware networking analysis [9–11]. The authors in [9] investigated the computation of Shapley value in the network centrality. In [10], the Shapley-value was used to enhance the networking Navigability in SIoT by measuring the influence of neighbor nodes. In [11], the authors adopted the Shapley-value to select the cluster head for each social community, and the selected seed user can improve the local offloading through multi-hop device-to-device (D2D) transmission. However, few work study how to utilize the Shapley-value to guide the task assignment in UAV networks.

To the best of our knowledge, the idea of leveraging social attributes to improve the task diffusion efficiency in Flying IoT has not been investigated yet. In this paper we intend to fill this gap, by designing a social coalition-aware task assignment method. The main contribution is the proposed novel social-aware scheme, which can select a set of suitable coalition leaders that can improve the efficiency of task coordination. In general, the proposed scheme aims to optimize the task diffusion speed through multi-hop delivery. To measure the collaborative effect of coalition leader, a social-aware coalition game is formulated. For this game, we adopt the Shapley value to capture the collaborative effect of one coalition member on the effectiveness of task diffusion.

The rest of this paper is organized as follows. In Sect. 2, we depict the social coalition-aware communication model with the framework of three different graph models and the optimization problem is formulated. In Sect. 3, the

Shapley value is adopted to measure the collaborative effect of coalition leader. The simulation results are conducted in Sect. 4 while conclusions are drawn in Sect. 5.

2 System Model and Problem Formulation

As shown in Fig. 1, we consider a multi-UAV networks with N UAVs and one ground controller. The sets of UAVs are denoted as $U = \{u_i\}_{i=1}^{N}$. Each UAV belongs to a coalition corresponding to a type of mission. Each UAV can directly receive the real-time task from a ground controller or from a coalition leader in a multi-hop delivery way. In this scenario, a UAV can directly deliver a B bits of data to the surrounding UAV, within one time slot t. We further assume that the span of each time slot is T seconds. Similar to assumption in [11], we focuses on delay-tolerant services, and thus the potential delay caused by multi-hop delivery can be ignored. Meanwhile, each drones is equipped with two antennas for transmit the mission information and control information, respectively [12]. The set of available service channel is denoted by $SC = \{sc_l\}_{l=1}^{L}$.

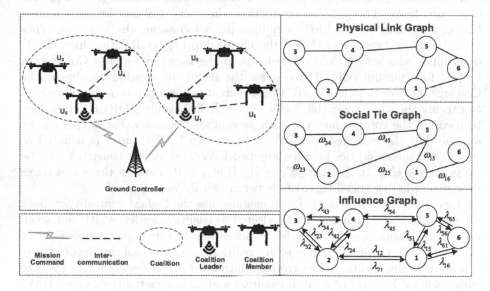

Fig. 1. The system model of coalition-based Flying IoT and its abstracted graph models.

Considering the dynamic 3D Cartesian coordinate model, the coordinate of u_i is denoted by $c_i = (x_i, y_i, z_i)$, and the distance between u_i and u_j can be represented as $d_{ij} = \|c_i - c_j\|$. Similar to the cooperative direct transmission in [12], the one-hop transmission rate for the link between u_i and u_j in channel sc_l is given by

$$R(i, j, l) = \frac{W}{2I_i(l)} \log_2(1 + \gamma_{i,j}) \tag{1}$$

where W is the normalized bandwidth block, $I_i(l)$ represents the sets of UAVs which select channel sc_l within the interference range, $\gamma_{i,j}$ represents the signal-to-noise ratio (SNR), and the channels between drones are dominated by line of sight (LoS) links [12]. Similar to [13], we further assume the fading process is constant within each time slot, and thus the data rate is constant during the time span of each one-hop link.

By establishing one-hop link among UAVs, we first introduce a physical link graph $G_p(U, E_p)$, where U represents the set of vertices (UAVs), and E_p represents the set of edges (available links). The edges can be constructed by satisfying the condition $E_p = \{(i,j)|B/R(i,j,l) \leq T\}$, which indicates that available link exists when u_i can entirely deliver the data packet to u_j within a given time slot. Considering that UAVs collaborate on different tasks, we assume that all the N UAVs form K coalitions. Let $C = \{c_k\}_{k=1}^{K}$ denotes the coalition set, where c_k denotes the set of UAVs in coalition k, and thus $\cup_{k=1}^{K} c_k = U$. In this paper, we assume that there exists no overlay part between any two coalitions.

In this paper, we innovatively consider the impact of social attributes on the performance of UAV task assignment. The context information such as position and mission target needs to be exchanged among drones to ensure the completion of various tasks. Through the interaction, the social ties can be built among UAVs. According to the SIoT paradigm [6], we consider the following types of social ties between two UAVs: the Ownership Object Relationship (OOR) determines whether two objects belongs to the same owner, the CoLocation Object Relationship (CLOR) measures the degree of location proximity; the Co-target Object Relationship (CTOR) is created when two objects collaborate to execute the same task; the Social Object Relationship (SOR) is established to measure the contact interval, such as contact frequency. For simplicity, we assume that the weighted social tie between two UAV i and j is denoted as $\omega_{i,j}$. Given the social ties between any two UAVs, we further construct a social tie graph $G_s(U, E_s, w_s)$, where $E_s = \{(i,j)|w_{i,j} > 0\}$ denotes the set of edges determine whether social tie exists between two UAVs.

Intuitively, in each coalition, the member has the highest centrality in both physical and social domains can be elected as coalition leader, which can accelerate the diffusion speed of task by broadcasting. In this paper, we consider K task data packets corresponding to K coalitions, and one coalition leader corresponding to one coalition. The ground controller first releases different tasks to each coalition leader. For a given coalition and its associated task, other UAVs in other coalitions can help to forward the task by multi-hop delivery. Hence, the successful delivery of task is influenced by both the social tie and physical condition between any two UAVs.

Next, we want to measure the influence according to both social tie graph and physical link graph. For instance, if $u_i \in c_k$ have to deliver the task information to another coalition member and the shortest multi-hop path containing $u_j \in c_{k'}$, there must exist a one-hop link passing u_j. Hence, the *directed influence* $\lambda_{i,j}$ is defined as the preference of $u_i \in c_k$ delivering task to its neighbor u_j, as follows:

$$\lambda_{i,j} = \sum_{u_{i'} \in c_k \setminus \{u_i\}, u_j \in SP_{i,i'}} \frac{\omega_{i,i'}}{|SP_{i,i'}|} \tag{2}$$

where $|SP_{i,i'}|$ denotes the number of UAVs within the shortest multi-hop path in $G_p(U, E_p)$.

Given the directed influence between any two UAVs, we further depict the influence graph $G_i(U, E_i, \lambda_i)$ to model the global influence. We aim to maximize the expected number of UAVs receiving the real-time task through multihop delivery by selecting suitable coalition leader for each coalition based on the influence graph. Let ρ_t be the UAVs set that have received task until time slot t, and $\Lambda = \{\rho_0, \rho_1, ... \rho_t\}$ represents the task diffusion process.

To accurately model the influence, we denote $\mathcal{N}_i$ as the one-hop neighbor set of u_i, and d_{ij}^s as the sum distance of shortest path between u_i and u_j. Next, we further define $\mathcal{N}_i^m = \{u_j | d_{ij}^s < d_{th}, u_j \in G_p\}$ as the multi-hop neighbor, whose distance from u_i is less than a given threshold. Next, we redefine the *undirected influence* based on closer neighbor set.

Definition 1. *The undirected influence of ρ_t on $u_i \in U \setminus \rho_t$ is defined as the expected number of UAVs on the multi-hop neighbor set $\mathcal{N}_i^m$ that will successfully receive the task data packet from the UAVs in ρ_t, which is expressed as*

$$\lambda_u(i, \rho_t) = \sum_{u_{i'} \in N_i^m} \left(1 - \prod_{u_j \in N_{i'}^m \cap \rho_t} (1 - \lambda_{i,j}) \right) \tag{3}$$

If $u_k \in \rho_t$ delivers the data packet to any UAV in the close neighbor set of u_i, we can say that u_k has an influence on u_i. If only u_k in ρ_t can affect u_i, the u_k is considered to has an *exclusive influence* on u_i, which is given by

$$
\begin{aligned}
\lambda_u(i, k) &= \lambda_u(i, \rho_t) - \lambda_u(i, \rho_t \setminus u_k) \\
&= \sum_{u_{i'} \in N_i^m} \left(1 - \prod_{u_j \in N_{i'}^m \cap \rho_t} (1 - \lambda_{i,j}) \right) - \sum_{u_{i'} \in N_i^m} \left(1 - \prod_{u_j \in N_{i'}^m \cap \rho_t \setminus u_k} (1 - \lambda_{i,j}) \right) \\
&= \sum_{u_{i'} \in N_i^m} \left(\prod_{u_j \in N_{i'}^m \cap \rho_t \setminus u_k} (1 - \lambda_{i,j}) - \prod_{u_j \in N_{i'}^m \cap \rho_t} (1 - \lambda_{i,j}) \right) \\
&= \sum_{u_{i'} \in N_i^m} \left(\left(1 - \prod_{u_j \in N_{i'}^m \cap \rho_t} (1 - \lambda_{i,j}) \right) \prod_{u_j \in N_{i'}^m \cap \rho_t \setminus u_k} (1 - \lambda_{i,j}) \right)
\end{aligned}
\tag{4}
$$

where the first term denotes the probability that u_k delivers the received data packet with at least one one-hop neighbor of u_j, and the second term is the probability that none of the members of $\rho_t \setminus u_k$ can delivers the received data packet with at least one one-hop neighbor of u_j. Hence, the *exclusive influence* can be expressed as the multiplication of these two terms.

Therefore, the optimization problem in this paper can be formulated as

$$\max_{\rho_0} \frac{1}{t} \sum_{x=1}^{t} |\rho_x| \tag{5}$$

It is obviously that we should select the initial coalition leader set ρ_0 to optimize the expected number of UAVs receiving the data packet in task diffusion process. There are some conventional methods in graph theory to solve problem (5) such as betweenness centrality and closeness centrality. However, these methods cannot accurately measure the relative importance of each coalition member. Hence, we need to adopt a suitable approach to accurately quantify the cumulative contributions for any single member.

3   Cooperative Game-Based Task Assignment

Game-theoretic solutions have recently been applied into the analysis network centrality. Particularly, the Shapley value in cooperative game can effectively measure the relative importance of single player to a group of players [14]. In general, if the characteristic function is defined as the influence of a node on other nodes over a graph, the Shapley value can be adopted to measure the influence of each node on other nodes. Consequently, the Shapley value in the a given game can be used to measure the centrality of a node over a graph, which has a high degree of flexibility to capture the relative importance in both social and physical domains. Now, we will formulate the optimization problem as a coalition game, and then adopt the Shapley value to measure the importance of each coalition member.

Based on the graph model above, the social-aware coalition game can be defined as $\mathcal{G} = (U, C, v)$, where U denotes all the players (UAVs), C represents the coalition structure of U, and v is a characteristic function assigning each coalition in the structure a value. Since the coalition formation problem can be solved by the some existing methods [8], we assume that the coalition structure is given by $C = \{c_k\}_{k=1}^{K}$, and mainly focus on the coalition leader selection based on the value (characteristic function) design. The value of a given coalition c_k is denoted as $v(c_k)$. Based on the aforementioned diffusion model, it is obvious that the diffusion process Λ is determined by the directed influence, one-hop and multi-hop neighbor set. Therefore, the value function for the coalition game $\mathcal{G}$, reflecting the cumulative *undirected influence* can be defined as

$$v(c_k) = \begin{cases} \sum_{u_i \in U \setminus c_k} \sum_{u_j \in c_k} \theta \lambda_u(i, j), & \text{if } c_k \neq \emptyset; \\ 0, & \text{Otherwise.} \end{cases} \tag{6}$$

where θ represents the price parameter to quantify influence. In this way, the value function in (6) will be a monetary value and of transferable utility (TU) [14]. The higher value function indicates a higher probability of delivering the data packet to other UAVs.

Now, we adopt the Shapley value of TU game $\mathcal{G}$ to measure the contribution of each coalition member to the value function of any subset s_k belonging to coalition c_k compared to UAVs in c_k, which is given by

$$\phi_i(c_k, \mathcal{G}) = \sum_{s_k \subseteq c_k \setminus u_i} \frac{(|c_k| - |s_k| - 1)!|s_k|!}{|c_k|!}(v(s_k \cup u_i) - v(s_k)) \tag{7}$$

The Shapley value can measure the degree of collaboration between two coalition members. Then, we illustrate the relationship between Shapley value of u_i and its exclusive influence on UAVs in other coalition, which corresponds to the case that some coalition member without the ability to communicate with its leader directly, so they need assistance from other coalitions to forward data packet.

Theorem 1. *The Shapley value of $u_i \in c_k$, can be calculated as the exclusive influence of u_i on other UAVs that are not in c_k, which is given by*

$$\phi_i(\mathcal{G}) = \sum_{\mathcal{N}_j^m \cap \mathcal{N}_i \neq \emptyset} \frac{\theta}{1 + |\mathcal{N}_j^m|} \tag{8}$$

Proof. Based on the expression of *exclusive influence*, the Eq. (7) can be rewritten as

$$\phi_i(c_k, \mathcal{G}) = \sum_{s_k \subseteq c_k \setminus u_i} \frac{(|c_k| - |s_k| - 1)!|s_k|!}{|c_k|!} \sum_{u_j \in U \setminus s_k} \lambda_u(i, j) \tag{9}$$

Given a coalition s_k and $u_i \notin s_k$, the u_i is viewed to have a exclusive influence another u_j. if and only if there is no intersection between the multi-hop neighbor set of u_j and the one-hop neighbor sets of all coalition members in s_k, satisfying that $\{\cup_{u_{i'} \in s_k} N_{i'}\} \cap N_j^m = \emptyset$. According to [9], the probability of satisfying this condition equals to $\frac{1}{1 + |\mathcal{N}_j^m|}$. Moreover, if u_i wants to deliver task to the multi-hop neighbor set of u_j, at least one multi-hop neighbor of u_j must belong to one-hop neighbor set of u_i, which implies $\mathcal{N}_j^m \cap \mathcal{N}_i \neq \emptyset$. Hence the Theorem 1.

Given the expression of Shapley value for each coalition member, the strategy of selecting coalition leader can be summarized as follows:

- All UAVs are divided into different coalition by mission types and they exchange context information in both social domain and physical domain. The ground controller constructs virtual graph models based on the physical link condition and social tie.
- In each coalition, the coalition member with highest Shapley value among the members of its coalition will be elected as the coalition leader. Then, the ground controller sends the data packets about different task to the coalition leaders. The coalition leaders first broadcast the data packet to its one-hop neighbor, and all coalition member will help other members within or without its coalition, by acting as the multi-hop relay node.

The main computational complexity consumption of proposed scheme comes form the computation of Shapley value. Firstly, the original Shapley value formula needs to consider $\mathcal{O}(2^{|N|})$ coalitions [9], which causes too large computational complexity consumption in the dense network. According to Theorem 1, the complexity can be simplified to only calculate the node degree and the number of shortest path in graph model. Herein, the complexity of calculating the node degree and calculating the shortest path can be represented as $\mathcal{O}(|N|)$ and $\mathcal{O}(|E_d| + |N| \log |N|)$, respectively.

4   Simulation Results and Performance Analysis

In this section, we compare our proposed social coalition-aware task assignment scheme, termed as "SCTA" with other benchmarks including betweenness centrality, closeness centrality, random selection, and our proposed scheme without considering the social tie. We conduct the simulation in a three-dimensional $(2 \times 2 \times 2 \, \mathrm{km}^3)$ space, where all UAVs are randomly distributed and in the relatively stationary state. The main simulation parameters are given as follows: bandwidth $B = 10 \, \mathrm{MHz}$ for each channel, the noise power density equals to $-174 \, \mathrm{dBm/Hz}$, the transmit powers of UAV are set to $30 \, \mathrm{dBm}$, the duration for each time-slot is $1 \, \mathrm{ms}$, the distance threshold to determine the close neighbor is

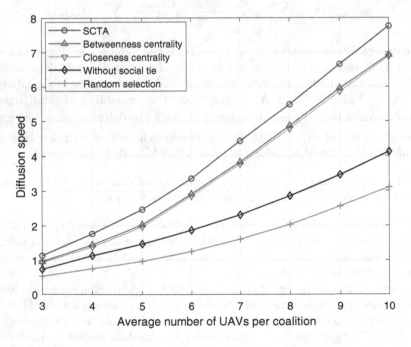

Fig. 2. The diffusion speed with varying number of UAVs per coalition.

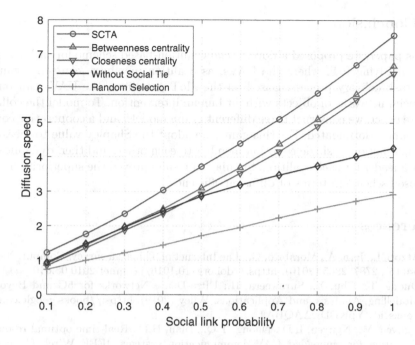

Fig. 3. The diffusion speed versus the social link probability.

400 m, and the size of each data packet is 1000 bits. To simplify, the internal of social tie is uniformly selected within [0, 1].

Figure 2 compares the diffusion speed attained by our proposed SCTA with other benchmarks. The diffusion speed of task represents the average difference between the number of UAVs that received the data packet during two consecutive time slots. The number coalition is set to 10. In this figure, it is obvious that the diffusion speed increases with the number of UAVs per coalition, and our proposed scheme can always achieve the optimal performance. This is because the betweenness centrality and closeness centrality do not comprehensively consider the relative importance in coalition leader selection. In general, the closeness centrality always select the coalition leader which is close with other coalition members, and betweenness centrality always select the coalition leader which are in the most of the shortest path sets.

In Fig. 3, the diffusion speed is shown with varying probability that the social tie exists, where the number of coalition is 10, and the number of UAVs is set to 10 per coalition. We aim to investigate the impact of social relationship on the task diffusion. It can be seen that the diffusion speed increase with the social link probability. It is worth mentioning that, as the social link probability increases, the advantage of our proposed algorithm is more obvious. Therefore, the rational utilization of social attributes can effectively enhance the performance of task assignment.

5 Conclusion

In this paper, we proposed a novel social coalition-based scheme for task assignment in Flying IoT, where the UAVs, as coalition members, can self-organize their task delivery process. Based on the SIoT paradigm, the UAVs can build the social networks of objects without human intervention. To model the collaborative effect, we construct three different graph models, and a cooperative coalition game is formulated. For this game, we adopt the Shapley value to measure the contribution of single coalition member to each other, and then dynamically select a leader for each coalition. Simulation results proved the superiority of our proposed scheme in terms of diffusion efficiency.

References

1. Atzori, L., Iera, A., Morabito, G.: The Internet of Things: a survey. Comput. Netw. **54**(15), 2787–2805 (2010). https://doi.org/10.1016/j.comnet.2010.05.010
2. Duong, T., Chu, X., Suraweera, H.: Ultra-Dense Networks for 5G and Beyond: Modelling, Analysis, and Applications. Wiley (2019). https://books.google.co.uk/books?id=YpKDDwAAQBAJ
3. Nguyen, M., Nguyen, L.D., Duong, T.Q., Tuan, H.D.: Real-time optimal resource allocation for embedded UAV communication systems. IEEE Wirel. Commun. Lett. **8**(1), 225–228 (2019). https://doi.org/10.1109/LWC.2018.2867775
4. Liu, D., Wang, J., Xu, Y., Ruan, L., Zhang, Y.: A coalition-based communication framework for intelligent flying ad-hoc networks. CoRR abs/1812.00896 (2018). http://arxiv.org/abs/1812.00896
5. Genc, H., Zu, Y., Chin, T., Halpern, M., Reddi, V.J.: Flying IoT: toward low-power vision in the sky. IEEE Micro **37**(6), 40–51 (2017). https://doi.org/10.1109/MM.2017.4241339
6. Atzori, L., Iera, A., Morabito, G., Nitti, M.: The Social Internet of Things (SIoT) - when social networks meet the internet of things: concept, architecture and network characterization. Comput. Netw. **56**(16), 3594–3608 (2012)
7. Han, Z., Niyato, D., Saad, W., Baar, T., Hjrungnes, A.: Game Theory in Wireless and Communication Networks: Theory, Models, and Applications, 1st edn. Cambridge University Press, New York (2012)
8. Saad, W., Han, Z., Debbah, M., Hjørungnes, A., Başar, T.: Coalitional game theory for communication networks. IEEE Signal Process. Mag. **26**(5), 77–97 (2009). https://doi.org/10.1109/MSP.2009.000000
9. Aadithya, K.V., Ravindran, B., Michalak, T.P., Jennings, N.R.: Efficient computation of the shapley value for centrality in networks. In: Saberi, A. (ed.) WINE 2010. LNCS, vol. 6484, pp. 1–13. Springer, Heidelberg (2010). https://doi.org/10.1007/978-3-642-17572-5_1
10. Militano, L., Nitti, M., Atzori, L., Iera, A.: Enhancing the navigability in a social network of smart objects: a Shapley-value based approach. Comput. Netw. **103**, 1–14 (2016). https://doi.org/10.1016/j.comnet.2016.03.007. http://www.sciencedirect.com/science/article/pii/S1389128616300743
11. Soorki, M.N., Saad, W., Manshaei, M.H., Saidi, H.: Social community-aware content placement in wireless device-to-device communication networks. IEEE Trans. Mob. Comput. **18**, 1938–1950 (2018). https://doi.org/10.1109/TMC.2018.2866100

12. Liu, D., Wang, J., Xu, Y., Xu, Y., Yang, Y., Wu, Q.: Opportunistic mobility utilization in flying Ad-Hoc networks: a dynamic matching approach. IEEE Commun. Lett. **23**, 728–731 (2019). https://doi.org/10.1109/LCOMM.2019.2899602
13. Yin, C., Garcia-Palacios, E., Vo, N., Duong, T.Q.: Cognitive heterogeneous networks with multiple primary users and unreliable backhaul connections. IEEE Access **7**, 3644–3655 (2019). https://doi.org/10.1109/ACCESS.2018.2887344
14. Serrano, R.: Cooperative games: Core and Shapley value. Working Paper 2007-11, Providence, RI (2007). http://hdl.handle.net/10419/80201

Industrial Networks and Applications

Ontology-Based Semantic Search for National Database of Natural Resources and Environment

Ngoc-Vu Nguyen[1,2(✉)], Hong-Son Bui[3], and Quang-Thuy Ha[1]

[1] University of Engineering and Technology (UET),
Vietnam National University, Hanoi (VNU),
144, Xuan Thuy, Cau Giay, Hanoi, Vietnam
thuyhq@vnu.edu.vn
[2] Department of Information Technology,
Ministry of Natural Resources and Environment,
28 Pham van Dong, Cau Giay, Hanoi, Vietnam
nnvu@monre.gov.vn
[3] Ho Chi Minh City Department of Natural Resources and Environment,
63 Ly Tu Trong, District 1, Ho Chi Minh City, Vietnam
son.ciren@gmail.com

Abstract. Semantic search helps the user queries to be understandable for electric agents searching. In this way, ontology plays the main role to define the semantic and the relations between user queries. The national database of natural resources and environment is a very large database system. Therefore, building a search engine software for the system with high accuracy and fast speed is very important for sharing information in the field of natural resources and environment. However, the existing search engine software in the national database needs to be improved to better meet user's needs. We proposed the architecture of ontology-based semantic search for the national database of natural resources and environment. Based on the proposed architecture, we have built semantic search software (VnNRESS) to demonstrate better results than the existing search software (NRESearch).

Keywords: Ontology · Semantic search · Search engine ·
Ontology-based semantic search

1 Introduction

In recent decades, the knowledge economy has been chosen as the development strategy in almost all countries. Studies have shown that ICT is the infrastructure to implement most of the activities of the knowledge economy, is an effective means of supporting and facilitating for creating and using knowledge.

The national database of natural resources and environment is a very large database system. Therefore, building a search engine software for the system with high accuracy and fast speed is very important for sharing information in the field of natural resources and environment.

T. Q. Duong et al. (Eds.): INISCOM 2019, LNICST 293, pp. 155–164, 2019.
https://doi.org/10.1007/978-3-030-30149-1_13

Keyword-based searching software was built in the project of building the national database of natural resources and environment. However, the software needed to improve some functions in order to better meet the needs of users: (i) keyword auto-suggestions and autocomplete input box, (ii) pre-processing in the case of Vietnamese without accents, (iii) word tokenizer and extracting a group of keywords from the user's query, (iv) processing with synonyms and related semantic words, (v) removing stop words and (vi) sorting search results. Thus, to reach improvement requirements with the above functions require the new searching software to analyze and "understand" the semantics of user query from which search results are the most relevant to user's demand.

Currently, semantic search and ontology-based semantic search are interested topics of the scientific community. There are works related ontology-based semantic search. Gupta et al. [2] proposed a novel architecture of ontology-based semantic search engine. Aghajani et al. [3] introduced a semantic search engine based on ontology with the name is Semoogle. Moawad et al. [4] proposed an ontology-based architecture for the semantic search engine for the Arabic language. Khan et al. [5] introduced an ontology-based semantic search engine in Holy Quran and an architecture of ontology-driven semantic search proposed by Bonino et al. [6].

In Vietnam, related to the field of research on semantic search, there are two typical research projects. The first is the knowledge-based search for water-related information system for the Mekong Delta, Vietnam by Tran Thai Binh (Universität Bonn University) and the second is classifying questions towards Vietnamese semantic search in the field of health by Nguyen Minh Tuan (University of Technology - VNU).

Therefore, building a semantic search software for the national database of natural resources and environment can solve the above-mentioned problems. In the current solutions to build semantic search software, ontology-based semantic search is a suitable and feasible solution.

This paper presents an ontology-based semantic search architecture for the national database of natural resources and environment. This paper has two main contributions: (i) propose an architecture of ontology-based semantic search for the national database of natural resources and environment, (ii) develop a semantic search software based the proposed architecture to evaluate the effectiveness of the proposed architecture.

The rest of this paper is organized as follows. In the next section, we briefly outline fundamental concepts of ontology building, semantic web and semantic search. Section 3 describes the proposed architecture of ontology-based semantic search for the national database of natural resources and environment. In Sect. 4, we present experimental results and comparing search results and performance of the semantic search software that was developed in our experiments with the existing search engine software. In the last section, we present conclusions and future work.

2 Ontology and Semantic Search

2.1 The Semantic Web

Semantic Web came up in 1998 by Tim Berners-Lee which was published on the road map to the Semantic Web on the home page of the WWW Consortium. According to Tim Berners-Lee, "the Semantic Web is an extension of the current web in which information is given well-defined meaning, better enabling computers and people to work in cooperation". The Semantic Web is considered a future generation of the Web, in which:

- Data could be processed by machine and automated agents.
- The meaning of information and services on the Web is defined for the Web to comprehend and satisfy the requests of people and machines to use its content.
- Explicit semantic information on the Web pages which can be used by intelligent agents to solve complex problems of information collection and Query-reply.

The final objective of the Semantic Web is to be able to keep the accountability of Web content and capability to semantically analyze. It needs a group of structures to model the knowledge and a link between the knowledge and contents. In this way, it depends on two basic components: ontology and semantic annotations. Ontology is used to interpret the textual content of a resource regardless of its format. There have been many fundamental approximations in the Semantic Web in which it is supposed that resources have been semantically annotated (Fig. 1).

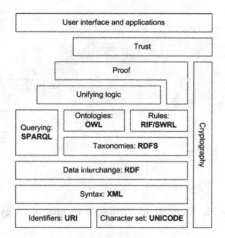

Fig. 1. The semantic web architecture

2.2 Ontology

Ontology term was initially used preferably by AI researchers and now it is one of the bases of the Semantic Web. It is impossible to envision the Semantic Web with no ontology because Semantic Web is the prime research project concerning ontology. The term Ontology is lent from philosophy. There are different definitions for the concept of ontology applied to information systems, each emphasizes a specific aspect. Gruber (1993) defines an ontology as a formal specification of conceptualization or, in other words, declarative representation of knowledge relevant to a particular domain. According to Uschold and Gruninger (1996) ontology is as a shared understanding of some domain of interest. Ontology provides "well-defined meaning" to the information enclosed in the Web also the benefit that different parties over the internet now have "shared" definitions about certain key concepts. The most important characteristic of ontology for the present research effort is their role as a structured form of knowledge representation. Ontology is used for the reason of interoperability among systems based on different schemas and comprehensively describing knowledge about a domain in a structured and sharable way, ideally in a readable format that is processable by a computer.

Noy and McGuinness [7] introduced the process of ontology building with four main steps. Firstly, define classes of the ontology. Secondly, reorganize classes in the class hierarchy of the ontology. Thirdly, define the properties of classes and finally, find instances of classes and value of properties of classes (Fig. 2).

In the TNMT.2015.08.06 project [1], we built an ontology in the field of natural resources and environment and this ontology will be used for the proposed architecture in Sect. 3.

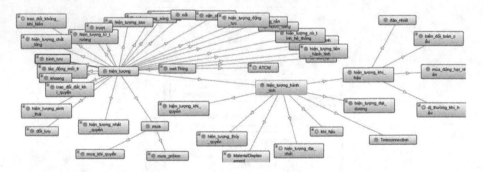

Fig. 2. Example of Vietnam's weather ontology

2.3 Semantic Search

Basically, a semantic search engine has a similar structure to a regular search engine that also includes two main components: the front-end component and the back-end component.

The front-end component has two main functions:

- Query interface: allows users to enter queries.
- Display searching results.

The back-end component plays an important role in the semantic search engine and it has three main functions:

- Analyzing user's queries.
- Searching in databases or document stores for user's queries.
- Document sets, search data/semantic network.

The difference in the structure of semantic search engines compared to a search engine is usually in the internal architecture part, specifically in two components: analyzing questions and searching data sets. Figure 3 shows the basic architecture of the semantic search system:

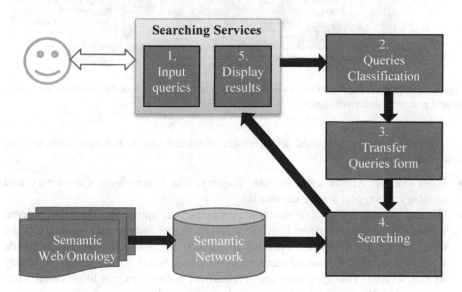

Fig. 3. The basic architecture of a semantic search system

3 The Proposed Architecture of Ontology-Based Semantic Search

Figure 4 shows the proposed architecture of ontology-based semantic search of the national database of natural resources and environment:

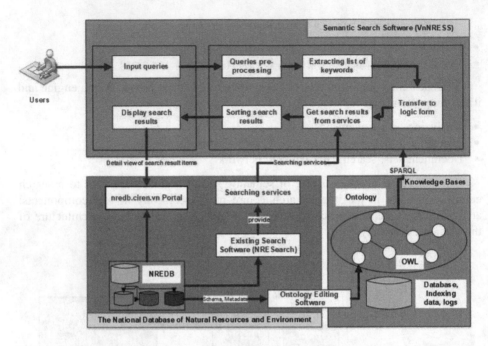

Fig. 4. The proposed architecture of ontology-based semantic search of the national database of natural resources and environment

According to the proposed architecture, semantic search software includes the following main functions:

- Input queries: Allows users to enter a query. The system uses a dictionary and ontology to suggest search keywords.
- Queries prepossessing: analysis and processing the query the user has entered includes the following steps: Firstly, removing spaces and special characters in the query except for the search operator characters. Secondly, removing stop words which are words that have no meaning in the query (using a Vietnamese dictionary of stop words). Thirdly, converting Vietnamese without accents to accented Vietnamese. Finally, separating the query into sentences by search operators.
- Extracting list of keywords: using Vietnamese natural language processing techniques with VietWordNet (a WordNet for the Vietnamese language) and ontology as additional knowledge to extract a list of keywords that match the meaning that the user desires.
- Transfer queries to logic form: building the query in the searchable query format in DBMS.
- Get searching results from the service: using the search service provided by the NRESearch software with input is the processed query to retrieve the search results.
- Sorting search results.
- Display search results.

In functions from 1st to 5th, the ontology of natural resources and environment domain is used as additional knowledge, to improve efficiency and accuracy.

4 Experimental Results

4.1 Datasets

Ontology

The ontology used in this research was built in the TNMT.2015.08.06 project for two domains of natural resources and environment field include geodesy and cartography domain and meteorological hydrological domain. Total classes in this ontology about 111.150 classes.

Databases

Data used for searching include text, tabled data, geospatial data were stored in database systems of the national database of natural resources and environment. The database systems were built in the project of building a national database of natural resources and environment by Ministry of Natural Resources and Environment.

4.2 Evaluation Method

We have created 300 different queries belonging to different types of queries such as Vietnamese-accented queries, Vietnamese without accents queries, and queries have Vietnamese language errors (extra space, contain special characters, contain stop words, etc.) and queries contain a combination of search operators (AND, OR, XOR, NOT).

We used these queries as input to the VnNRESS software and the NRESearch search software. Then for each query, we evaluate the search results of two search softwares based on a list of evaluation criteria such as:

- The semantic relevance of the search results for the user's question.
- The number of search results.
- The sort order of search results.
- Total processing time to produce search results.

4.3 Experiment Results and Analysis

- With most of the queries, the VnNRESS software gave better results than NRE-Search software on the following issues:
    - Keyword suggestions: suggestion of search keywords in NRESearch software is not correct. VnNRESS software not only suggests exact search keywords with the Vietnamese-accented query but also exactly with the Vietnamese without accents query. In addition, VnNRESS software also supports a suggestion of search keywords with synonyms and near meanings in the conceptual hierarchy of the ontology.

- Handling Vietnamese without accents: Since using the ontology has established the equivalent meaning of the Vietnamese word with accents and the Vietnamese word without accents corresponding, so the VnNRESS software handled very well with Vietnamese without accents queries, while the NRESearch software did not yet have this function.
- Tokenizer and extracting a list of search keywords: it is very simple with the NRESearch software because of separation of the query by space into a list of single words. Meanwhile, the VnNRESS software performs separation by meaningful words using the dictionary and the ontology (e.g. "Ho Chi Minh city" can be understood as three keywords "Ho Chi Minh City", "City" and "Ho Chi Minh". With the separation of keywords into the above meaning, the search results will be more appropriate for the searching needs of users.
- Elimination of stop words: using a dictionary of Vietnamese stop words inherited from previous studies, eliminating stop words of the VnNRESS software will help to get more suitable search results.
- Sorting search results: The order of sorting the search results of results display function of the VnNRESS software is more appropriate than the NRESearch software.
- Handling search operators: The VnNRESS software supports most of the basic search operators and gives more accuracy than the NRESearch software.
- Accuracy of the search results: Through the above tests, the VnNRESS software has made a list of results that are more suitable for search requests in the user's search content.
- In addition, the VnNRESS software has been added some following functions:
  - Support suggests other search content, more suitable than the content that the user has entered.
  - Having the function of identifying the name in the content of the user's search by using information about place names and information about administrative units: provinces, districts and communes.
  - Supporting the standardization of key words through the relationship between words (for example, geographic names often come together; provincial/district/ commune information often goes together,…).
  - Support the identification of information about place names, give correct answers about names, geographical locations, locations on the map of places that appear in the user's search sentence.
  - Supporting search by synonyms, related words (upper level concepts, lower levels in the ontology).
- Performance: Although it takes a lot of time for analyzing users' questions, extracting a list of appropriate keywords but search execution time of the VnNRESS software is acceptable and guaranteed (Figs. 5 and 6).

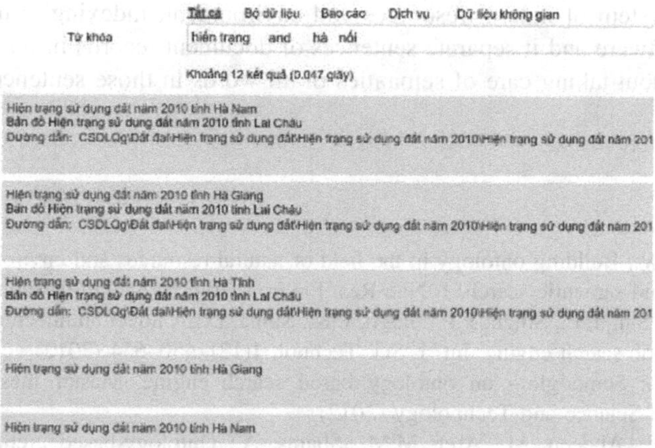

Fig. 5. The search results of the NRESearch software with input query is "hiển trạng and hà nội"

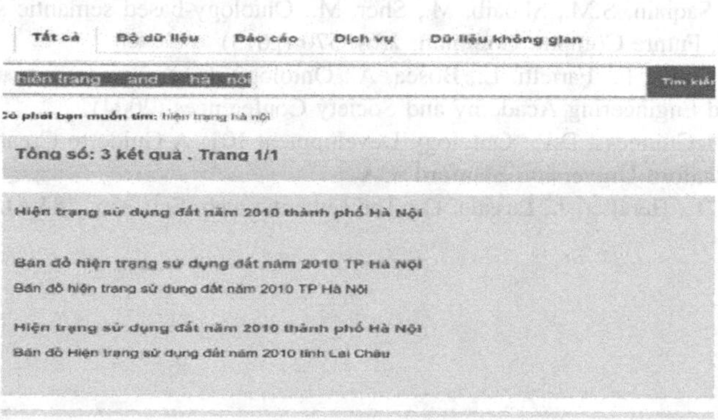

Fig. 6. The search results of the VnNRESS software with input query is "hiển trạng and hà nội"

5 Conclusions and Future Work

Through the above experimental and assessment results, the VnNRESS software has been proven effective and suitable, more accurate than the NRESearch software. At the same time, the empirical results also confirm the proposed architecture of ontology-based semantic search for the national database of natural resources and environment is appropriate and feasible.

However, in order to achieve greater efficiency, our future works need focus on two main issues:

- Firstly, we need to build a completed ontology for all domains of natural resources and environment field using semi-automatic and automated techniques based on web resources to reduce the effort and cost of ontology building.
- Secondly, we will have to use the complete ontology for all phases of the data search process. Especially in the search indexing phase because in the national

database system of natural resources and environment, indexing is using Apache Lucene software and it separate sentences of document records into a list of single words without taking care of semantics of all words in those sentences.

References

1. Nguyen, N.-V.: Building ontology in the field of natural resources and environment for data integration and semantic search. J. Nat. Res. Environ. 30–32 (2017). (in Vietnamese)
2. Gupta, P.N., Singh, P., Singh, P.P., Singh, P.K., Sinha, D.: A novel architecture of ontology-based semantic search engine. Int. J. Sci. Technol. 1(12), 650–654 (2012)
3. Aghajani, N.: Semoogle - an ontology based search engine, Master thesis, Norwegian University of Science and Technology (2012)
4. Moawad, I.F., Abdeen, M., Aref, M.M., Shams, A.: Ontology-based architecture for an Arabic semantic search engine. In: The Tenth Conference on Language Engineering (ESOLEC 2010) (2010)
5. Khan, H.U., Saqlain, S.M., Shoaib, M., Sher, M.: Ontology-based semantic search in Holy Quran. Int. J. Future Comput. Commun. 2(6), 570 (2013)
6. Bonino, D., Corno, F., Farietti, L., Bosca, A.: Ontology driven semantic search. In: World Scientific and Engineering Academy and Society Conferences (2004)
7. Noy, N.F., McGuinness, D.L.: Ontology Development 101: A Guide to Creating Your First Ontology. Stanford University, Stanford, CA
8. Berners-Lee, T., Hendler, J., Lassila, O.: The semantic web. Sci. Am. 284(5), 34–43 (2001)

Optimized Multi-cascade Fuzzy Model
for Ship Dynamic Positioning System
Based on Genetic Algorithm

Viet-Dung Do[1,2(✉)], Xuan-Kien Dang[1], Leminh-Thien Huynh[3],
and Van-Cuu Ho[3]

[1] Ho Chi Minh City University of Transport, Ho Chi Minh City, Vietnam
[2] Dong An Polytechnic, Di An, Vietnam
vietdung@dongan.edu.vn
[3] SaiGon University, Ho Chi Minh City, Vietnam

Abstract. In this paper, we aim to develop a multi-cascade fuzzy model for the ship dynamic positioning system influenced by environment to enhance its quality. The cascades of fuzzy model are selected corresponding to the level of output feedback error. The optimized tuning of the structure parameter for fuzzy-case 2 and fuzzy-case 4 is realized by the genetic algorithm. Then, the simulation studies which compare our proposed control strategy with fuzzy control strategy using Matlab are carried out, and the simulation result proves the effectiveness of the multi-cascade fuzzy model.

Keywords: Dynamic positioning system · Environmental impact · Multi-cascade fuzzy · Genetic algorithm

1 Introduction

Offshore exploration and exploitation of ocean resources have led to the increasing demands for ship dynamic positioning (DP) system. In a DP system, a computer coordinates the propulsion system consisting of thrusters and propellers to keep the vessel in position. The environmental factors, which have an effect on a hull, are regularly changing. So the vessel moves under various conditions that make the object highly nonlinear. In the early 1960s, the first ship dynamic positioning system uses traditional proportional-integral-derivative control [1]. Subsequently, the modern theories are employed for improving the quality of controller are applied. Among them, the fuzzy algorithm is one of the most widely used theories in DP control because it has robust characteristics for the nonlinear object. Chang et al. apply an application of Takagi-Sugeno (TS) fuzzy to represent the nonlinear DP system [2]. Via parallel distributed compensation rule, the DP control system can be merged by the linear controllers of all rules. However, it is necessary to take practical test cases such as environmental impacts for verifying the advantage of proposed solution. In order to improve the control quality, Chen et al. provide an approximation technique based on adaptive control in combination with type-2 fuzzy model [3]. The proposed adaptive type-2 fuzzy model is able to reduce the hydrodynamic disturbances. But the rigorous

© ICST Institute for Computer Sciences, Social Informatics and Telecommunications Engineering 2019
Published by Springer Nature Switzerland AG 2019. All Rights Reserved
T. Q. Duong et al. (Eds.): INISCOM 2019, LNICST 293, pp. 165–180, 2019.
https://doi.org/10.1007/978-3-030-30149-1_14

theoretical examination is necessary. Ho et al. combine the orthogonal function approach and the hybrid Taguchi-genetic algorithm to define quadratic finite horizon optimal controller design fuzzy problems [4]. The standard algebraic process in suggestion approach allows easy calculation. Therefore, the process of designing finite quadratic optimal controllers for DP is much simpler. Hu et al. present an adaptive fuzzy controller for the DP system under unexpected impacts from environmental operation [5]. The unexpected impacts are approximated by the adaptive fuzzy structure. The proposed solution scheme does not require knowledge of vessel dynamic model parameters and time-varying environmental disturbances. Fang et al. apply a Neural-Fuzzy algorithm into practice to find out the best of ship propulsion systems [6]. In this case, the environmental disturbances are estimated and reduced by the neural structure. However, the real-time signal treatment of sensor with a Kalman suggests and the thruster power lags are critical aspects that should be carefully analysis. The fuzzy method has been widely used for identifying nonlinear system, only simple structure tests. For an uncertain DP system structure, it requires more work to determine the optimal fuzzy rules. On the other side, the goal of control should be confirmed under the operating condition of different seas to enhance the quality structure of controller.

This paper provides an optimized multi-cascade fuzzy model for the nonlinear DP system. The designed controller is composed of four fuzzy cascades which are selected from simple to complex edition which correspond to the level of output error, due to environmental impact. The unknown parameters caused by environmental factors that are estimated by the fuzzy sets. Subsequently, a real-coded genetic algorithm (GA) is applied in an iterative fashion together with a rule base of fuzzy algorithm in order to optimize and simplify the model, respectively. The proposed results is demonstrated for system identification and a classification problem.

The paper is organized as follows. In the next section, the problem formulation and preliminaries are discussed. The multi-cascade fuzzy model controller is proposed in Sect. 3. Section 4 presents the fuzzy genetic algorithm design controller. The supervisor design is given in Sect. 5. The simulation results are provided to validate the designed controller in Sect. 6, followed by the conclusions of the paper in Sect. 7.

2 Problem Formulation and Preliminaries

The DP motion of vessel is described by three degrees of freedom [1]. Two separate coordinate systems presented by Fig. 1 include: one is a vessel fixed non-inertial frame $O - XYZ$; and the other is the inertial system approximated to the earth $O_0 - X_0Y_0Z_0$. Model representation of the DP system with three degrees of freedom, namely, surge, sway, yaw and external force acting are shown in equations as below:

$$\dot{\eta} = J(\eta)v \tag{1}$$

$$M\dot{v} + Dv = \tau + J^T(\eta)\tau_e \tag{2}$$

where position (x, y) and heading (ψ) of the absolute coordinate system $X_0 Y_0 Z_0$ are denoted as a vector from $\eta = (x, y, \psi)^T$. The vector $v = (u, v, r)^T$ describes velocities of the vessel motion in the relative frame of reference. The control vector τ produced by propeller and thruster systems. Vector τ_e represents the impact forces from environmental factors, including wave, wind and current.

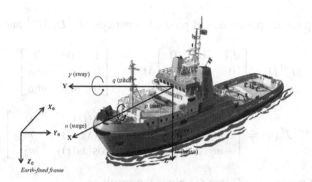

Fig. 1. Definition of the earth-fixed and the vessel-fixed reference frames.

The vertical centering of the relative coordinate system XYZ is placed at the roll axis of vessel, x_G denotes the longitudinal position of the gravity centre of the vessel towards the relative frame of reference. The transformation matrix $J(\psi)$ and $M \in R^{3 \times 3}$ and $D \in R^{3 \times 3}$ are the inertia and damping matrix, respectively. Such matrixes are taken as

$$J(\psi) = \begin{bmatrix} \cos(\psi) & -\sin(\psi) & 0 \\ \sin(\psi) & \cos(\psi) & 0 \\ 0 & 0 & 1 \end{bmatrix} \tag{3}$$

$$M = \begin{bmatrix} m - X_{\dot{u}} & 0 & 0 \\ 0 & m - Y_{\dot{v}} & mx_G - Y_{\dot{r}} \\ 0 & mx_G - N_{\dot{v}} & I_z - N_{\dot{r}} \end{bmatrix}; D = \begin{bmatrix} -X_u & 0 & 0 \\ 0 & -Y_v & mu_0 - Y_r \\ 0 & -N_v & mx_G u_0 - N_r \end{bmatrix} \tag{4}$$

where m is the vessel mass, I_z is the moment of inertia about the body-fixed Z-axis, x_G represents the location of G in x-axis direction, u_0 is velocity component at mid-vessel. Rewriting (1) and (2) in matrix form [7], yields the following

$$\begin{bmatrix} \dot{\eta} \\ \dot{v} \end{bmatrix} = \begin{bmatrix} 0 & J(\eta) \\ 0 & -M^{-1}D \end{bmatrix} \begin{bmatrix} \eta \\ v \end{bmatrix} + \begin{bmatrix} 0 \\ M^{-1} \end{bmatrix} \tau + \begin{bmatrix} 0 \\ M^{-1} J^T(\eta) \end{bmatrix} \tau_e \tag{5}$$

We defined the following vatiables

$$x(t) = [\eta, v]^T = [x_1, x_2, x_3, x_4, x_5, x_6]^T$$
$$u(t) = \tau = [u_1, u_2, u_3]^T \tag{6}$$
$$\tau_e(t) = [\tau_{e1}, \tau_{e2}, \tau_{e3}]^T$$

Let us compute the parameters of products matrices $M^{-1}D$, M^{-1} and $M^{-1}J^T(\eta)$ as

$$M^{-1}D = \begin{bmatrix} d_{11} & 0 & 0 \\ 0 & a_{22} & a_{23} \\ 0 & a_{32} & a_{33} \end{bmatrix} M^{-1} = \begin{bmatrix} 1 & 0 & 0 \\ 0 & m_{01} & m_{02} \\ 0 & m_{20} & m_{10} \end{bmatrix} \tag{7}$$

$$M^{-1}J^T(\eta) = \begin{bmatrix} \cos(x_3(t)) & \sin(x_3(t)) & 0 \\ -m_{33}\sin(x_3(t)) & m_{33}\cos(x_3(t)) & -m_{32} \\ -m_{23}\sin(x_3(t)) & -m_{23}\cos(x_3(t)) & m_{22} \end{bmatrix} \tag{8}$$

Next, substituting (7)–(8) into (5) gives the result as

$$\begin{bmatrix} \dot{x}_1 \\ \dot{x}_2 \\ \dot{x}_3 \\ \dot{x}_4 \\ \dot{x}_5 \\ \dot{x}_6 \end{bmatrix} = \begin{bmatrix} 0 & 0 & 0 & \cos x_3 & -\sin x_3 & 0 \\ 0 & 0 & 0 & \sin x_3 & \cos x_3 & 0 \\ 0 & 0 & 0 & 0 & 0 & 1 \\ 0 & 0 & 0 & -d_{11} & 0 & 0 \\ 0 & 0 & 0 & 0 & -a_{22} & -a_{23} \\ 0 & 0 & 0 & 0 & -a_{32} & -a_{32} \end{bmatrix} \begin{bmatrix} x_1 \\ x_2 \\ x_3 \\ x_4 \\ x_5 \\ x_6 \end{bmatrix} + \begin{bmatrix} 0 & 0 & 0 \\ 0 & 0 & 0 \\ 0 & 0 & 0 \\ 1 & 0 & 0 \\ 0 & m_{01} & m_{02} \\ 0 & m_{20} & m_{10} \end{bmatrix} \begin{bmatrix} 0 \\ 0 \\ 0 \\ u_1 \\ u_2 \\ u_3 \end{bmatrix}$$

$$+ \begin{bmatrix} 0 & 0 & 0 \\ 0 & 0 & 0 \\ 0 & 0 & 0 \\ \cos(x_3(t)) & \sin(x_3(t)) & 0 \\ -m_{33}\sin(x_3(t)) & -m_{33}\cos(x_3(t)) & -m_{33} \\ -m_{23}\sin(x_3(t)) & -m_{23}\cos(x_3(t)) & -m_{23} \end{bmatrix} \begin{bmatrix} 0 \\ 0 \\ 0 \\ \tau_{e1} \\ \tau_{e1} \\ \tau_{e1} \end{bmatrix} \tag{9}$$

The state equation of DP system can be rewritten in more compact form as

$$\dot{x} = Ax(t) + Bu(t) + E\tau_e(t) \tag{10}$$

In this paper we make the following assumptions:

Assumption 1: The force of environmental factors τ_e are time-varying and unknown. On the other hand, the structure of controllers are built with a fixed status. Therefore, DP system of vessel with unknown parameters and fixed status becomes the significant challenge of the DP control design.

Remark 1: The vessel motion operates in a practical case under environment impacts, so the object parameter is highly nonlinear underlying physical processes. Thereby carrying out the DP control with fixed status on the nonlinear object does not give high precision. As such, the Assumption 1 is reasonable and more practical.

In this paper, the controller is to design a multi-cascade fuzzy model for the DP system of vessel (1) and (2) under the Assumption 1 such that the vessel is maintained at the desired values of its position and heading with arbitrary accuracy, while GA suggestion is adapted to calibrate the fuzzy-case 2 and fuzzy-case 4. Thereby enhancing the quality of system and optimizing the controller structure for vessel motion.

3   Multi-cascade Fuzzy Model

In this study, we design a multi-cascade controller which comprises four fuzzy cases for the DP system which constitutes a nonlinear object. Structurally, the designed controllers form a cascade architecture. A fuzzy-TS dynamic model has been proposed by Do et al. to represent local linear input/output relations of DP systems [8]. The inference process system combines membership functions (MFs) with if-then rules and the fuzzy logic operators. The TS model consists of rules where the rule consequents are often taken to be linear functions of the inputs given as follow [9]:

Plan rule R_i

$$\text{If } z_1 \text{ is } F_{k1}^i \text{ and} \dots \text{ and } z_n \text{ is } F_{kn}^i \text{ Then } \dot{x}(t) = A_i x(t) + B_i u(t) + E\tau_e(t) \qquad (11)$$

where $F_{k1}^i, F_{k2}^i, ..F_{kn}^i$ are the fuzzy sets [10], for $i = 1, 2, \dots, n$, $A_i \in R^{n \times n}$ and $B_i \in R^{n \times m}$ are the input and state matrix, n is the number of If-Then rules, and $z_1, z_2, \dots, z_n$ are the premise variables. The overall fuzzy system is given by

$$\dot{x}(t) = \frac{\sum_{i=1}^{n} \mu_i(z(t))(A_i x(t) + B_i u(t))}{\sum_{i=1}^{n} \mu_i(z(t))} + E\tau_e(t)$$

$$= \sum_{i=1}^{n} h_i(z(t))(A_i x(t) + B_i u(t)) + E\tau_e(t) \qquad (12)$$

$$\mu_i(z(t)) = \prod_{j=1}^{n} F_{ij}(z_j(t))$$

$$h_i(z(t)) = \frac{\mu_i(z(t))}{\sum_{l=1}^{n} \mu_l(z(t))} \qquad (13)$$

$$z(t) = [z_1(t), z_2(t), \dots, z_n(t)]$$

for $F_{ij}(z_j(t))$ is the rank of MFs of $z_j(t)$ in F_j. In this paper, we assumes that $\mu_i(z(t)) \geq 0$ with $i = 1, 2, \ldots, n$ and

$$\sum_{i=1}^{n} \mu_i(z(t)) \geq 0, \bigvee t \tag{14}$$

We get $h_i(z(t)) \geq 0$, for $i = 1, 2, \ldots, n$ and

$$\sum_{i=1}^{n} h_i(z(t)) = 1 \tag{15}$$

Therefore, from (5) we have the following

$$\dot{x}(t) = Fx(t) + Gu(t) + E\tau_e(t) \tag{16}$$
$$= \sum_{i=1}^{n} h_i(z(t))(A_ix(t) + B_iu(t)) + \left\{ \left(F(x) - \sum_{i=1}^{n} h_i(z(t))A_ix(t) \right) \right.$$
$$\left. + \left(G(x) - \sum_{i=1}^{n} h_i(z(t))B_ix(t) \right) u(t) \right\} + E\tau_e(t)$$

where $\left\{ \left(F(x) - \sum_{i=1}^{n} h_i(z(t))A_ix(t) \right) + \left(G(x) - \sum_{i=1}^{n} h_i(z(t))B_ix(t) \right) u(t) \right\}$ indicates the estimation error between the DP system (5) and the fuzzy system (11). Let us assume that the following fuzzy system is carried out to control the internal system (15) design as:

Plan rule R_i:

If $z_1(t)$ is F_{k1}^i and ... and $z_n(t)$ is F_{kn}^i, Then $u(t) = K_jx(t)$, for $j = 1, 2, \ldots, r$ (17)

where K_j is the control adjustment. Hence, the following overall fuzzy controller is represented as:

$$u(t) = \frac{\sum_{j=1}^{r} \mu_j(z(t))K_jx(t)}{\sum_{j=1}^{n} \mu_j(z(t))} = \sum_{j=1}^{r} h_j(z(t))K_jx(t) \tag{18}$$

where $h_j(z(t))$ is defined in (13) (for $j = 1, 2, \ldots, r$). Next, substituting (18) into (16) yields, the closed-loop DP nonlinear control system as follows:

$$\dot{x}(t) = Fx(t) + Gu(t) + E\tau_e(t)$$

$$= \sum_{i=1}^{n}\sum_{j=1}^{r} h_i(z(t))h_j(z(t))(A_i + B_iK_j)x(t) + \left(F(x) - \sum_{i=1}^{n} h_i(z(t))A_ix(t)\right)$$

$$+ \sum_{i=1}^{n} h_i(z(t))\sum_{j=1}^{r} h_j(z(t))(G - B_ix(t))K_jx(t) + E\tau_e(t)$$

$$= \sum_{i=1}^{n}\sum_{j=1}^{r} h_i(z(t))h_j(z(t))(A_i + B_iK_j)x(t) + \Delta f + \Delta g + E\tau_e(t) \tag{19}$$

Defining and applying

$$\Delta f = \left(F(x) - \sum_{i=1}^{n} h_i(z(t))A_ix(t)\right) \tag{20}$$

and

$$\Delta g = \sum_{i=1}^{n} h_i(z(t))\sum_{j=1}^{r} h_j(z(t))(G - B_ix(t))K_jx(t) \tag{21}$$

The environmental factors make the control signal to be erroneous. So the building of MFs is an important thing to ensure the quality of fuzzy controller. If the passive parameter of MFs are used, the system performance will be lowered, and the object will be even out of balance. The GA is an effective tool for structure optimization of the controllers. Hence, integration and synthesis of fuzzy schematic and GA have been proposed to reduce the nonlinear characteristic. However, if the vessel operates under normal environmental conditions, the complex fuzzy controller (optimized by GA) is not effective, leading to slow response time. In the paper, we propose a multi-cascade fuzzy model which is divided into four fuzzy control cases corresponding to the level of environmental impact. In the no environmental impact case, the first fuzzy model is defined by a simple architecture which is described by 3×3 memberships function (fuzzy-case 1). So the closed-loop DP nonlinear control is rewritten as

$$\dot{x}_1(t) = K_s \sum_{i=1}^{3}\sum_{j=1}^{3} h_i(z(t))h_j(z(t))(A_i + B_iK_j)x(t) + \Delta f + \Delta g + E\tau_e(t) \tag{22}$$

where K_s is the fuzzy model selection, is defined by the supervisor module. In the low environment impact, the structure of second fuzzy is the same in first case. However, the value of fuzzy sets is optimized by GA (fuzzy-case 2). The DP control is given by

$$\dot{x}_2(t) = K_s \sum_{i=1}^{3}\sum_{j=1}^{3} h_i(z(t))h_j(z(t))(A_i + B_iK_j\lambda)x(t) + \Delta f + \Delta g + E\tau_e(t) \tag{23}$$

where λ is the adjusting fuzzy structure coefficient. In the case of medium environment impact, we propose the architecture of third fuzzy model which is defined by 5×3 memberships function (fuzzy-case 3). The DP control is rewritten as

$$\dot{x}_3(t) = K_s \sum_{i=1}^{5} \sum_{j=1}^{3} h_i(z(t)) h_j(z(t)) (A_i + B_i K_j) x(t) + \Delta f + \Delta g + E\tau_e(t) \qquad (24)$$

As in the case of high level environmental impact, the structure of fourth fuzzy model would be optimized by GA in the same as the second fuzzy model (fuzzy-case 4). So the DP control is expressed as follow:

$$\dot{x}_4(t) = K_s \sum_{i=1}^{5} \sum_{j=1}^{3} h_i(z(t)) h_j(z(t)) (A_i + B_i K_j \lambda) x(t) + \Delta f + \Delta g + E\tau_e(t) \qquad (25)$$

In the next section, the design of fuzzy structure which is optimized by GA is discussed in detail. Since the fuzzy sets are set up by experience according to the Remark 1. Thereby, the object membership function is fixed to the control signal for the nonlinear DP system is not optimized with time-varying object.

4   Fuzzy Genetic Algorithm Design

In the cascade structure, the second fuzzy controller case (3×3) and fourth fuzzy controller case (5×3) will be fixed by the optimal algorithm. This paper presents a combined genetic algorithm and fuzzy logic method, a fuzzy genetic algorithm (fuzzy-GA) for the DP system is designed to achieve the control goal stated in Sect. 2. So Fig. 2 represents the flowchart of optimizing fuzzy system. Thereby solving problems presented by Remark 1. Building control consists of a two-stage process.

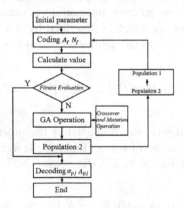

Fig. 2. Optimization flowchart of fuzzy system uses the GA.

Process 1: Determine the fuzzy system with optimized parameter. The fuzzy modulator has a double-input, $e_x(t)$, $d_{e_x(t)}/d(t)$ and a single-output, $\tau(t)$ [11]. These fuzzy sets adjusted flexibly by λ coefficient to optimize the control structure. In the fuzzy-GA controller, the value and overlap degree of MFs, which are optimized by the GA, are used. Following Eq. 18, the fuzzy output can be redescribed as below:

$$u_{(t)} = \sum_{j=1}^{r} h_j(z(t))K_jx(t)\lambda \tag{26}$$

Process 2: Adjusting the optimized fuzzy parameter. The GA optimization module completes the calibrating of fuzzy sets with vector $\lambda(\lambda_1, \lambda_2, \lambda_3, \lambda_4)$ adjusting coefficients that synthesise the optimum control structure. Adjusting a structure of fuzzy system can be viewed as an optimization subject in 4-search space as many options of the controller could be achieving better response. The adjustment of normal solutions do not ensure an optimal goal and in the adjusted structure controller needs demonstrate through the erroneous. The GA includes a group of solutions named population and continuity modifies to them. At every stage, the GA chooses an λ individuals from the current population to become parents and uses these individuals to make children for the next pedigree [12]. The detail of coding and decoding system, crossover and mutation operation and fitness evaluation are given in the next sections.

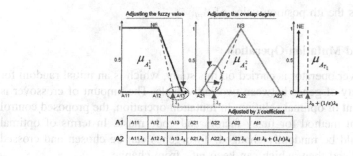

Fig. 3. Describing the coding values of MFs.

4.1 Coding and Decoding

Coding is the genetic representation of solutions [13]. The nominate solutions are displayed by strings of determine length, which named chromosomes. Figure 3 presents a method of genotype for individuals. Code mask A_f is a binary vector with a length of N_f, where a 0 or 1 at the ith position represents the absence or presence of the ith feature. The value of ith position is given by

$$A_i = \begin{cases} 1, & P\left(A_f^i = 1\right) = mf_{se}/mf, i = 1, 2 \ldots N_f \\ 0, & P\left(A_f^i = 1\right) = 1 - mf_{se}/mf \end{cases} \tag{27}$$

where mf_{se} is an original factor and expresses the amount of selected characteristic. This suggestion make the characteristics of genetic evolution in lower cases. The relationship between code mask A_f and its phenotype representation is

$$\tilde{D} = D.diag(A_f) \tag{28}$$

where D and $\tilde{D}$ present the preset parameter and the parameter after characteristic selection, respectively. The N_{pj} expresses the genotypic length of ith parameter for the fuzzy set. N_{pj} is given by

$$N_{pj} = round\left[log_2\left(\frac{\sigma_{pj,up} - \sigma_{pj,low} + \Delta}{\Delta_{pj}}\right)\right] + 1 \tag{29}$$

where $\sigma_{pj,up}$ and $\sigma_{pj,low}$ show the upper and lower searching bound, respectively. Δ_{pj} is an original factor and display an accurate appraisal. The genotype A_{pj} of parameters j should be decoded into phenotype σ_{pj} by

$$\sigma_{pj} = \sigma_{pj,low} + \left(\sigma_{pj,up} - \sigma_{pj,low}\right)\left(\frac{\sum_{i=1}^{N_{pj}}\left(A_{pj}^{(i)}\right)^{N_{pj}-1}}{2^{N_{pj}}}\right) \tag{30}$$

where $A_{pj}^{(i)}$ represents the ith position value of A_{pj}.

4.2   Crossover and Mutation Operation

Homologous crossover operator is carried on this study, which is an initial random for creating a new variety of genes between two individuals. The amount of crossover is defined by the amount of optimal solution. In mutation operation, the proposed control realizes two different method for fuzzy factors and code mask. In terms of optimal solution, a gene could be randomly changed. Two genes can be chosen and crossed their parameter now and then, which can keep mf_{se} from change.

4.3   Fitness Evaluation

The minimum initial values that caused by the error between response value and referent value is the optimal goal. So the fitness function is used for iterations to value the quality of all the proposed solutions to the problem in the current population. The fuzzy-GA control structure for DP system shows in Fig. 4. In this paper, the fitness function is chosen by the ITAE criteria [15] as follows:

$$ITAE = \int_0^\infty t|e(t)|dt \tag{31}$$

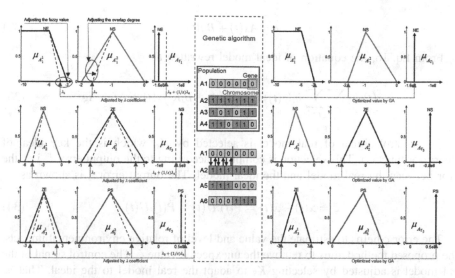

Fig. 4. The fuzzy set values of MFs are calibrated by GA [14].

5 Supervisor Design

The supervisor consists of two subsystems: Estimator, and a switch logic. Figure 5 shows the structure of the supervisor using two different process models [16]. Inputs to the supervisor are the process input, τ, and measured process states, η. The supervisor output is the switching signal, K_S.

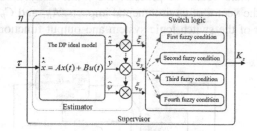

Fig. 5. The supervisor structure for cascade multi fuzzy model.

In the proposed method, namely multi-cascade fuzzy model, control signal for real model (environmental impact) is adjusted by the K_S factor. This adjustment is based on the output error ξ between ideal model (without peripheral factors) and real model. The response of real model is controlled to adapt to the ideal model. Therefore, the error caused by peripheral factors is minimized [17]. The structure of multi-cascade fuzzy model is shown in Fig. 6. Ideally, we eliminate the impact of environmental factors, it is mean $E\tau_e(t) = 0$. So the equation of ideal model are as

$$\dot{x} = Ax(t) + Bu(t) \tag{32}$$

From Eq. 16, the equation of ideal model rewritten as

$$\dot{x}(t) = \sum_{i=1}^{n} \sum_{j=1}^{r} h_i(z(t)) h_j(z(t)) (A_i + B_i K_j) x(t) + \Delta f + \Delta g \tag{33}$$

The control signal of real model is selected by K_s, which is the key ideal of supervisor function. This selection is computed according to the output error ξ, i.e., the error between the real model and the ideal model. The output error ξ is shown as

$$\xi = \dot{x}(t) - \dot{\hat{x}}(t) = P(t).U(t) - \hat{P}(t).U(t) \tag{34}$$

The error ξ helps to estimate the value and level of relative environmental impacts. The proposed method aims to remove the unexpected value ξ. The control signal of the real model is adjusted by selecting K_S to adapt the real model to the ideal. That is, $\xi \to 0$ when $t \to \infty$, and then

$$P(t).U(t) = \hat{P}(t).U(t) \tag{35}$$

The switch logic [18] signal of the real model can be presented by

$$\dot{x} = A_{K_s}(x, \tau_e) \tag{36}$$

$$e_p = C_p(x, \tau_e), p \in P \tag{37}$$

where x defines the state vector of process, the multi-fuzzy model and the multi-estimator, and τ_e is the vector of environmental impacts. A_{K_s} and C_p are functions that denote the dynamics of the switch logic system and output functions, respectively.

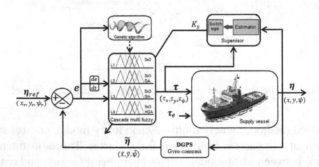

Fig. 6. The optimized multi-cascade fuzzy model structure for DP system of vessel.

6 Simulation Studies

6.1 Configuration Parameter

Performance comparisons between the proposed optimized multi-cascade fuzzy model and the fuzzy controller are conducted to assess the precision and effectiveness of the proposed controller. Multi-cascade fuzzy model (Sect. 3) with the optimal goal (Eq. 31) is tested on the supply vessel with length overall 52.08 m, length between perpendiculars 46 m, beam 12.07 m, design draft 5.52 m and design speed 11 knots [1].

$$D \begin{bmatrix} 5.024e4 & 0 & 0 \\ 0 & 2.722e5 & -4.393e6 \\ 0 & -4.393e6 & 4.189e8 \end{bmatrix}; M = \begin{bmatrix} 5.312e6 & 0 & 0 \\ 0 & 8.283e6 & 0 \\ 0 & 0 & 3.745e9 \end{bmatrix}$$

Case 1: The optimized multi-cascade fuzzy model and fuzzy force the vessel to arrive at the desired value [3 m, 7 m, 20°] in around 200 s from reference [0 m, 0 m, 0°]. Case 2: These controllers are proposed to keep vessel routine for achieving desired trajectory under environmental impacts. Figures 7(a) and 8(c) reveal that the optimized multi-cascade fuzzy model can make the vessel motion to aim at the expected position in simulation cases. The real position (x, y) and heading are kept at the target value illustrated by Figs. 7(b) and 8(a). On the other hand, Figs. 7(d) and 8(b) show that the control forces and moment by the optimized multi-cascade fuzzy model and fuzzy controller are glossy and justice. The environmental impacts are presented by Fig. 7(c) [19]. The environment impacts include wave, wind and currents τ_e which expressed by

$$\tau_e = \tau_{wave} + \tau_{wind} + \tau_{current} \tag{38}$$

here, the wave impact is described as follow [1]:

$$\tau_{wave} = \zeta_{qr}(x, y, t) = \zeta_{aqr} \sin \left(\omega_q t + \phi_{qr} - k_q (x \cos \psi_r + y \sin \psi_r) \right) \tag{39}$$

where wave height $H_s = 0.8$ m, wave spectrum peak frequency $\omega_p = 0$ rad/s, wave direction $\psi_0 = -30°$, spreading factor $s = 2$, number of frequencies $N = 20$, number of directions $M = 10$, cutoff frequency factor $\zeta = 3$, wave component energy limit $k = 0.005$ and wave direction limit $\psi_{lim} = 0$. The wind forces are performed by

$$V_R = V_w$$
$$g_R = \beta_w - \psi_L - \psi_H \tag{40}$$

The wind simulation parameters are sorted as follows: $A_L = 2.4$, $A_T = 9.34$, wind speed $V_\omega = 2$ m/s and the angle of impact wind $\beta_\omega = 20°$.

$$u_c = V_c \cos(\beta_c - \psi_L - \psi_H)$$
$$v_c = V_c \sin(\beta_c - \psi_L - \psi_H) \tag{41}$$
$$\tau_{current} = [u_c, v_c, 0]^T$$

Besides that, the simulation parameters for current factor are set to their default values accept as follows: $V_C = 2$ m/s, vessel direction $\beta_C = 30^0$, low frequency and high frequency of rotation are ignored $\psi_L = \psi_H = 0$.

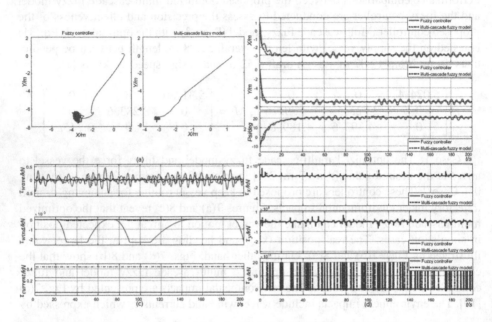

Fig. 7. The simulation of cases 1 consist of two controllers. (a) Trajectory of the vessel position in xy-plane. (b) The real position (x, y) of vessel and the heading ψ of vessel. (c) Environmental impacts (τ_{wave}, τ_{wind} and $\tau_{current}$) have an effect on the vessel. (d) Surge control force τ_x, sway control force τ_y and yaw control force τ_ψ.

6.2   Simulation Results

The DP controllers of the position error vectors in cases are given by Figs. 7(a) and 8 (c), which illustrate that the optimized multi-cascade fuzzy model has a good stable performance in each of simulation case under environmental impacts acting on the vessel in case, respectively. Having done so, it dealt with the question of causation according to Assumption 1 and Remark 1. Only using a fuzzy controller for keeping balance of the DP system, the vessel position will be stable from the low-impact case and vibratile at the higher impact cases. Besides that, the vessel heading fluctuates strongly according to the level of environmental impacts. The satisfactory results prove that the optimized multi-cascade fuzzy model has the adaptability to nonlinear systems of vessel motion and against time-varying environmental impacts. Thereby improving the quality of control signal, that make amplitude of surge, sway and yaw fluctuation at low-level and keep the vessel balance.

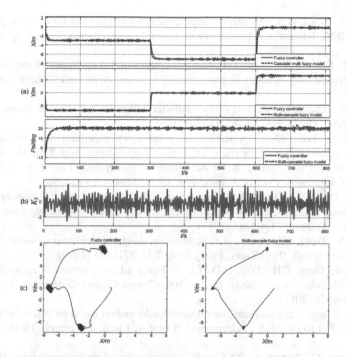

Fig. 8. The simulation of cases 2 consist of two controllers. (a) Real position (x, y) and the vessel heading ψ. (b) Output supervisor K_s which adjust multi-cascade fuzzy model structure. (c) Trajectory of the vessel position in xy-plane.

7 Conclusions

In this paper, an optimized multi-cascade fuzzy model has been developed for the DP system in the presence of environmental impacts. Our proposed control strategy is able to keep the vessel at the desired values of its position and heading with accuracy. Proposed algorithm optimize structure parameters fuzzy and reduce the nonlinear characteristics of the DP system which caused by environmental impacts. Calibrating control structure of fuzzy system was intuitive to perform. The main advantage of the proposed controller, compared to the conventional fuzzy control, it is optimal to variations in its displacement and the environmental conditions. This study can be extended by using the robust algorithm to improve the DP control quality when the vessel operates in unstable state for constantly.

References

1. Fossen, T.I.: Marine control systems – Guidance, navigation and control of ship, rigs and underwater vehicles. Marine Cybernetics, Trondheim, Norway (2002)
2. Chang, W.F., Chen, G.J., Yeh, Y.L.: Fuzzy control of dynamic positioning systems for ships. J. Mar. Technol. **10**(1), 47–53 (2002)

3. Chen, X.T., Woei, W.T.: An adaptive type-2 fuzzy logic controller for dynamic positioning. In: 2011 IEEE International Conference on Fuzzy Systems, Taipei, Taiwan, pp. 2147–2154 (2011)

4. Ho, W.H., Chen, S.H., Chou, J.H.: Optimal control of Takagi-Sugeno fuzzy-model-based systems representing dynamic ship positioning systems. Appl. Soft Comput. **13**(7), 3197–3210 (2013)

5. Hu, X., Du, J., Shi, J.: Adaptive fuzzy controller design for dynamic positioning system of vessels. Appl. Ocean Res. **53**, 46–53 (2015)

6. Fang, M.C., Lee, Z.L.: Application of neuro-fuzzy algorithm to portable dynamic positioning control system for ships. Int. J. Nav. Arch. Ocean. Eng. **8**(1), 38–52 (2016)

7. Ngongi, W.E., Du, J., Wang, R.: Robust fuzzy controller design for dynamic positioning system of ship. Int. J. Control Autom. Syst. **13**(5), 1294–1305 (2015)

8. Do, V.D., Dang, X.K., Ho, L.A.H.: Enhancing quality of the dynamic positioning system for supply vessel under unexpected impact based on fuzzy Takagi-Sugeno algorithm. Vietnam. J. Mar. Sci. Technol. **51**, 92–95 (2017)

9. Chen, B.S., Tseng, C.S., Uang, H.J.: Robustness design of nonlinear dynamic systems via fuzzy linear control. IEEE Trans. Fuzzy Syst. **7**(5), 571–585 (1999)

10. Dang, X.K., Guan, Z.H., Tran, H.D., Li, T.: Fuzzy adaptive control of networked control system with unknown time-delay. In: The 30th Chinese Control Conference, Yantai, China, pp. 4622–4626 (2011)

11. Do, V.D., Dang, X.K.: Optimal control for torpedo motion based on fuzzy-PSO advantage technical. TELKOMNIKA (Telecommun. Comput. Electron. Control) **15**(4), 2999–3007 (2018)

12. Kuppusamy, M., Natarajan, R.: Genetic algorithm based proportional integral controller design for induction motor. J. Comput. Sci. **7**, 416–420 (2011)

13. Liu, T., Zhang, W., McLean, C., UeLand, M., Forbes, S.L., Su, S.W.: Electronic nose-based odor classification using genetic algorithms and fuzzy support vector machines. Int. J. Fuzzy Syst. **20**, 1309–1320 (2018)

14. Do, V.D., Dang, X.K., Ho, L.A.H., Dong, V.H.: Optimization of control parameter for dynamic positioning system based on genetic algorithm advantage technique. In: The 17th Asia Maritime & Fisheries Universities Forum (AMFUF 2018), Guangdong, China, pp. 117–129 (2018)

15. Lu, Q., Peng, Z., Chu, F., Huang, J.: Design of fuzzy controller for smart structures using genetic algorithms. Smart Mater. Struct. J. **12**, 979–986 (2003)

16. Nguyen, T.D., Sørbø, A.H., Sørensen, A.J.: Modelling and control for dynamic positioned vessels in level ice. In: The 8th IFAC International Conference on Manoeuving and Control of Marine Craft, Guarujá, Brazil, pp. 229–236 (2009)

17. Do, V.D., Dang, X.K., Le, A.T.: Fuzzy adaptive interactive algorithm for rig balancing optimization. In: International Conference on Recent Advances in Signal Processing, Telecommunication and Computing, Danang, Vietnam, pp. 143–148 (2017)

18. Lin, X., Li, H., Liang, K., Nie, J., Li, J.: Fault-tolerant supervisory control for dynamic positioning of ships. Math. Probl. Eng. **2019**(6), 1–11 (2019)

19. Dang, X.K., Ho, L.A.H., Do, V.D.: Analyzing the sea weather effects to the ship maneuvering in Vietnam's sea from BinhThuan province to Ca Mau province based on fuzzy control method. TELKOMNIKA (Telecommun. Comput. Electron. Control) **16**(2), 533–543 (2018)

UiTiOt-Vlab: A Low Cost Physical IoTs Testbed Based on Over-The-Air Programming Approach

Hung Le-Viet, Dang Huynh-Van, and Quan Le-Trung$^{(\boxtimes)}$

Department of Computer Networks, University of Information Technology,
Viet Nam National University, Ho Chi Minh City, Vietnam
14520339@gm.uit.edu.vn, {danghv, quanlt}@uit.edu.vn

Abstract. In the plethora of technologies in wireless networks and embedded systems, Internet of Things (IoTs) technology plays a vital role, be applicable in many different domains, such as agriculture, industry, education, transportation, just to name a few. The deployment of IoTs applications in these application domains normally requires complicated phases not only in the analysis, design, implementation but also in the actual operations, e.g., to control the devices and application remotely via the wireless protocols. Therefore, the development of a low cost physical IoTs-based reconfigurable testbed supporting well-known hardware and software platforms is significant and will be presented in this paper. By enabling over-the-air programming (OTAP) with a proper enhancement on Arduino ESP266 device, the proposed testbed system can be used for testing real-time IoTs application, as well as for (re)programming to change or adapt the operations of IoTs devices and IoTs applications remotely, through the wireless network protocols. In addition, the testbed system also provides a virtual laboratory for studying IoTs technology, be a cost-effective approach for students to share IoTs devices and practice IoTs programming remotely.

Keywords: OTA (re)programming · Arduino ESP8266 ·
OTA upgrade firmware · IoTs reconfiguration

1 Introduction

The Internet of Things (IoTs) connects everything through the Internet, bringing many benefits to users such as more convenient services, utilities for users; better resource allocation; better control of devices and applications remotely via the wireless network protocols. Today, IoTs technologies are indispensable for users in ambient assisted living, transportation, and healthcare [1]. More and more IoTs products have been launched to serve consumers in many different fields such as agriculture [2], industry [3], transportation [4], etc. The release of the breakthrough technologies as 5G technology [5] providing faster speeds, reducing the power consumption and improving the response time and bandwidth. Thus, the appearance of IoTs/5G technologies is necessary for billions of IoTs devices to communicate easily and effectively. These improvements are expected to not only improve the user experience but also pave the way for further improvements in the development of IoTs applications and solutions. In

T. Q. Duong et al. (Eds.): INISCOM 2019, LNICST 293, pp. 181–192, 2019.
https://doi.org/10.1007/978-3-030-30149-1_15

this approach, IoTs testbeds have been developed in order to evaluate IoTs applications and protocols before deploying such IoTs applications and protocols into the real world.

At present, some IoTs-based systems have been integrated with the reconfiguration and reprogramming features to manage IoTs devices and applications remotely. This approach allows the users to reconfigure behaviors of IoTs devices to adapt upon the dynamic changes or unpredictable events in the environment. The remote reprogramming of IoTs devices and IoTs applications through wireless network protocols, aka Over-The-Air-Programming (OTAP), is a potential research topic which has been attracted the attention of many famous research centers in the world. However, most of the existing testbeds are expensive and not to be popular with beginners. In this paper, we propose UiTiOt-VLab, which uses cost-effective Arduino-based IoTs devices based on the OTA programing approach.

The UiTiOt-VLab has been deployed on the IoTs lab and integrated into the cloud infrastructure at the University of Information Technology - VNUHCM. Because the UiTiOt-VLab can be remotely accessed and controlled through a web interface, different users can use this testbed for doing their own researches and experiments in the areas of OTAP/reprogramming, reconfiguration/re-adaptation of both the operations in IoTs devices and the behaviors in IoTs applications.

The paper structure is organized as follows. Section 1 introduces the topic and shares researches and systems deployed in the world as well as contributions in our research. Section 2 lists similar works implemented in the world, summarizing the nature, purpose, and mode of operation of each project. Section 3 focuses on clarifying the process of system deployment. This section contains 4 sub-sections, each of which presents the core work when building the system. Part 3 is the result of our deployment in a realistic scenario. Finally, Sect. 4 concludes the research paper with the conclusions and future work.

2 Related Work

Nowadays, there are many existing IoT device management systems which released in the modern industry and bring positive effects.

PlanetLab [6] is a global coverage network for developing and delivering wide area network services. The goal of PlanetLab is to grow up to 1000 geographically distributed sensor nodes, which are connected by diverse links. PlanetLab allows multiple services to run simultaneously and continuously, each service in a separate sensor node. Worldwide network services have appeared such as network-embedded storage [7], peer-to-peer file sharing [8], content distribution networks [9], robust routing overlays [10], scalable object location [11], scalable event propagation [11]. All of these applications have one common: take advantage of the wide connectivity with network protocols. To support the design and evaluation of the applications, PlanetLab has been developed to form a global network of networks which is initially deployed over 100 nodes distributed on 42 websites. In the near future, PlanetLab will be a microcosm for the next generation Internet.

MoteLab [12] includes a lot of sensor nodes connected to a central server that handles reprogramming and logs generated data by experiments by a continuous database. The user can access data via a web interface or directly from the database. MoteLab also allows users to interact directly with individual sensor node during testing using the Web interface. MoteLab helps everyone can deploy quickly by providing connections into real-sensor network device network. Additionally, MoteLab speeds up error correction and application development by automatically data logging, allowing the software to evaluate the performance of sensor networks operate offline. In addition, by providing a web interface, MoteLab allows both internal and external users to access the testbed, the system limits access time and limits bandwidth to ensure uniform distribution. The project has great significance for research as well as teaching. MoteLab's source is shared free, easy to install and has been used in some research institutes. The widespread use of MoteLab will accelerate and improve the research of wireless sensor networks.

FIT/IoT-LAB [13] is part of the FIT experimental platform, providing a method for testing IoT with mobile wireless communication devices in both network and application layer, thereby speeding up design and research advanced network technologies for the Internet. There are more than 1500 wireless sensors nodes spread over six different sites in Grenoble, Lille, Saclay, Strasbourg, Paris, Lyon (France) to create a heterogeneous platform. The project provides fixed nodes or mobile nodes moving on the ground to serve mobility in testing. The remote users access the Web interface to register necessary entities, with direct command line connection to the platform.

EU IoT Lab [14] is a large-scale research project in Europe, aiming to study the potential of providing IoTs equipment in society, creating a multidisciplinary research environment, experiment with interaction from multiple users. EU IoT Lab also provides Testbed service, a platform to create a combination of participant groups (workers, end-users, and researchers) to solve practical and collaborative challenges at work.

Indiana IoT Lab [15] provides pioneers in this IoTs industry with resources in a collaborative environment. Indiana IoT Lab builds a connection between industries established in India with technology companies specializing in IoT. The area is 24,000 feet wide.

Almost systems in the world use the expensive boards with the ability to operate in the large project; however, these types of board are not popular with students, who start to study and approach to IoTs technologies. The re-programming research uses Arduino ESP8266 platform, which is low cost and easy to program, is suited to the current research needs. To sum up, the proposed solution solves problems in remote (re)programming low-cost Arduino-based devices with many utilities such as managing IoTs equipment remotely, providing an environment for users to test their code, connecting with each other to support research and work. The equipment in the project can be provided for a large number of users to conduct their experiments simultaneously with various scenarios.

3   Implementation

The IoTiOt-VLab is a system which comprises a set of software tools for managing an IoTs testbed of connected sensor network nodes. A central server handles the authorized connection between user and device, reprogramming nodes, logging data, and providing a web interface for users. The UiTiOt-VLab consists of three main software components include:

- Mongo Database Backend: stores data collected during experiments, information used to generate web content, and state driving testbed operation.
- Web Interface: ReactJS generate pages which present a user interface for deployment, granting, and data collection as well as an administrative interface to testbed control functionality.
- Job Daemon: Python script run as a cron job to set up and tear down jobs.

Throughout this section, we refer to a "job" running on UiTiOt-VLab. A UiTiOt-VLab job includes some number of executables nodes, a description mapping each pair of nodes used to executable, and several Javascript files used for data logging. To create a job, the user uploads the sketch onto board via a web interface. Once the job is created, UiTiOt-VLab executes the script on the real device and send data to the user. The UiTiOt-VLab limits usage time of each device so that other users can use the testbed to run their own code.

The system architecture is described in detail by listing all components of the model. The Arduino device block includes a master and slave board. The master is preconfigured to receive and transmit code to Slave. The sensor/actuator devices directly connect to the Slave board to run the author's programs. The receiver data, parameters are distributed to Master, then forwarded to the server.

In Fig. 1, describes the overview architecture of the system, in which:

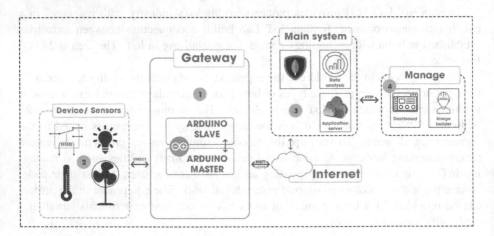

Fig. 1. Architecture of testbed system

- IoTs gateways are Arduino-based devices which are remotely (re) programmable.
- Sensors connected directly to IoTs gateway via the Slave board.
- Application system, including rental service, management, and access devices.
- The manage block is a web-based user interface which is used to upload code into the board.

According to the above architecture, our IoTs system consists of three main components: *(i)* sensors, actuators control block, *(ii)* firmware update mechanism for the master and slave Arduino boards, and *(iii)* user interface for remote deployment.

3.1   Sensors and Actuator Control Block

Depend on the using purposes, there are many types of sensor nodes for monitoring environmental parameters. For example, the sensors array can include pH sensor, temperature sensor, humidity sensor, module relay to control the on-off of the light or the machine, which is connected to the Arduino circuit directly. In this proposed system, all sensors, actuators have to be connected in advanced to provide the sensor lists and connection ports for users on the website.

3.2   OTA Update Firmware Mechanism for Arduino Boards

Arduino [16] is a microprocessor circuit board used to build applications that interact with each other or with a favorable environment. The Arduino hardware includes a power circuit board designed on the Atmel 8bit AVR processor platform or 32-bit ARM Atmel. The current models are equipped with 1 USB interface port, 6 analog input pins, 14 digital I/O ports compatible with many different expansion boards.

It is reported that Arduino-based programming is easy, inexpensive for scientists, students, and professionals. It is easy to design a simple IoT-based application that can interacts with the environment through sensors and actuators such as simple robots, environmental parameters collection or motion detection. In addition, the Arduino community has also provided an integrated development environment (IDE) that runs on regular personal computers and allows users to write Arduino programs in code C/C++.

The Arduino supports both digital and analog input. In terms of output, we can use PWM to simulate an analog output. Digital I/O pins on Arduino can be configured for input or output. Only analog input pins are used for Input.

3.2.1   The Limitation of the Existing OTA Update on ESP8266

To conduct the basic OTA update firmware process on ESP8266, it is efficient to use an embedded computer such as Raspberry Pi works as a local server. Raspberry Pi (RPi) is a small-sized computer with powerful hardware built-in capable of running the operating system and installing many applications. RPi is suitable for applications that need powerful processing, multitasking, entertainment, and especially low cost. RPi features built around the Broadcom BCM2835 SoC processor (is a powerful mobile processor with small size, or used in mobile phones) including CPU, GPU, audio/video processor, and other features... integrated inside this low-power chip (Fig. 2).

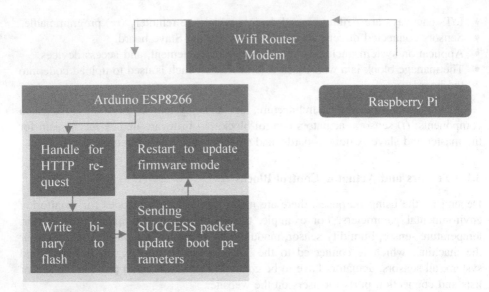

Fig. 2. Diagram of loading an Arduino ESP8266 with basic OTA library

In this library, RPi works as a local server, to control and deploy program onto Arduino ESP8266. The detailed operation of the reprogramming via the RPi follows steps below:

1. RPi checks and compiles sketch into a .bin file, send a signal via Wifi router to Arduino start the process of reconfiguring.
2. The packet is received by the Arduino, processed and written to the last section of the flash memory.
3. After receiving the last packet of the binary file, Arduino returns the SUCCESS packet to RPi and installs boot params to enable the built-in OTA mode in the bootloader.
4. The OTA mode available after ESP restart, copy the entire binary to 0x000000 in flash and restart after the copy has been completed.

However, this approach has some disadvantages which are inappropriate to be used in a reusable IoT testbed. These drawbacks include: (1) the user's code must be reprocessed, insert the code to call the OTA library to make sure OTA is active after being uploaded; (2) the OTA library won't work properly if there is an infinite loop or delay too long in the user code; (3) all Serial.print() functions can't run.

3.2.2 The Proposed Method for OTA Programming the Running Board

Due to the physical design limitation, Arduino only handles single threads per task, so we cannot run the received signal to begin the reconfiguration process and user code execution tasks. In this case, we suggest using an ESP8266 board to be able to handle the two tasks simultaneously (Fig. 3).

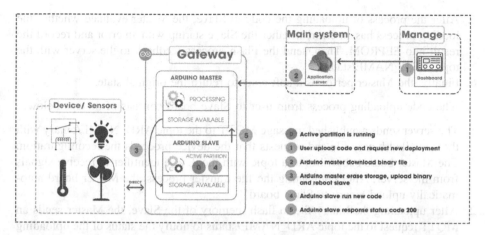

Fig. 3. The diagram of uploading firmware uses an additional ESP as a virtual COM port

The code on the ESP8266 master simulates the operation of the upload code with the virtual COM port set up by sending and receiving HTTP request. After receiving the binary from the Server via Ethernet port and writing to flash, the ESP8266 Master sends restart signal to Slave and then switches to the upload mode and sends it to SYNC to establish the connection. The flash section used to store the binary file is blanked by writing 0xff. The Master board continues to transmit binary packets to write to flash starting at 0x000000 address. Finally, the Slave board is restarted into the executive mode.

The detailed operation of the Master board sequentially follows steps below:

1. Every 10 s, the Master board sends a ping signal to the server with a device name and password to inform the status of the operation.
2. Every 10 ms, the Master board reads packets from Slave's serial port; The Master sends information from the serial port to the MQTT server with topic ARD_NAME/com every 100 ms, so that this information is available to the user promptly.
3. If there is an MQTT packet to topic ARD_NAME/binary, then receive the file, write it to memory and stop reading the data from the Serial, stop sending the Ping packet to set device status to offline.
4. After receiving the binary file completely, the master begins waiting for the server sends the signal to start the process of uploading code on the device. From the time of the last packet of the binary file is completed, the master waits for a 10 s timeout period, if there is no request, responding to the server status with the topic ARD_NAME/status.

5. After the process of uploading the code to Slave, the Master evaluate whether the loading process has an error, whether the Slave startup with an error and record the result into EEPROM. Then, send the Flash process feedback to the server with the topic ARD_NAME/status.
6. Finally, the Master performs a soft reset to return the original state.

The code uploading process from user to Slave board consists of 3 steps below:

1. The server sends a published message MQTT to the topic ARD_NAME/binary with the compiled binary file and requests that the device proceeds the reconfiguration. The Master board subscribes to a topic with the device identifier to receive signals from the server. After completing the file transfer process, the Master board automatically upload binary to Slave board.
2. After uploading the binary file to flash memory of the Slave, the Master sends an MQTT request to the topic ARD_NAME/status to notify the status of the uploading process to the server (Failed, successfully uploaded or no file). If the timeout expires, report the error reported from the server to the user.
3. The server continues to send requests include the data received from Slave, which are processed by the server and forward to users.

3.3 The Web-Based Application for Remote Deployment

The main function of the web-app is the user interface to manage, deploy the reconfiguration, and activate the reprogramming process. The main components of the web-app include: (i) Displaying ready-to-use devices, separating each platform, (ii) Managing lists, modifying information of all devices in the system, (iii) Managing user list, edit profile, system management permissions for each object, (iv) managing content of practice tutorials, (v) Interacting with Arduino devices (Fig. 4).

Fig. 4. Manage devices in use interface

Device management system with functions for management level (admin, technical staff, lecturers): add, delete, get device description, information access equipment, edit the description, show device's status; functions for all role, including show device lists, borrow and return devices, load code onto Arduino board (Fig. 5).

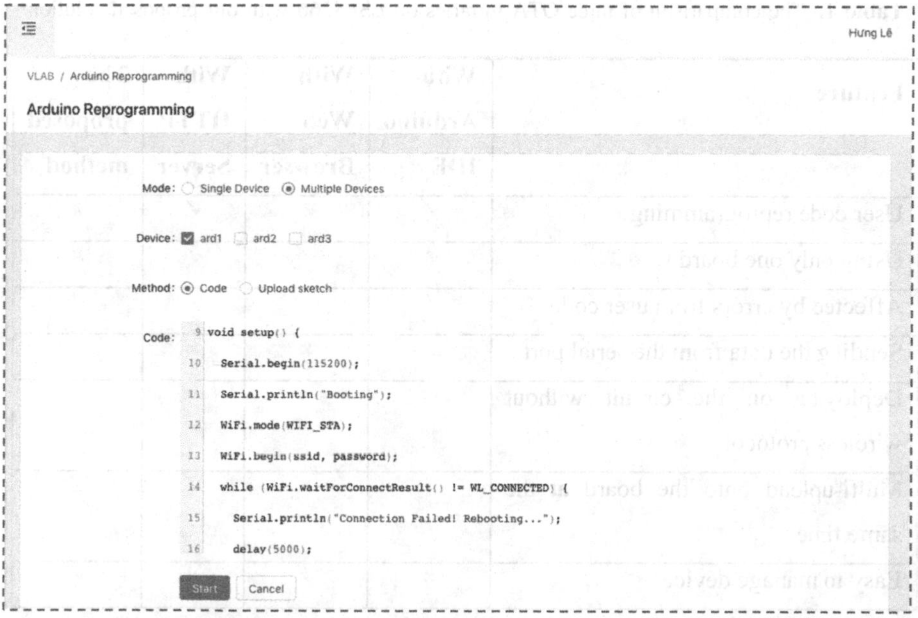

Fig. 5. Upload code onto the Arduino board interface

The editing source code page provides 2 modes: compile and upload code onto one device or multiple devices. The upload to single device mode is to upload and deploy on the device that the user presses the "Load" button on the management site. The upload to multi-device mode is to upload and deploy one array of user-selected devices below. In addition, we provide two more methods: type the code directly on the website or upload the sketch file to the system. However, the method of processing code and feedback for users is similar.

4 Experiment Results

In this section, the experimental results of the proposed system are illustrated. The results provide information on various metrics related to the use of algorithms, analysis of packet transmission performance of wireless sensor nodes.

Compared to the three OTA libraries provided by Arduino Dev, our solution has some remarkable betterment, which is suitable to be used in practical implementation. The Table 1 below indicates the advantages and disadvantages of our proposed solution with the existing OTA approaches in ESP8266.

Table 1. The comparison of three OTA libraries on ESP8266 with our proposed solution

Feature	With Arduino IDE	With Web Browser	With HTTP Server	The proposed method
User code reprogramming	✓	✓	✓	✗
Using only one board	✓	✓	✓	✗
Affected by errors from user code	✓	✓	✓	✓
Sending the data from the serial port	✗	✗	✗	✓
Deploying on the circuit without wireless protocol	✗	✗	✗	✓
Multi-upload onto the board at the same time	✗	✗	✓	✓
Easy to manage device	✗	✗	✗	✓

By embedding a timer counter on the device, the time of distribution code from the server system to the device clearly, accurately. We have collected some useful information such as the code size before compilation, the code size after compiling (before loading onto the device), disseminating time, number of packets to be transmitted as well as the average size of the package. The diagram below shows the results in the distribution process of source code with 05 different applications, each application tested 05 times (Table 2).

Table 2. The experimental results of the distribution process with 5 applications

- Application name - File .ino length (kilobytes) - File .bin length (kilobytes)	No.	Download time (s)	Time of deployment from master to slave (s)	Time of deployment with the system (s)	Time of deployment with IDE (s)
Blink	1	5,500	31,604	37,104	32,80
0,627	2	5,394	31,605	36,999	33,13
261.824	3	5,273	31,575	36,848	32,71
	4	5,505	31,581	37,086	32,41
	5	5,369	31,575	36,944	32,10
Basic HTTP Client	1	5,685	33,341	39,026	35,52
1,712	2	5,603	33,345	38,948	36,34
280.752	3	5,475	33,338	38,813	36,54
	4	5,550	33,339	38,889	36,13
	5	5,583	33,340	38,923	36,05
Web updater	1	6,025	35,635	41,660	39,13
0,989	2	6,000	35,634	41,634	39,31
306.352	3	6,061	35,642	41,703	39,80
	4	5,990	35,639	41,629	39,33
	5	6,014	35,633	41,647	39,04
Chat server	1	5,938	34,535	44,473	43,06
2,026	2	5,676	34,542	44,218	42,92
294.768	3	5,785	34,538	44,323	42,95
	4	5,674	34,542	44,216	42,88
	5	5,784	34,534	44,318	43,10
Basic OTA	1	5,845	33,986	39,831	37,29
1,540	2	5,719	33,987	39,706	37,51
287.792	3	5,840	33,985	39,825	37,54
	4	5,684	33,959	39,643	38,34
	5	5,739	33,985	39,724	37,99

5 Conclusion and Future Work

In summary, this paper introduces the development of UiTiOt-VLab, which is low cost, physical IoTs testbed, working as an IoTs virtual laboratory for testing and learning IoTs programming remotely via the wireless protocols. Our enhancement on OTA upgrade firmware mechanism in ESP8266 board allows the running devices can receive new firmware version simultaneously. The proposed solution is not only suitable to build a physical IoTs testbed but also helpful for real world IoT-based deployments.

In the future, we will do more research in the development of reconfiguration and reprogramming solutions on other microchip platforms and integrate such developed solutions in the actual IoTs-based projects.

Acknowledgement. This research is funded by University of Information Technology-Vietnam National University HoChiMinh City under grant number D1-2019-12.

References

1. Building A Digital World Consumers Can Trust: Proposed recommendations from the consumer movement to the G20 member states. Consumers International and The Federation of German Consumer Organizations (2017)
2. TongKe, F.: Smart agriculture based on cloud computing and IOT. J. Convergence Inf. Technol. (JCIT) **8**(2) (2013)
3. Li, R., Song, T., Capurso, N., Yu, J., Couture, J., Cheng, X.: IoT applications on secure smart shopping system. IEEE Internet Things J. **4**(6), 1945–1954 (2017)
4. Song, T., Capurso, N., Cheng, X., Yu, J., Chen, B., Zhao, W.: Enhancing GPS with lane-level navigation to facilitate highway driving. IEEE Trans. Veh. Technol. **66**(6), 4579–4591 (2017)
5. Hattachi, R.E., Erfanian, J.: 5G White Paper. NGMN Board (2015)
6. Chun, B., et al.: PlanetLab: an overlay testbed for broad-coverage services. ACM SIGCOMM Comput. Commun. Rev. **33**(3), 3–12 (2003)
7. Kubiatowicz, J., et al.: OceanStore: an architecture for global-scale persistent storage. In: ACM SIGARCH Computer Architecture News - Special Issue: Proceedings of the Ninth International Conference on Architectural Support for Programming Languages and Operating Systems (ASPLOS 2000), vol. 28, no. 5, pp. 190–201 (2000)
8. Rowstron, A., Druschel, P.: Storage management and caching in PAST, a large-scale, persistent peer-to-peer storage utility. In: SOSP 2001 Proceedings of the Eighteenth ACM Symposium on Operating Systems Principles, pp. 188–201 (2001)
9. Wang, L., Pai, V., Peterson, L.: The effectiveness of request redirection on CDN robustness. In: ACM SIGOPS Operating Systems Review - OSDI 2002: Proceedings of the 5th Symposium on Operating Systems Design and Implementation, vol. 36, pp. 345–360 (2002)
10. Andersen, D., Balakrishnan, H., Kaashoek, F., Morris, R.: Resilient overlay networks. In: SOSP 2001 Proceedings of the Eighteenth ACM Symposium on Operating Systems Principles, vol. 35, no. 5, pp. 131–145 (2001)
11. Ratnasamy, S., Handly, M., Karp, R., Shenker, S.: Topologically-aware overlay construction and server selection. In: Proceedings of the IEEE INFOCOM Conference, pp. 1190–1199, 2002
12. Werner-Allen, G., Swieskowski, P., Welsh, M.: MoteLab: a wireless sensor network testbed. In: IPSN 2005 Proceedings of the 4th International Symposium on Information Processing in Sensor Networks, vol. 68 (2005)
13. Adjih, C., et al.: FIT IoT-LAB: a large scale open experimental IoT testbed. In: 2015 IEEE 2nd World Forum on Internet of Things (WF-IoT), Milan, Italy (2016)
14. Ziegler, S., Rolim, J., Nikoletsea, S.: Internet of Things, crowdsourcing and systemic risk management for smart cities and nations: initial insight from IoT Lab European research project. In: 2016 30th International Conference on Advanced Information Networking and Applications Workshops (WAINA), Crans-Montana, Switzerland (2016)
15. Velis, M.: 'Smart City' trends and opportunities for collaboration. In: Consulate General of the Kingdom of the Netherlands, Chicago (2017)
16. Sarik, J., Kymissis, I.: Lab kits using the Arduino prototyping platform. In: 2010 IEEE Frontiers in Education Conference (FIE), Washington, DC, USA (2010)

Performance Analysis on Wireless Power Transfer Wireless Sensor Network with Best AF Relay Selection over Nakagami-m Fading

Duy-Hung Ha[1], Dac-Binh Ha[2($\boxtimes$)], Van-An Vo[3], and Miroslav Voznak[1]

[1] Faculty of Electrical Engineering and Computer Science,
VSB - Technical University of Ostrava,
17. lisltopadu 2172/15, 708 00 Ostrava, Czechia
haduyhung@tdtu.edu.vn, miroslav.voznak@vsb.cz
[2] Faculty of Electrical and Electronics Engineering, Duy Tan University,
Danang, Vietnam
hadacbinh@duytan.edu.vn
[3] Faculty of Engineering Technology,
Binh Duong Economics and Technology University,
Binh Duong, Vietnam
vvan@ktkt.edu.vn

Abstract. In this paper, we present the performance analysis of energy harvesting amplify-and-forward (AF) relaying wireless sensor network with best relay selection scheme over Nakagami-m fading. Specifically, this considered network consists of one sink, multiple energy-constrained relays, and one destination sensor node. The best relay is chosen to amplify and forward the message to the destination after powered by the sink. In order to evaluate the performance, the closed-form expression of outage probability and throughput are derived by applying the discrete optimal power splitting ratio. Based on this expression, we investigate the behavior of this network according to the key parameters such as transmit power, number of relays, time switching ratio and the distance.

Keywords: Wireless sensor network · Wireless power ·
Relaying network · Amplify and forward · Time switching ·
Power splitting · Outage probability

1 Introduction

Nowadays, wireless sensor networks (WSNs) present extensive potential in human life, i.e., health monitoring, security and tactical surveillance, intrusion detection, manufacturing control, disaster management, weather monitoring, traceability, farming monitoring, and safety services. The next generation network (i.e., 5G), the platform of the Internet of Things (IoT), is developing

© ICST Institute for Computer Sciences, Social Informatics and Telecommunications Engineering 2019
Published by Springer Nature Switzerland AG 2019. All Rights Reserved
T. Q. Duong et al. (Eds.): INISCOM 2019, LNICST 293, pp. 193–204, 2019.
https://doi.org/10.1007/978-3-030-30149-1_16

to deploy in the near future. This drives the wide application of WSNs into real life. However, there are a number of challenges that WSNs faced when they are deployed widely, such as the limitations of the resource (i.e., processing ability, memory capacity, antenna, and battery capacity). The limited energy of sensor nodes degrade the coverage of WSN, reduce the processing ability of the sensor node and decrease the lifetime of the network.

A novel technology trend in energy harvesting that can solve the limited energy problem in WSNs is wireless power transfer, namely RF energy harvesting (EH). RF EH in WSNs (RF EH-WSNs) refers to the sensor nodes (SNs) harvest energy from RF energy sources (TV/radio broadcasts, mobile base stations, and handheld radios) and converting it into electrical energy for information transmission. Although the harvested energy from the above environmental sources is dependent on the presence of the energy sources and must consider their unstable natures, the RF energy harvesting approach may represent a practical trend for future energy-aware systems because of it's ready available in the form of transmitted energy and in small form factor implementations, and low cost [1–4]. The relaying and cooperative techniques are applied in the modern WSNs to improve the performance and extend the coverage area of wireless networks and reduce the energy consumption of SNs [5–10]. Naturally, the embedding EH into relaying WSNs or cooperative WSNs have attracted a lot of attention from the academia and industry in the recent decade [11–21]. There are two models of RF EH architecture: time switching (TS) and power splitting (PS). In the TS architecture, the SNs switch and use either the RF energy harvesting circuit or the information receiver circuit for the received RF signals, for example, a part of the time for energy harvesting and remain time for information receiving. Meanwhile, in the PS architecture, the received RF signals are split into two streams for the RF energy harvester circuit or information receiver circuit according to PS ratios. One problem introduced is how to find the optimal TS or PS ratio to enhance the performance of WSNs. In [13], the authors introduced the block-wise TS-based protocol for EH AF relaying network in which the relay can implement EH and information processing. By derivation of analytical expressions of the achievable throughput, the performance of this considered networks with two modes of continuous and discrete time EH at the relay was analyzed and evaluated. The PS-based protocol was applied at each relay of simultaneous wireless information and power transfer (SWIPT) multi-relay network to coordinate the received signal energy for information decoding and EH in [15]. By using the interior-point method, the solutions for the optimization problems of PS ratios at the relays were provided for both basic relay schemes, i.e., decode-and-forward (DF) and AF. In the work of [19], the authors proposed a hybrid TS-APS protocol for EH DF relay networks with the discrete-level battery of relays. The results of this work have shown that the proposed TS-APS scheme can achieve better effective transmission rate than the previous works. The authors in the work [20] analyzed the performance of an energy harvesting relay-aided cooperative network with proposed on-off relay-aided cooperative protocol. By the help of the Markov property of energy buffer status, the analytical closed-form expression of outage probability was derived for Nakagami-m fading channels. This paper concluded that this approach can improve the system outage performance when EH relays were employed and The studied results have also shown that the more relays the better system performance. A TSAPS-OBR protocol based on optimal capacity for energy harvesting AF relaying network was proposed in [21]. The optimal PS ratio was found to maximum the end-to-end SNR and the simulation results confirmed that the performance of this considered system is improved by applied this optimal value. However, they have not derived the expression of outage probability for performance analysis.

Motivate by the work of [21], in this paper we consider the AF multi-relay wireless sensor networks with RF energy harvesting over Nakagami-m fading channels. The main contributions of our paper are as follows.

1. Deriving the closed-form expressions of outage probability and throughput of this considered WSN in two cases: fixed and adaptive PS ratio.
2. Evaluating the performance of the considered system in different key system parameters, such as transmit SNR, EH time, relay location, and number of relays in two cases: fixed and adaptive PS ratio.

The remain of this paper is organized as follows. Section 2 presents the system and channel models. The closed-form expressions of outage probability and throughput are derived in Sect. 3. The numerical results and discussion are shown in Sect. 4. Finally, the conclusion of the paper is provided in Sect. 5.

Notation: P_0 is the transmit power of S; n represents additive Gaussian noise, i.e., $n \sim \mathcal{CN}(0, N_0)$; d_1 and d_2 are the distances of $S - R$ and $R - D$, respectively; θ is the path loss exponent, for simplicity, the path loss exponent is assumed the same for all relays; η $(0 \leq \eta \leq 1)$ is the energy conversion efficiency; $*$ is denoted as the best relay chosen according to TSAPS-ORS protocol [21].

2 System and Channel Models

The Fig. 1 depicts a RF EH AF wireless sensor network, where a sink node (S) communicates with destination sensor node (D) via the assistance of K energy-constrained relays R_k $(1 \leq k \leq K)$.

The operation scenario of this considered system is assumed as [21]. The dual-phase protocol for this considered system as follows:

(i) In the first phase, S broadcasts information/energy signal to the energy-constrained relays in the time of αT (α is the fraction of the block time, $0 \leq \alpha \leq 1$) based on TS scheme. The relays split the received signal into two parts with the splitting ratio ρ $(0 \leq \rho \leq 1)$ based on PS scheme: one part is for information signal and other parts for EH leaving for amplifying and retransmitting;
(ii) In the second phase of the remaining duration of $(1 - \alpha)T$, the best relay selected among K relays amplifies the information part of RF signal received from S and retransmit to D by using the harvested energy in the first phase.

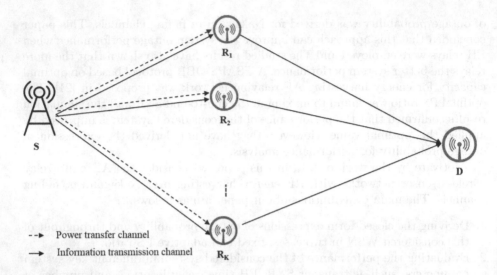

Fig. 1. System and channel models for EH AF WSNs

Note that, the PS ratio can be dynamically adjusted according to the variation of channel coefficient to maximum the end-to-end signal-to-noise ratio (SNR). Note that the optimal relay is selected among N AF relays based on the criteria of optimal system capacity. The relays can estimate the instantaneous channel gain based on the algorithm of channel estimation in request-to-send (RTS)/clear-to-send (CTS) transmission from the source and the destination [21].

According to [21], in high SNR region, the end-to-end SNR of this considered system is given by

$$\gamma_{e2e}^* \sim \frac{c\gamma_0\gamma_1^*\gamma_2^*(1-\rho^*)\rho^*}{c(1-\rho^*)\gamma_2^* + \rho^*} , \qquad (1)$$

where $c = \dfrac{\eta\alpha}{1-\alpha}$, $\gamma_0 = \dfrac{P_0}{N_0}$, $\gamma_1^* = \dfrac{|h_1|^2}{d_1^\theta}$, $\gamma_2^* = \dfrac{|h_2|^2}{d_2^\theta}$.

In order to maximum γ_{e2e}^*, the optimal ρ^* was introduced in [21] as follows

$$\rho^* = \frac{\sqrt{c\gamma_2^*}}{\sqrt{c\gamma_2^*} + 1}. \qquad (2)$$

Due to the relay node can only split the received signal into two power parts based on a finite discrete set of PS ratios in practice, we design that ρ_i^* can only pick the value from the following set:

$$\rho_l \in \left\{ \frac{1}{L}, \frac{2}{L}, \cdots, \frac{L-1}{L} \right\}, \qquad (3)$$

where L is the number of PS ratio levels and $1 \leq l \leq L-1$. Note that, ρ_l^* cannot is selected as zero (no information part) or one (no energy part). According to (2) and (3), we assign $\rho_l^* = \dfrac{l}{L}$ when the channel gain of R - D link satisfies the following condition:

$$b_l = \frac{l^2}{(L-l)^2 c} < \gamma_2^* < b_{l+1} = \frac{(l+1)^2}{(L-l-1)^2 c}, \tag{4}$$

where $l \in \{1, 2, \cdots, L-1\}$.

Note that due to the links of S - R and R - D undergo the Nakagami-m fading, the cumulative distribution function (CDF) and probability density function (PDF) of random variable (RV) SNRs, i.e., γ_n, $n \in \{1, 2\}$ are respectively given by

$$F_{\gamma_n}(x) = 1 - e^{-\frac{m_n}{\lambda_n} x} \sum_{k=0}^{m_n-1} \frac{1}{k!} \left(\frac{m_n}{\lambda_n} x\right)^k, \tag{5}$$

$$f_{\gamma_n}(x) = \frac{x^{m_n-1}}{(m_n-1)!} \left(\frac{m_n}{\lambda_n}\right)^{m_n} e^{-\frac{m_n}{\lambda_n} x}, \tag{6}$$

where $\lambda_n = \mathbf{E}(\gamma_n)$, $m_n \geq 1/2$ is the fading severity factor, in which $m_n = 1$ corresponds to Rayleigh fading and $m_n = (V+1)^2/(2V+1)$ approximates Rician fading with parameter V.

3   Performance Analysis

This Section presents the derivation of the expression of outage probability and throughput of this considered WSN system.

System Outage Probability. In order to characterize the performance of a wireless communication system, the outage probability is used as an important performance metric. It is defined as the probability that the instantaneous capacity (C) falls below a predetermined rate threshold $R > 0$, which is expressed as

$$OP = \Pr(C < R). \tag{7}$$

In this considered system with TSAPS-ORS scheme [21], the overall outage probability can be obtained as

$$OP^* \overset{(a)}{=} (P_{out}^*)^K = \left[\Pr\left(C_{opt}^* < R\right)\right]^K = \left[\Pr(\gamma_{e2e}^* < 2^{\frac{2R}{1-\alpha}} - 1)\right]^K, \tag{8}$$

where C_{opt}^* is the optimal instantaneous capacity for best relay. Note that step (a) is obtained by assuming the channels are modeled as i.i.d over different relaying channels [19].

To analyze the performance of this system, we obtain the following theorems.

Theorem 1. Under Nakagami-m fading, the outage probability of overall system is obtained as

$$OP^* = \{1 - \sum_{l=1}^{L-1} \sum_{j=0}^{m_1-1} \sum_{i=0}^{j} \frac{e^{-\frac{m_1\gamma_{th}}{\lambda_1\rho_l\gamma_0}}}{i!(j-i)!(m_2-1)!c^i(1-\rho_l)^i\rho_l^{j-i}} \left(\frac{m_1\gamma_{th}}{\lambda_1\gamma_0}\right)^j \left(\frac{m_2}{\lambda_2}\right)^{m_2}$$

$$\times \sum_{p=0}^{\infty} \frac{(-1)^p a_1^p}{p!a_2^{m_2-i-p}} \left[\Gamma(m_2-i-p, a_2b_l) - \Gamma(m_2-i-p, a_2b_{l+1})\right]\}^K, \quad (9)$$

where $a_1 = \frac{m_1\gamma_{th}}{\lambda_1 c(1-\rho_l)\gamma_0}$, $a_2 = \frac{m_2}{\lambda_2}$.

Proof. See Appendix.

System Throughput. The second important and related performance metric is the throughput (τ) at the destination under the delay-limited transmission mode. This metric is found by evaluating the outage probability at a fixed source transmission rate $-$ R bps/Hz. Taking into account the fixed source transmission rate R bps/Hz, the effective communication time from the source node to the destination node in the total block time T is $\frac{(1-\alpha)T}{2}$. We obtain the following theorem.

Theorem 2. Under Nakagami-m fading, the overall throughput of this considered system is written as

$$\tau = \frac{1}{2}(1-\alpha)R\left\{1 - \left\{1 - \sum_{l=1}^{L-1} \sum_{j=0}^{m_1-1} \sum_{i=0}^{j} \frac{e^{-\frac{m_1\gamma_{th}}{\lambda_1\rho_l\gamma_0}}}{i!(j-i)!(m_2-1)!c^i(1-\rho_l)^i\rho_l^{j-i}} \left(\frac{m_1\gamma_{th}}{\lambda_1\gamma_0}\right)^j \right.\right.$$

$$\times \left. \left(\frac{m_2}{\lambda_2}\right)^{m_2} \sum_{p=0}^{\infty} \frac{(-1)^p a_1^p}{p!a_2^{m_2-i-p}} \left[\Gamma(m_2-i-p, a_2b_l) - \Gamma(m_2-i-p, a_2b_{l+1})\right] \right\}^K \right\}. (10)$$

Proof. According to the definition of throughput [14], we have

$$\tau = (1 - OP^*) R\frac{(1-\alpha)T/2}{T} = \frac{1}{2}(1-\alpha)R(1-OP^*). \quad (11)$$

Substituting (9) into (11), we obtained the throughput of this considered system as (10). This is the end of our proof.

4   Numerical Results and Discussion

In this section, we provide the simulation and analysis results in terms of OP^* and τ to reveal the impact of key system parameters on system performance, such as average transmit SNR (γ_0), number of relays (N), TS ratio (α), EH efficiency (η), relay location (d_1) and fading severity factor (m).

4.1 Impact of Average Transmit SNR and Number of Relays

The impact of average transmit SNR γ_0 and the number of relays K on system performance are shown in Figs. 2 and 3. According to these figures, the performance gets better with increasing γ_0 and K. This is because the higher transmit power the better signal and the more energy harvested, leading to the higher power to amplify the retransmit signal in the second phase. However, when γ_0

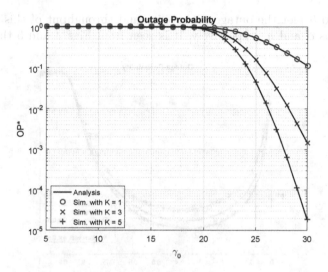

Fig. 2. OP^* vs. average transmit SNR γ_0 with $d_1 = d_2 = 1$, $m_1 = 2$, $m_2 = 25$, $R = 1\,\mathrm{bps/Hz}$, $\theta = 2$, $\alpha = 0.7$, $\eta = 1$, $L = 20$.

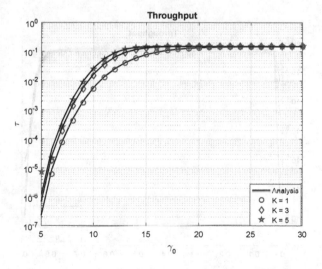

Fig. 3. τ vs. average transmit SNR γ_0 with $d_1 = d_2 = 1$, $m_1 = 2$, $m_2 = 25$, $R = 1\,\mathrm{bps/Hz}$, $\theta = 2$, $\alpha = 0.7$, $\eta = 1$, $L = 20$.

is large enough, $OP^* \to 0$ and $\tau \to \frac{(1-\alpha)R}{2}$. From these figures, we can also understand that increasing the number of relays can improve the performance of this system because we have more choices to select the best relay to forward the information to the destination sensor node.

4.2 Impact of Energy Harvesting Time and Energy Harvesting Efficiency

Figures 4 and 5 plot the outage probability and throughput of this considered system versus α and η, respectively. It is seen from Figs. 4 and 5 that when α

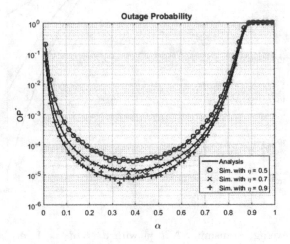

Fig. 4. OP^* vs. α and η with $\gamma_0 = 20\,\text{dB}$, $d_1 = d_2 = 1$, $m_1 = 2$, $m_2 = 15$, $R = 1\,\text{bps/Hz}$, $\theta = 2$, $L = 20$.

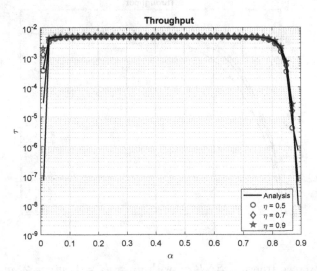

Fig. 5. τ vs. α and η with $\gamma_0 = 20\,\text{dB}$, $d_1 = d_2 = 1$, $m_1 = 2$, $m_2 = 15$, $R = 1\,\text{bps/Hz}$, $\theta = 2$, $L = 20$.

grow up, OP^* decreases, τ scales up and the performance is upgraded. This is explained that due to more time spent on energy harvesting as α grows leads to higher transmission power, hence better performance results. However, when α continues to increase, OP^* gets larger, τ scales down and the performance is degraded. That is because the information time is reduced, which leads to the increasing of real data rate. Overall, there is an optimal value of α that can minimize OP^*. This is clearly seen by the single bottom of the OP^* curve plotted in Fig. 4 as a function of α.

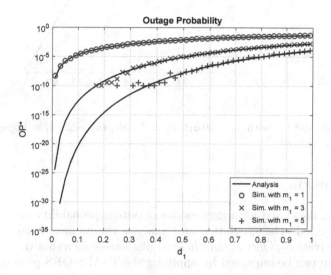

Fig. 6. OP^* vs. d_1 and m_1 with $\gamma_0 = 20\,\text{dB}$, $d_2 = 2 - d_1$, $m_2 = 25$, $R = 1\,\text{bps/Hz}$, $\theta = 2$, $\alpha = 0.7$, $\eta = 1$, $L = 20$.

4.3   Impact of Relay Location and Fading Severity Parameters

The impact of relay location (d_1) and fading severity parameters (m_1) on outage probability and throughput are illustrated in Figs. 6 and 7, respectively. In these two figures, increasing d_1 makes OP^* and τ worse. This is because of the higher values of d_1^θ lead to the smaller values of energy collected as well as poorer received signal strength at the relay nodes. Similarly, we can understand that OP^* decreases and τ increases with increasing m_1. This is because the channel qualities are better with larger fading severity parameters.

In general, from above figures we can see that the superior match between analytical and simulation results occurs in the high average transmit SNR or in large P_0 region. For a more clear explanation, when γ_0 holds low values, the analytical and simulation results do not match well because we use the approximated expression of the end-to-end SNR at the destination node as (1) [21] and using finite terms of (1.211-1) in [22].

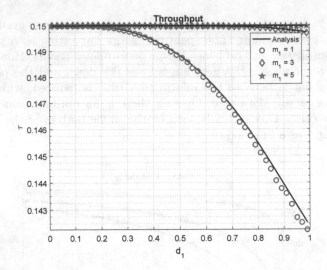

Fig. 7. τ vs. d_1 and m_1 with $\gamma_0 = 20\,$dB, $d_2 = 2 - d_1$, $m_2 = 25$, $R = 1\,$bps/Hz, $\theta = 2$, $\alpha = 0.7$, $\eta = 1$, $L = 20$.

5   Conclusion

In this paper, the closed-form expressions of outage probability and throughput have been derived. The simulation and analysis results have been presented to verify our derivations. Once again, these results have shown that the performance of this system can be improved by applying the TSAPS-ORS protocol.

Appendix

Here, we derive the expression of P_{out}^* as (12) on the top of next page. Substituting (12) into (8), we obtain the closed-form expression of outage probability for this system.

$$P_{out}^* = \Pr(\gamma_{e2e}^* < 2^{\frac{2R}{1-\alpha}} - 1)$$

$$= 1 - \sum_{l=1}^{L-1} \Pr\left(\frac{c\gamma_0\gamma_1\gamma_2(1-\rho_l)\rho_l}{c(1-\rho_l)\gamma_2 + \rho_l} > \gamma_{th}, b_l \leq \gamma_2 < b_{l+1} \right)$$

$$= 1 - \sum_{l=1}^{L-1} \int_{b_l}^{b_{l+1}} \left[1 - F_{\gamma_1}\left(\frac{c(1-\rho_l)\gamma_{th}x + \rho_l\gamma_{th}}{c(1-\rho_l)\rho_l\gamma_0 x} \right) \right] f_{\gamma_2}(x)dx$$

$$= 1 - \sum_{l=1}^{L-1}\sum_{j=0}^{m_1-1}\sum_{i=0}^{j} \frac{e^{-\frac{m_1\gamma_{th}}{\lambda_1\rho_l\gamma_0}}}{i!(j-i)!(m_2-1)!c^i(1-\rho_l)^i\rho_l^{j-i}} \left(\frac{m_1\gamma_{th}}{\lambda_1\gamma_0} \right)^j \left(\frac{m_2}{\lambda_2} \right)^{m_2}$$

$$\times \int_{b_l}^{b_{l+1}} x^{m_2-i-1}e^{-\frac{m_1\gamma_{th}}{\lambda_1 c(1-\rho_l)\gamma_0 x} - \frac{m_2 x}{\lambda_2}} dx$$

$$= 1 - \sum_{l=1}^{L-1} \sum_{j=0}^{m_1-1} \sum_{i=0}^{j} \frac{e^{-\frac{m_1\gamma_{th}}{\lambda_1\rho_l\gamma_0}}}{i!(j-i)!(m_2-1)!c^i(1-\rho_l)^i\rho_l^{j-i}} \left(\frac{m_1\gamma_{th}}{\lambda_1\gamma_0}\right)^j \left(\frac{m_2}{\lambda_2}\right)^{m_2}$$

$$\times \left[\int_{b_l}^{\infty} x^{m_2-i-1} e^{-\frac{m_1\gamma_{th}}{\lambda_1 c(1-\rho_l)\gamma_0 x} - \frac{m_2 x}{\lambda_2}} dx - \int_{b_{l+1}}^{\infty} x^{m_2-i-1} e^{-\frac{m_1\gamma_{th}}{\lambda_1 c(1-\rho_l)\gamma_0 x} - \frac{m_2 x}{\lambda_2}} dx \right]$$

$$\overset{(b)}{=} 1 - \sum_{l=1}^{L-1} \sum_{j=0}^{m_1-1} \sum_{i=0}^{j} \frac{e^{-\frac{m_1\gamma_{th}}{\lambda_1\rho_l\gamma_0}}}{i!(j-i)!(m_2-1)!c^i(1-\rho_l)^i\rho_l^{j-i}} \left(\frac{m_1\gamma_{th}}{\lambda_1\gamma_0}\right)^j \left(\frac{m_2}{\lambda_2}\right)^{m_2}$$

$$\times \left[\sum_{p=0}^{\infty} \frac{(-1)^p a_1^p}{p!} \int_{b_l}^{\infty} x^{m_2-i-p-1} e^{-a_2 x} dx - \sum_{q=0}^{\infty} \frac{(-1)^q a_1^q}{q!} \int_{b_{l+1}}^{\infty} x^{m_2-i-q-1} e^{-a_2 x} dx \right]$$

$$\overset{(c)}{=} 1 - \sum_{l=1}^{L-1} \sum_{j=0}^{m_1-1} \sum_{i=0}^{j} \frac{e^{-\frac{m_1\gamma_{th}}{\lambda_1\rho_l\gamma_0}}}{i!(j-i)!(m_2-1)!c^i(1-\rho_l)^i\rho_l^{j-i}} \left(\frac{m_1\gamma_{th}}{\lambda_1\gamma_0}\right)^j \left(\frac{m_2}{\lambda_2}\right)^{m_2}$$

$$\times \left[\sum_{p=0}^{\infty} \frac{(-1)^p a_1^p}{p! a_2^{m_2-i-p}} \Gamma(m_2-i-p, a_2 b_l) - \sum_{q=0}^{\infty} \frac{(-1)^q a_1^q}{q! a_2^{m_2-i-q}} \Gamma(m_2-i-q, a_2 b_{l+1}) \right], \quad (12)$$

where $\gamma_{th} = 2^{\frac{2R}{(1-\alpha)}} - 1$. Note that step (b) and (c) are obtained by the help of (1.211-1) and (3.381-3), respectively, in [22].

This concludes our proof.

References

1. Chen, X., Zhang, Z., Chen, H.H., Zhang, H.: Enhancing wireless information and power transfer by exploiting multi-antenna techniques. IEEE Commun. Mag. **53**(4), 133–141 (2015)
2. Ha, D.B., Tran, D.D., Truong, T.V., Vo, N.V.: Physical layer secrecy performance of energy harvesting networks with power transfer station selection. In: IEEE International Conference on Communications and Electronics (ICCE), pp. 451–456 (2016)
3. Vo, V.N., Nguyen, T.G., So-In, C., Ha, D.B.: Secrecy performance analysis of energy harvesting wireless sensor networks with a friendly jammer. IEEE Access (2017)
4. Xu, K., Shen, Z., Wang, Y., Xia, X.: Beam-domain hybrid time-switching and power splitting SWIPT in full-duplex massive MIMO system. EURASIP J. Wirel. Commun. Netw., 1–21 (2018)
5. Suraweera, H.A., Karagiannidis, G.K., Smith, P.J.: Performance analysis of the dual hop asymmetric fading channel. IEEE Trans. Wirel. Commun. **8**(6), 2783–2788 (2009)
6. Gurung, A.K., Al-Qahtani, F.S., Hussain, Z.M., Alnuweiri, H.: Performance analysis of amplify-forward relay in mixed Nakagami-m and Rician fading channels. In: The 2010 International Conference on Advanced Technologies for Communications, Ho Chi Minh City, Vietnam, 20–22 October 2010, pp. 321–326 (2010)

7. Haghighat, J., Eslami, M., Hamouda, W.: Relay pre-selection for reducing CSI transmission in wireless sensor networks. IEEE Commun. Lett. **20**(9), 1828–1831 (2016)
8. Mousavi, S.H., Haghighat, J., Hamouda, W., Dastbasteh, R.: Analysis of a subset selection scheme for wireless sensor networks in timevarying fading channels. IEEE Trans. Signal Process. **64**(9), 2193–2208 (2016)
9. Kim, J.B., Song, M.S., Lee, I.H.: Achievable rate of best relay selection for non-orthogonal multiple access-based cooperative relaying systems. In: International Conference on Information and Communication Technology Convergence (ICTC), Jeju, South Korea, pp. 960–962. IEEE (2016)
10. Gendia, A.H., Elsabrouty, M., Emran, A.A.: Cooperative multi-relay non-orthogonal multiple access for downlink transmission in 5G communication systems. In: 2017 Wireless Days, Porto, Portugal. IEEE (2017)
11. Luo, Y., Zhang, J., Letaief, K.B.: Relay selection for energy harvesting cooperative communication systems. In: IEEE Global Communications Conference (GLOBE-COM), pp. 2514–2519 (2013)
12. Ishibashi, K.: Dynamic harvest-and-forward: new cooperative diversity with RF energy harvesting. In: 2014 Sixth International Conference on Wireless Communications and Signal Processing (WCSP), 23–25 October 2014, pp. 1–5 (2014)
13. Nasir, A.A., Zhou, X., Durrani, S., Kennedy, R.A.: Block-wise time-switching energy harvesting protocol for wireless-powered AF relays. In: 2015 IEEE International Conference on Communications (ICC), 8–12 June 2015, pp. 80–85 (2015)
14. Ha, D.B., Tran, D.D., Tran-Ha, V., Hong, E.K.: Performance of amplify-and-forward relaying with wireless power transfer over dissimilar channels. Elektronika ir Elektrotechnika J. **21**(5), 90–95 (2015)
15. Liu, Y.: Wireless information and power transfer for multi-relay assisted cooperative communication. IEEE Commun. Lett. **20**(4), 784–787 (2016)
16. Nguyen, H.S., Do, D.T., Nguyen, T.S., Voznak, M.: Exploiting hybrid time switching-based and power splitting-based relaying protocol in wireless powered communication networks with outdated channel state information. J. Control Meas. Electron. Comput. Commun. **58**(1), 111–118 (2017)
17. Ha, D.B., Nguyen, Q.S.: Outage performance of energy harvesting DF relaying NOMA networks. Mobile Netw. Appl. (2017)
18. Cvetkovic, A., Blagojevic, V., Ivanis, P.: Performance analysis of nonlinear energy-harvesting DF relay system in interference-limited Nakagami-m fading environment. ETRI J. **39**(6), 803–812 (2017)
19. Singh, V., Ochiai, H.: A efficient time switching protocol with adaptive power splitting for wireless energy harvesting relay networks. In: IEEE 85th Vehicular Technology Conference (VTC Spring) (2017)
20. Zhong, S., Huang, H., Li, R.: Performance analysis of energy-harvesting-aware multi-relay networks in Nakagami-m fading. EURASIP J. Wirel. Commun. Netw. **2018**, 63 (2018)
21. Ha, D.H., Ha, D.B., Zdralek, J., Voznak, M.: A new protocol based on optimal capacity for energy harvesting amplify-and-forward relaying networks. In: 5th NAFOSTED Conference on Information and Computer Science (NICS), HCMC, Vietnam (2018)
22. Gradshteyn, I., Ryzhik, I.: Table of Integrals, Series, and Products. Elsevier Academic Press, Cambridge (2007)

Hardware and Software Design, Information Processing and Data Analysis

Generating Test Data for Blackbox Testing from UML-Based Web Engineering Content and Presentation Models

Quyet-Thang Huynh[1], Dinh-Dien Tran[1], Duc-Man Nguyen[2],
Nhu-Hang Ha[2], Thi-Mai-Anh Bui[1], and Phi-Le Nguyen[1(✉)]

[1] School of Information and Communication Technology,
Hanoi University of Science and Technology, Hanoi, Vietnam
{thanghq, anhbtm}@hust.soict.edu.vn,
trandinhdien@gmail.com, lenp@soict.hust.edu.vn
[2] Duy Tan University, Da Nang, Vietnam
{mannd, hatnhuhang}@duytan.edu.vn

Abstract. Software testing is a process that produces and consumes huge amounts of data. Thus, the test data is usually either gathered manually by the testers or randomly generated by tools. The manual method consumes lot of time and highly depends on the testers' experience while the random approach faces the problem of redundant test data caused by identical use cases. By leveraging the concept of Model-based testing, this paper provides a novel method of testing to save the cost of manual testing and to increase the reliability of the testing processes. In Model-based testing, test cases and test data can be derived from different models. In this paper, we present a technique to generate test data from UML-based Web Engineering (UWE) presentation model for web application testing by using formal specification and Z3 SMT solver. We also build a model-based testing Eclipse Plug-in tool called TESTGER-UWE that generates test data based on the model of UWE for the web application. We evaluate the proposed methods by applying them to generate test data for an Address Book project of UWE. Experimental results show that our proposed methods can reduce the time significantly when generating test data for automation test tools such as Selenium, Katalon, Unit test, etc.

Keywords: Web application testing · Model-based testing · Test case generation · UML-based Web Engineering

1 Introduction

The UWE is an object-oriented approach, which was presented by the end of the 90s [1, 2]. This concept aims to find a standard for building models of analyzing and designing web systems based on object-oriented hypermedia design method (OOHDM) [3], relationship management methodology (RMM) [4], and web search and data mining (WSDM) [5]. These models are built at the different phases of the software development process and represent different views of the Web application corresponding to the different concerns. UWE follows a strict separation of concerns in the

© ICST Institute for Computer Sciences, Social Informatics and Telecommunications Engineering 2019
Published by Springer Nature Switzerland AG 2019. All Rights Reserved
T. Q. Duong et al. (Eds.): INISCOM 2019, LNICST 293, pp. 207–219, 2019.
https://doi.org/10.1007/978-3-030-30149-1_17

early stages of the development and implements a model-driven development process. In UWE, UML diagrams are exploited to visualize the models. By such ways, UWE can represent the structural aspects of the different views and provide support for model-driven web applications development [5]. Model-driven software development stresses the use of models at all levels of the software development process. This result changes the way software is designed, maintained, and tested [6]. According to [7]: "Testing often accounts for more than 50% of the required effort during system development".

A testing cycle encompasses three main parts: (i) Test case generation, (ii) Test execution, and (iii) Test evaluation. Test case and test data generation is perhaps the most complex and challenging part. Testing is a cumbersome task which needs to assure customer satisfaction and product safety. Automated test data generation is one of the main factors that contribute to the quality of automated testing. It is used for automatically generating test data for software under test (SUT) during software testing. Automation in the test data generation process could reduce testing expenses and increase the reliability of the entire testing process. For these reasons, automated test data generation has remained a topic of interest for the past four decades [9]. In many cases, model-based testing is more efficient than other testing techniques due to its possibility in generating large test suites and test data to provide evidence for accurate system implementation [8, 10].

In this paper, we aim at proposing a technique to generate test cases and test data from the UWE presentation model for Web application testing. We used the results of the previous study of Nguyen et al. [29] on the generation of test data using the formal specification and Z3 SMT Solver. From UWE presentation model and content models, they are transformed to XMI file, from which the formal specification file (called myDSL-Domain Specific Language defined by [29]) is generated for each Presentation class. Then the generation test engine invokes Z3 SMT Solvers to generate test data. We also build a plug-in tool called TESTGER-UWE that supports to transform XMI to myDSL and generate test data. The proposed method applies to the Address Book project of UWE.

The rest of this paper is organizing as follows: Sect. 2 provides the related studies; the proposed method is presenting in Sect. 3; In Sect. 4, we apply the proposed approach to generate test data for an Address book case study; Sect. 5 presents the results and discussion; Sect. 6 concludes the paper and describes our future works.

2   Related Works

UML is most generally used to provide a standard way to visualize the design of a system and also widely used for test case generations. There are many types of research in recent years about various techniques for the generation of test cases from UML diagrams.

Wang, et al. [11] proposed use case modeling for system tests generation. The proposed technique uses use case specifications, a domain model, a class diagram and constraints to generate executable system test cases. Oluwagbemi and Asmuni [12] presented an enhanced method for generating test cases from various UML diagrams.

A robust scheme for collecting artifacts from the underlying diagrams of the software under test was proposed. The intermediate representation of the artifacts is in the form of a tree over which the traversal of contents generates test cases. Anbunathan and Anirban [13] developed a method using basis path testing approach for test cases generation from UML class diagrams. From the class diagrams, a corresponding state chart diagram has been drawn and then is converted into a control flow graph. Test cases are generated manually as well as automatically, and the effectiveness of test cases has been performed using mutation analysis. Results show a significant decrease in cost as compared to other existing techniques.

Vinaya and Ketan [14] have presented a method for the generation of test cases from UML use case diagrams, class diagrams, and sequence diagrams, and then transform it into a Sequence Diagram Graph. A data dictionary is presented in the form of Object Constrained Language (OCL). The UML diagrams are drawn with the help of magic draw tool and then exported to XML format. The XML file has been parsed in java for extracting different nodes of the graph and generates all set of the scenarios from start node to end nodes.

Papadopoulos and Walkinshaw [15] presented a model-inference driven testing framework that is designed to support the inference-driven test generation for programs that are not sequential. The framework is designed to be modular; it is not necessarily tied to a specific model inference or test generation framework and can be in principle applied to any executable program, without the need for access to the source code.

The framework is deliberately flexible and uses C4.5 algorithm to infer decision trees from program executions and uses the Z3 solver to generate and execute tests from it [15]. The authors provided an openly-available Java implementation that can be extended to handle different types of programs, models, and test generators. The authors developed an evaluation of three openly-available programs and indicated that inference-driven testing could produce better test sets more efficiently than random testing.

According to Jain and Porwal survey, most of the researches proposed an approach to test data generation based on actually executing the program, analyzing the dynamic data flow, and using a function minimization method. These researches use popular heuristic approaches like Genetic Algorithm, Simulating Annealing, Particle Swarm Optimization, Ant Colony Optimization has been widely applied to search for effective test data. These approaches are verified to be more optimized than the random technique [16–22].

Currently, on the market, there are some tools to support test data generation online and offline for Data Driven Testing such as Generate data, Mockaroo, Yan Data Ellan [26–28] but these tools only let testers create data generation fields, select predefined data types and generate data. Their limitation is not to modify data size constraints and other user-defined constraints.

From the literature review, we could observe that there are different methods available to generate test cases and test data using different techniques. Studies focused on test case generation from UML diagrams by converting UML models to intermediate graphs and generating test cases. Other studies on test data generation are mostly based on the program code using meta-heuristics algorithms and optimization techniques.

3 A Novel Approach for Test Data Generation from UWE Content and Presentation Models

A metamodel represents these model elements and their relationships. UWE is compatible with the MOF exchange metamodel and therefore with XMI-based XML exchange format tools. The advantage of UML CASE tools which support for UML profiles or UML extension mechanisms can be used to create UWE models of web applications such as automatic model generation.

The presentation model provides an abstract view of the user interface (UI) of a web application. It is based on the navigation model. It describes the basic structure of the user interface, such as UI components (e.g., text, images, anchors, forms) used to represent navigation nodes. The UI components do not represent specific components of any presentation technology, but only describe what functionality is required at that particular point in the user interface. The basic elements of the presentation model are presentation classes, which are directly based on the nodes from the navigation model. Presentation classes may contain other presentation elements. In the case of UI components, such as text or images, the presentation properties associated with the navigation attribute containing the content will be displayed.

Based on our previous research on model transformation with OCL integration [23, 24] and development of rules and algorithms for code in UWE [30], we continue to expand and develop test case/test data generation technique for generating code, as well as test data based on UWE metamodels. In this context, we propose a new approach to generate test data using XMI file, domain specific language and Z3 SMT Solvers from the study of Nguyen and et al. [29]. The proposed framework is shown on Fig. 1.

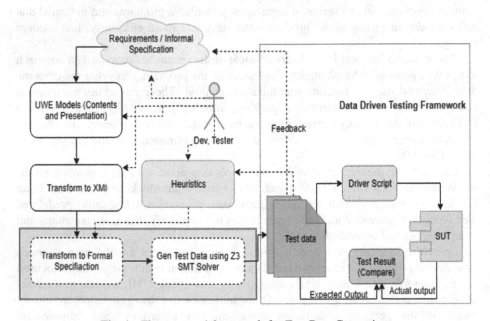

Fig. 1. The proposed framework for Test Data Generation.

The process of developing Web Application with model transformation techniques, using MagicUWE will include 5 steps as follows. We propose the technique for test data generation from UWE Content and Presentation Models to be implemented in step 3, step 4 and step 5, according to the framework shown in Fig. 1.

Step 1. Build up the charts in term of Content, Navigation, Presentation, Process structure and/or Process flow.

Step 2. Use MagicDraw to transform Content, Presentation Models to XMI file (e.g. Fig. 2)

```
  </packagedElement>
- <packagedElement xmi:id="AAAAAAFpreBuJ4/4Ip4=" name="Contact" xmi:type="uml:Class" visibility="public"
  isActive="false" isLeaf="false" isFinalSpecialization="false" isAbstract="false">
    + <ownedMember xmi:id="AAAAAAFpreZmqZC/lAw=" xmi:type="uml:Association" visibility="public"
      isDerived="false">
    <ownedAttribute xmi:id="AAAAAAFpreCo95Akclg=" name="name" xmi:type="uml:Property" type="String_id"
      visibility="public" isLeaf="false" isID="false" isDerived="false" aggregation="none" isUnique="false"
      isOrdered="false" isReadOnly="false" isStatic="false"/>
    <ownedAttribute xmi:id="AAAAAAFpreDKqJArW+E=" name="email" xmi:type="uml:Property" type="String_id"
      visibility="public" isLeaf="false" isID="false" isDerived="false" aggregation="none" isUnique="false"
      isOrdered="false" isReadOnly="false" isStatic="false"/>
  </packagedElement>
- <packagedElement xmi:id="AAAAAAFpreEukJAym04=" name="Address" xmi:type="uml:Class" visibility="public"
```

Fig. 2. An example of XMI file of Content model

Step 3. Leverage suggested functions to transfer XMI to formal specifications DSL. We utilize the results of previous research, proposed by Nguyen et al. [29] regarding the use of formal language to specify software requirements, the myDSL edit tool is built upon xText and DSL (Java environment), Fig. 3 shows the formal specification (called myDSL) of the Contact Class in Presentation model.

```
1    enum Contact {SUCCEESS FAIL INVALID}
2    function Contact (String name, String email)
3    define VALIDname {name.length > 8 && name.length <30}
4    define VALIDemail {email.contain(@) && email.contain(.)}
5    define conFail {!VALIDname || !VALIDemail}
6    precondition {name.length > 0 && name.length <255}
7    precondition {email.length > 0 && email.length <255}
8    testcase {
9        "SUCCESS" VALIDname && VALIDemail
10       "FAIL" conFail
11       "INVALID"
12   }
13   run
14
```

Fig. 3. myDSL of Contact class.

Step 4. Then, the engine generates data by transferring specification from the 3rd step to Z3 SMT language and call Z3 SMT solvers to find a set of solutions (test cases and test inputs). Z3 offers a compelling match for software analysis and verification tools since several common software constructs map directly into supported theories. It is best used as a component in the context of other tools that require solving logical formulas. Figure 4 show the screenshot of generating test data.

Step 5. The results provide the test data in the format of XLS or CSV that shown in Table 1 (an example data), Tester and Developer can use these results for different testing purposes, especially for Data Driven Testing method of web applications. They also can use the results of test data to do some heuristics to optimize the data results by adding other constraints to the myDSL file and performing test generation again.

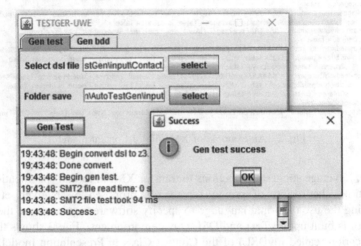

Fig. 4. Screenshot of generating test data tool

In Black Box Testing the code is not visible to the tester, functional test cases can have test data meeting following criteria:

- No data: Check system response when no data is submitted
- Valid data: Check system response when Valid test data is submitted
- Invalid data: Check system response when InValid test data is submitted
- Illegal data format: Check system response when test data is in an invalid format
- Boundary Condition Dataset (BVA): Test data meeting boundary value conditions
- Equivalence Partition Data Set (EPC): Test data qualifying your equivalence partitions.

4   Case Study: Address Book Web Application

Address Book with Searches is a typical example, used as a case study in UWE engineering research [25]. In this study, we also use this example to illustrate the proposed technique given in Sect. 3. This is an address book of contacts. Each contact will contain a name, two phone numbers (main and alternative), two postal addresses (main and alternative), an e-mail address and a picture. The page will publish the details of the contact(s) matching a filtering condition. Users can create new, edit, update and search the contacts.

Figure 5 shows the content model of the Address Book with Searches, with the classes defined for Address-Book, Contact, Address, and Phone [25]. Figure 6 indicates the presentation model. The address book page contains the Introduction section and the Contacts list. For each contact, the corresponding email, phone and address fields are displayed [25].

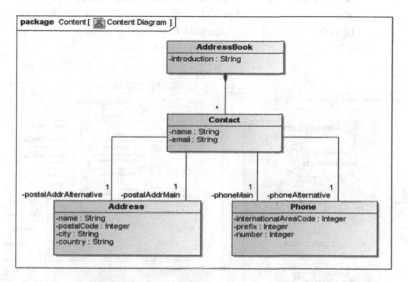

Fig. 5. UWE content model of the Address Book with Searches [25]

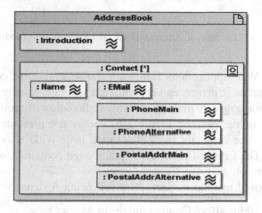

Fig. 6. UWE presentation model of a simple Address Book [25]

Table 1. Example test data for Contact form

Name	Email	Result
oJRNjn	dmC3Ih	Fail
hhC4fCkA	0ZQ6kf@qe9Y.ca	Success
m	x	Fail
6de	FeX	Fail
yrw9bobAMpLyOrAfw	6elcYlQV	Fail
5u	j7	Fail
jfyK8Yi	SsQtPWd	Fail
5GhPr0LXw	oJ3WWV@NShJV.bMN	Success
ntZt	y13 M	Fail
Hj%rD	G6amh@CO.Va	Fail

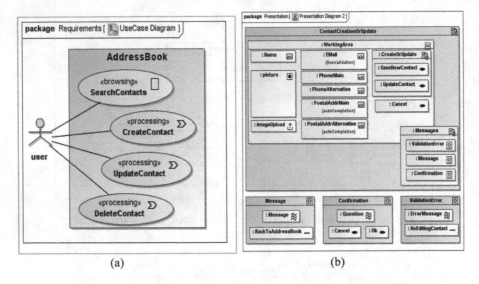

(a)   (b)

Fig. 7. (a) Use case of Address Book, (b) Presentation models [25]

UWE specifies Web applications following the separation of concerns, i.e. modeling content, navigation structure, and presentation separately. Increasing functionality of the web application suggests making a detailed elicitation of requirements. Figure 7 (a) is the use case of the project. Figure 7(b) shows the presentation model of the running case study. The container form is selected to provide a more intuitive representation of pages. The ContactCreationOnUpdate page contains Contact's fields to be updated, buttons, image, message forms.

Apply the proposed method in Sect. 3 by following 6 steps:

Step 1 – Open AddressBookContent.mdzip in MagicDraw.
Step 2 – Export UWE content and presentation models to XMI format.

Step 3 – Transform XMI to DSL.

Step 4 – Call gentest engine to generate test data from DSL input file.

Step 5 – Validate the output and add some Heuristics and/or modify DSL file and re-generate test data.

Steps 3 to 5 are features of TESTGER-UWE plug-in tool.

Use the output to different purposes such as Selenium, Katalon, Testcomplete or Data Driven Testing framework.

5 Results and Discussion

Experimental results for Address Book application (including Simple Address Book, Address Book with Search feature and Address Book with Update Content), depending on the number of input fields on each form, or attribute fields in classes of Content models, the data type of each field as well as their size that are adapted from XMI to each respective DSL specification. For data fields at Presentation model, it must be based on the class diagram of a Content model to determine the type and size (or data constraints). The data generated for the fields is described in Table 2.

Table 2. The result of generating test data for content and presentation models

Class of content/presentation	Data type	Coverage of test data	No. Rows generated
AddressBook	String	Random values; minlength, maxlength, BVA, EPC (len), Valid data, Invalid data	100
Contact	String	Random values; minlength, maxlength, BVA, EPC (len), Valid data, Invalid data	200
Address	String	Random values; minlength, maxlength, BVA, EPC (len), Valid data, Invalid data	200
	Integer	Random values; min, maxint, BVA, EPC, valid data, invalid data	200
Phone	Integer	Random values; min, maxint, BVA, EPC, valid data, invalid data	200–1000
Picture	Integer	Random values; min, maxint, BVA, EPC, valid data, invalid data	200
SearchForm	String	Random values; minlength, maxlength, BVA, EPC (len), Valid data, Invalid data.	100
WorkingAreaForm	String	Random values; minlength, maxlength, BVA, EPC (len), Valid data, Invalid data.	200
	Number	Random values; min, maxint, BVA, EPC, valid data, invalid data	200

In this study, we only tested about 100 records data for each class/form. Data may be generated more by adjusting the data generation parameters. The data types supported in this study are Numeric, String (string length), Boolean type. Other data types will be studied in the future. The generated data corresponding to each data type is declared/bound to valid data, invalid data, invalid format, using BVA and EPC to optimize output and detect the corner error.

Table 3. The comparison of 3 online tools with TESTGER-UWE

	TESTGER-UWE	Generatedata	Mockaroo	Yan Data Ellan
Format output data	CSV, XLS	CSV, XLS, JSON, SQL, XML	CSV, XLS, JSON, SQL	CSV, XLS, JSON, SQL, XML
Data coverage	**Valid, invalid data, invalid format, BVA, EPC**	Valid, invalid data	Valid, invalid data	Valid, invalid data
Generate type	Random	Random	Random	Random
Change/add constraints	**Yes**	No	No	No
Number row of data at a time	**5000+**	5000	5000	10000
Meaningful of data	No	Yes	Yes	Yes
Ease of use	**Plug-in/standalone**	Online	Online	Online
Expected result	**Yes**	No	No	No

Bold values in this table are advanced features of TESTGER-UWE compared to other tools.

Fig. 8. Generatedata.com screenshot

Comparing experimental results with the Generate data [26], Mocka-roo [27], Yan Data Ellan [28] shows that these tools have many advanced features such as generating data into SQL, XML, Firebase, JSON, and more meaningful data. However, most of them have fees (to pay), do not generate data to cover cases of invalid data, data sizes, wrong format, additional constraints to optimize, adjust generated data as TESTGER-UWE tool. Table 3 presents a comparison of 3 online tools with TESTGER-UWE. Figures 8 and 9 are a screenshot of Generatedata.com and Mockaroo.com and generated data.

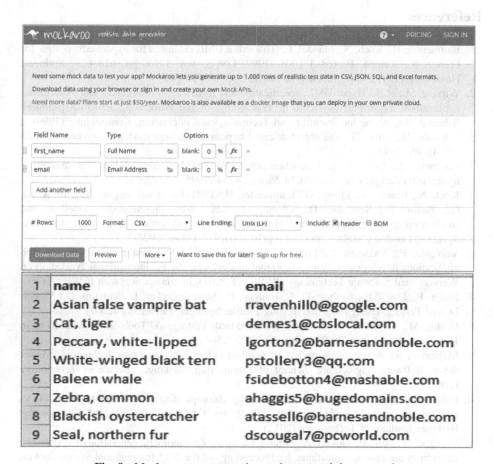

Fig. 9. Mockaroo.com screenshot and generated data example

6 Conclusion and Future Work

In this paper, we proposed a method to generate test data from UWE Content and Presentation. Specifically, our method first converts the content of the UWE Content and Presentation into XM. We also developed a tool named TESTGER-UWE tool, which generates the specification DSL and call Z3 SMT solver to generate test data. Our proposed approach is not only suitable for unit testing and Data Driven Testing but also can be applied to various testing purposes.

In the future, we will expand our approach to cover other types of data and generate test scripts for unit test or automated tests using Selenium, Katalon.

Acknowledgments. This research is funded by Hanoi University of Science and Technology under Grant number T2018-PC-015.

References

1. Baumeister, H., Koch, N., Mandel, L.: Towards a UML extension for hypermedia design. In: France, R., Rumpe, B. (eds.) UML 1999. LNCS, vol. 1723, pp. 614–629. Springer, Heidelberg (1999). https://doi.org/10.1007/3-540-46852-8_43
2. Wirsing, M., et al.: Hyper-UML: specification and modeling of multimedia and hypermedia applications in distributed systems. In: Proceedings of 2nd Workshop. German-Argentinian Bilateral Programme for Scientific and Technological Cooperation, Konigswinter (1999)
3. Schwabe, D., Rossi, G.: The object-oriented hypermedia design model. Commun. ACM **38**(8), 45–46 (1995)
4. Isakowitz, T., Stohr, E.A., Balasubramanian, P.: RMM: a methodology for structuring hypermedia design. Commun. ACM **38**(8), 34–44 (1995)
5. Koch, N., Knapp, A., Zhang, G., Baumeister, H.: UML-based web engineering. In: Rossi, G., Pastor, O., Schwabe, D., Olsina, L. (eds.) Web Engineering: Modelling and Implementing Web Applications. Human-Computer Interaction Series, pp. 157–191. Springer, London (2008). https://doi.org/10.1007/978-1-84628-923-1_7
6. Valverde, F., Valderas, P., Fons, J., Pastor, O.: A MDA-based environment for web applications development: from conceptual models to code. In: International Workshop on Web-oriented Software Technology (IWWOST 2007), in conjunction with ICWE (2007)
7. Baker, P., Dai, Z.R., Grabowski, J., Haugen, P., Schieferdecker, I., Williams, C.: Model-Driven Testing: Using the UML Testing Profile. Springer, Heidelberg (2007)
8. Utting, M., Legeard, B.: Practical Model-Based Testing: A Tools Approach. Morgan Kaufmann, Burlington (2006). ISBN-10: 0123725011
9. Mahmood, S.: A systematic review of automated test data generation techniques. Master thesis, Software Engineering, School of Engineering, Blekinge Institute of Technology (2007)
10. Polamreddy, R.R., Irtaza, S.A.: Software testing: a comparative study model-based testing vs test case-based testing. Master thesis, Software Engineering, School of Engineering, Blekinge Institute of Technology (2012)
11. Wang, C., Pastore, F., Goknil, A., Briand, L., Iqbal, Z.: Automatic generation of system test cases from use case specifications. In: Proceedings of the 2015 International Symposium on Software Testing and Analysis, ISSTA 2015 (2015). https://doi.org/10.1145/2771783.2771812
12. Oluwagbemi, O., Asmuni, H.: An approach for automatic generation of test cases from UML diagrams. Int. J. Softw. Eng. Appl. **9**(8), 87–106 (2015)
13. Anbunathan, R., Anirban, B.: Dataflow test case generation from UML Class diagrams. In: 2013 IEEE International Conference on Computational Intelligence and Computing Research (2013). https://doi.org/10.1109/iccic.2013.6724144

14. Vinaya, S., Ketan, S.: Automatic generation of test cases from UML models. In: International Conference on Technology Systems and Management (ICTSM) (2011)
15. Papadopoulos, P., Walkinshaw, N.: Black-box test generation from inferred models. In: 2015 IEEE/ACM 4th International Workshop on Realizing Artificial Intelligence Synergies in Software Engineering (2015). https://doi.org/10.1109/raise.2015.11
16. Jain, N., Porwal, R.: Automated test data generation applying heuristic approaches—a survey. In: Hoda, M.N., Chauhan, N., Quadri, S.M.K., Srivastava, P.R. (eds.) Software Engineering. AISC, vol. 731, pp. 699–708. Springer, Singapore (2019). https://doi.org/10.1007/978-981-10-8848-3_68
17. Mahadik, P.P., Thakore, D.M.: Survey on automatic test data generation tools and techniques for object-oriented code. Int. J. Innov. Res. Comput. Commun. Eng. 4, 357–364 (2016)
18. Korel, B.: Dynamic method for software test data generation. Softw. Test. Verif. Reliab. 2 (4), 203–213 (1992)
19. Latiu, G.I., Cret, O.A., Vacariu, L.: Automatic test data generation for software path testing using evolutionary algorithms. In: Third International Conference on Emerging Intelligent Data and Web Technologies (2012)
20. Varshney, S., Mehrotra, M.: Search based software test data generation for structural. ACM SIGSOFT Softw. Eng. Notes 38(4), 1–6 (2013)
21. Nayak, N., Mohapatra, D.P.: Automatic test data generation for data flow testing using particle swarm optimization. In: Ranka, S., et al. (eds.) IC3 2010. CCIS, vol. 95, pp. 1–12. Springer, Heidelberg (2010). https://doi.org/10.1007/978-3-642-14825-5_1
22. Jiang, S., Zhang, Y., Yi, D.: Test data generation approach for basis path coverage. ACM SIGSOFT Softw. Eng. Notes 37(3), 1–7 (2012)
23. Nguyen, T.T.L., Tran, D.D., Bui, Q.T., Huynh, Q.T.: Integration MDA techniques in solving a class of web application with similar structure. In: 2015 ANU/SEED-Net Regional Conference for Computer and Information Engineering, Hanoi, 1–2 October 2015, pp. 78–83 (2015). ISBN 978-604-938-689-3
24. Tran, D.D., Huynh, Q.T., Tran, Q.K.: Model transformation with OCL integration in UWE. In: FICTA2018: 7th International Conference on Frontiers of Intelligent Computing: Theory and Applications, 29–30 November 2018
25. http://uwe.pst.ifi.lmu.de/exampleAddressBookWithSearches.html
26. http://generatedata.com/
27. Mockaroo, Random realistic test data generation in CSV, JSON, SQL, and Excel formats. https://mockaroo.com/
28. Yan Data Ellan. http://www.yandataellan.com/
29. Nguyen, D.M., Huynh, Q.T., Nguyen, T.H., Ha, N.H.: Automated test input generation via model inference based on user story and acceptance criteria for mobile application development. Int. J. Softw. Eng. Knowl. Eng. (2019). ISSN 1793-6403
30. Tran, D.-D., Huynh, Q.-T., Bui, T.-M.-A., Nguyen, P.-L.: Development of rules and algorithms for model-driven code generator with UWE. In: SOMET (2019)

Linearizing RF Power Amplifiers
Using Adaptive RPEM Algorithm

Han Le Duc[1], Minh Hong Nguyen[1], Van-Phuc Hoang[1(✉)], Hien M. Nguyen[2],
and Duc Minh Nguyen[3]

[1] Le Quy Don Technical University,
No. 236 Hoang Quoc Viet Street, Hanoi, Vietnam
phuchv@lqdtu.edu.vn
[2] Duy Tan University, Da Nang, Vietnam
[3] Hanoi University of Science and Technology,
No. 1 Dai Co Viet Street, Hanoi, Vietnam

Abstract. This paper proposes the adaptive indirect learning architecture (ILA) based digital predistortion (DPD) technique using a recursive prediction error minimization (RPEM) algorithm for linearizing radio frequency (RF) power amplifiers (PAs). The RPEM algorithm allows the forgetting factor to vary with time, which makes the predistorter (PD) parameter estimates more consistent and accurate in steady state, and hence reduces mean square errors. The proposed DPD technique is evaluated with respect to the error vector magnitude (EVM) and the adjacent channel power ratio (ACPR). The simulated PA Wiener model is used to validate the efficiency of the proposed algorithms. The simulation results have confirmed the improvement of the proposed adaptive RPEM ILA based DPD in terms of EVM and ACPR.

Keywords: Power amplifier · RPEM algorithm · Linearizing

1 Introduction

The development of future wireless communication systems, e.g., the fifth generation (5G) or beyond, continuously demands higher data rates and larger user capacities, which faces significant challenges. It requires not only wideband transceiver architecture, but also higher-order modulation schemes. The signals of these systems characterized by non-constant envelopes and high peak-to-average power ratio (PAPR), leading to stringent linearity requirements for signal amplification. In the meantime, the power dissipation of the future communication systems must be remained as low as possible [1]. To cope with these challenges, high efficiency and linear radio frequency (RF) power amplifiers (PAs) are indispensable components. Unfortunately, due to the inherent

This research is funded by Vietnam National Foundation for Science and Technology Development (NAFOSTED) under grant number 102.02-2016.12.

T. Q. Duong et al. (Eds.): INISCOM 2019, LNICST 293, pp. 220–231, 2019.
https://doi.org/10.1007/978-3-030-30149-1_18

nonlinear behavior of PAs, efficiency and linearity requirements often conflict each other. In order to provide highly-efficient power conversion, PAs should be driven into the saturation region. However, the saturated PAs produce not only in-band distortion but also result in spectral regrowth that interferes the adjacent frequency band channels. Consequently, the spectra utilization efficiency is reduced. In contrast, the nonlinear distortion can be mitigated by a traditional back-off approach, but this generates low power efficiency due to high PAPR of the transmitted signals. In order to maintain a low level of distortion without sacrificing the system energy efficiency requirement, PA linearization techniques are often used [2]. Thanks to its flexibility and excellent linearization performance, baseband digital predistortion (DPD) has been recognized as one of the most cost-effective linearization techniques [3–9], and it also tends to be popularly and widely used in wireless transmitters for the next generation wireless communication systems. In this scheme, a predistorter (PD) block is placed in front of a PA. The PA input signal is pre-distorted by the PD whose transfer function is the inverse of that of the PA. Ideally, the cascade of the PD and PA behaves as a linear amplification system and the original input is amplified by a constant gain.

In practice, the PA characteristics change with time due to process, supply voltage, and temperature (PVT) variations. In order to track time-varying change in the PA characteristics, an adaptive DPD using cost-effective learning architectures has become one of the most preferred choices. There are two commonly and widely used learning architectures for PD parameter identification: indirect learning architecture (ILA) [10–12] and direct learning architecture (DLA) [8,9,13,14]. Although DLA is more robust than ILA in terms of noise at the PA output and can provide unbiased parameter estimates, it is more complex identification process since the adaptive algorithms used in DLA require many iterations to find a set of parameters that minimizes the optimization criterion [3]. For these reasons, the adaptive ILA is most often used for identifying the PD parameters in RF PAs [3]. The adaptive ILA using least mean squares (LMS) for linearizing PAs was developed in [15]. The advantage of LMS is its simple implementation. However, it provides inaccurate estimation and has slow convergence since increasing the step size parameter leads instability problems. Moreover, it is also sensitive to the scaling of the input signal, making it very hard to choose a proper step size [15]. In order to obtain faster convergence of the adaptation, authors in [10,12] proposed the adaptive ILA using recursive least squares (RLS). It is worth noting that the choice of forgetting factor λ is often essential to make a good trade-off between the convergence and accuracy. For RLS, a decrease in the forgetting factor λ leads to its sensitivity to noise and a larger fluctuation of parameter estimates [16], resulting in inefficiency linearization performance.

In this paper, we propose an adaptive ILA using recursive prediction error minimization (RPEM) algorithm to linearize PAs, which allows time-varying forgetting factor λ. Thus, the RPEM algorithm reduces the fluctuation of the PD parameter estimates, speeds up the convergence, mitigates the steady-state mean square error and hence minimizes the total nonlinear distortion at the PA output. As a result, the adaptive ILA with RPEM effectively compensate the nonlinear distortion of the PA even if the PA characteristics changes due to PVT drift and other factors such as type of signals, high-order modulation schemes, input power levels etc. The rest of the paper is organized as follows. Section 2 proposes the one using RPEM. Simulation results are presented in Sect. 3. Conclusions are finally drawn in Sect. 4.

2   Proposed Adaptive ILA Using RPEM for Linearizing RF Power Amplifiers

Figure 1 shows the block diagram of the ILA-based DPD technique, where a post-distorter (or training) block is used to identify the postinverse of the PA. The baseband signal $u(n)$ is fed to the predistorter, which generates a signal $x(n)$ that is a PA input. The PA output signal is normalized by a linear gain G_0, producing the normalized output $z(n)$, i.e., $z(n) = \frac{y(n)}{G_0}$. The postdistorter model has the input $z(n)$ and the output $z_p(n)$. Its parameters are identified by minimizing the error signal $e(n) = x(n) - z_p(n)$ using the adaptive algorithms. Note that both the PD and postdistorter models are identical. Thus, when the coefficients of the postdistorter are identified, they are directly copied to the PD model. This process is repeated iteratively until the ILA linearization has converged. At convergence, the cascaded PD and PA system behaves linearly. Since the MP models have owned low computational cost, satisfactory accuracy, and easy hardware implementation, they have become promising choices and been widely applied for behavioral modeling and predistortion of PAs exhibiting nonlinear memory effects [2,4,12,17]. Therefore, both the PD and postdistorter

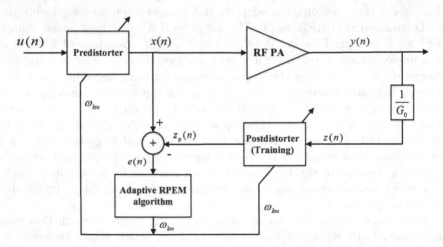

Fig. 1. Block diagram of Indirect learning architecture (ILA) using the proposed RPEM adaptive algorithm.

are modeled by the same MP model that has Q as the nonlinearity order and P as memory depth, and ω_{km} as coefficients. The input and output relation of the PD model is given by

$$x(n) = \sum_{k=1}^{Q} \sum_{m=0}^{P} \omega_{km} u(n-m)|u(n-m)|^{k-1} = \omega^{\mathrm{T}} \phi(n), \tag{1}$$

where

$$\omega = [\omega_{10}, \ldots, \omega_{Q0}, \ldots, \omega_{1P}, \ldots, \omega_{QP}]^{\mathrm{T}}, \tag{2}$$

and

$$\phi(n) = [\phi_{10}(n), \ldots, \phi_{Q0}(n), \ldots, \phi_{1P}(n), \ldots, \phi_{QP}(n)]^{\mathrm{T}} \tag{3}$$

with

$$\phi_{km}(n) = u(n-m)|u(n-m)|^{k-1}. \tag{4}$$

The symbol T indicates the matrix transpose.

The input and output of the postdistorter model can be expressed by

$$z_{\mathrm{p}}(n) = \sum_{k=1}^{Q} \sum_{m=0}^{P} \omega_{km} z(n-m)|z(n-m)|^{k-1} = \omega^{\mathrm{T}} \mathbf{z}(n), \tag{5}$$

where ω is defined as in (2) and

$$\mathbf{z}(n) = [z_{10}(n), \ldots, z_{Q0}(n), \ldots, z_{1P}(n), \ldots, z_{QP}(n)]^{\mathrm{T}} \tag{6}$$

with

$$z_{km}(n) = z(n-m)|z(n-m)|^{k-1}. \tag{7}$$

The prediction error $e(n, \omega)$ is defined by

$$e(n, \omega) = x(n) - z_{\mathrm{p}}(n) = x(n) - \omega^{\mathrm{T}} \mathbf{z}(n). \tag{8}$$

The adaptive algorithms are derived by minimizing corresponding lost functions that refer to scalar-valued functions of all the prediction errors $e(n, \omega)$.

The coefficient vector ω of the predistorter is estimated by using the Gauss-Newton RPEM algorithm in [16] that minimizes the following cost function.

$$f_L(\omega) = \lim_{L \to \infty} \frac{1}{L} \sum_{l=1}^{L} \mathrm{E}\left\{e^2(l, \omega)\right\}, \tag{9}$$

where $e(l, \omega)$ is given as in (8).

The formulation of the RPEM algorithm is derived in [16], which requires the negative gradient of $e(l, \omega)$ with respect to ω. From (8), the negative gradient is given by

$$-\frac{\partial e(n, \omega)}{\partial \omega} = \mathbf{z}^{\mathrm{T}}(n). \tag{10}$$

When applying the RPEM algorithm [16] for PA linearization, the adaptive ILA-based DPD using RPEM algorithm is described in Algorithm 1, where ρ also is

Algorithm 1. The proposed adaptive ILA-based DPD technique using RPEM.

1: Initialize: $n = 0, \lambda_0, \lambda(0), \mathbf{P}(0) = \rho\mathbf{I}$.

2: **for** $n = 1$ to $L - 1$ **do**

3:   $x(n) = \omega^{\mathrm{T}}(n-1)\phi(n)$

4:   $y(n) = F_{\mathrm{PA}}\{x(n)\}$.

5:   $z(n) = \frac{y(n)}{G_0}$

6:   $z_p = \omega^{\mathrm{T}}(n-1)\mathbf{z}(n)$

7:   $e(n) = x(n) - z_p(n)$

8:   $\lambda(n) = \lambda_0\lambda(n-1) + 1 - \lambda_0$

9:   $\mathbf{k}(n) = \frac{\mathbf{P}(n-1)\mathbf{z}(n)}{\lambda(n)+\mathbf{z}^{\mathrm{T}}(n)\mathbf{P}(n-1)\mathbf{z}(n)}$

10:   $\mathbf{P}(n) = \frac{1}{\lambda(n)}\left[\mathbf{P}(n-1) - \mathbf{k}(n)\mathbf{z}^{\mathrm{T}}(n)\mathbf{P}(n-1)\right]$

11:   $\omega(n) = \omega(n-1) + \mathbf{k}(n)e(n)$.

12: **End For**

---

a positive constant and $\lambda(n)$ is a forgetting factor that tends exponentially to 1 as $n \to \infty$. λ_0, $\lambda(0)$ and $\mathbf{P}(0)$ are initial variables designed by users. Typically chosen values for λ_0 and $\lambda(0)$ are $\lambda_0 = 0.99$ and $\lambda(0) = 0.95$ [16].

It is crucial that the evaluation criteria should be adopted to clearly validate the performance of PA behavioral modeling and DPDs. Therefore, this part defines the figures of merit for performance evaluation. The most commonly used criteria are normalized mean square error (NMSE) in time domain, adjacent channel power ratio (ACPR) in frequency domain, and error vector magnitude (EVM) that are defined as in [3,18].

Firstly, NMSE is an estimator of the overall difference between the predicted and measured signals in time domain. It is often defined in decibels as

$$\mathrm{NMSE} = 10\log_{10}\left(\frac{\sum\limits_{n=1}^{N}(|y[n] - x[n]|)^2}{\sum\limits_{n=1}^{N}(|x[n]|)^2}\right), \tag{11}$$

where $x(n)$ is the experimental output (or desired output) of the DUT, and $y(n)$ is the output obtained from the model.

Moreover, ACPR is the ratio between the total adjacent channels' powers to the main channel signal power. It describes the degree of the signal regrowth into neighbouring channels. Since the ACPR characterizes the maximum power allowed to be radiated outside the allocated band, it plays a very important role in wireless radio standards. The ACPR is often expressed in decibels as

$$\mathrm{ACPR} = 10\log_{10}\left(\frac{\int_{\mathrm{B_{adj}}}|Y(f)|^2}{\int_{\mathrm{B_{ch}}}|Y(f)|^2}\right) \tag{12}$$

where $|Y(f)|$ denotes the power spectrum of the measured output signal $y(n)$, $\mathrm{B_{adj}}$ and $\mathrm{B_{ch}}$ refer to the bandwidth of the adjacent and main channels, respectively.

The EVM is a measure criterion that quantifies the imperfection to the output signal when compared to the input one. It describes the in-band distortion of the PA and is defined as

$$
\text{EVM} = \sqrt{\frac{\sum_{j=0}^{L} \left[\left(I_j - \hat{I}_j\right)^2 + \left(Q_j - \hat{Q}_j\right)^2 \right]}{\sum_{j=0}^{L} \left[I_j^2 + Q_j^2\right]}} \tag{13}
$$

where I_j and Q_j are the ideal output signal in-phase and quadrature components, and $\hat{I}_j$ and $\hat{Q}_j$ are their output measured counterparts, respectively.

3 Simulation Results

In order to demonstrate the proposed DPD linearization method, we tested a simulated PA that is modeled by a Wiener model consisting of a FIR filter followed by memoryless nonlinearity model. The coefficients of the FIR filter are as in [19–21]

$$
h_0 = 0.7692, h_1 = 0.1538, h_2 = 0.0769. \tag{14}
$$

For the memoryless nonlinearity model, we use Saleh's model [22], which is defined by

$$
y(n) = \frac{\alpha_a \left|v(n)\right|}{1 + \beta_a |v(n)|^2} e^{j\angle \left[v(n) + \frac{\alpha_\varphi |v(n)|^2}{1 + \beta_\varphi |v(n)|^2}\right]}, \tag{15}
$$

with

$$
v(n) = h_0 x(n) + h_1 x(n-1) + h_2 x(n-2), \tag{16}
$$

where $x(n)$ and $y(n)$ are the input and output of the simulated PA, respectively, and $v(n)$ is the input of Saleh model. The parameters of Saleh model are as in [19]

$$
\alpha_a = 20, \beta_a = 2.2, \alpha_\varphi = 2, \beta_\varphi = 1. \tag{17}
$$

The transmitted symbols are modulated by 16-QAM with 3.84 MHz bandwidth. The input modulated signal is filtered by a raised cosine pulse shaping filter with the roll-off factor of 0.22.

The AM/AM and AM/PM characteristics computed at the instantaneous samples of the PA input and output, are shown in Fig. 2. It is clear that the simulated PA suffers from the nonlinearity and memory effects. Figure 3 shows the gain performance of the simulated PA with the average input power. One can observe that the gain in linear region is about 26 dB. The average input power at 1 dB compression point and at 3 dB are around −1 dBm and 4.3 dBm, respectively.

The MP model is used to model nonlinear behavior of the PA. In order to reduce the computational complexity, the orders (N and M) of the MP model are optimized by using a performance-based sweeping method [17]. Figure 4 shows

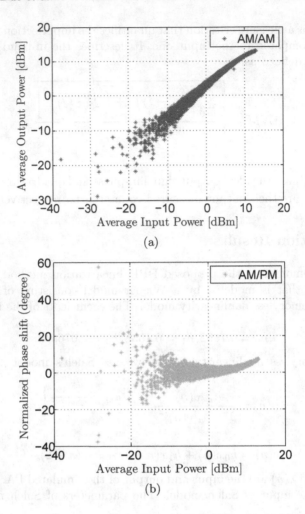

Fig. 2. The PA characteristics. (a) AM/AM. (b) AM/PM.

the NMSE performance versus the orders of the PA model. From this figure, we can see that the optimal values of N and M are $N = 5$ and $M = 2$, respectively, in order to achieve a good trade-off between the best NMSE and computational complexity.

In order to validate the proposed DPD, the RPEM algorithm is initialized when $\lambda_0 = 0.99$, $\lambda(0) = 0.95$ and the initial weight vectors $\omega(0)$ have a first element as 1 and the others as 0. In this simulation, the ACPR values are measured at the upper adjacent channels, corresponding to frequency offsets of 5 MHz.

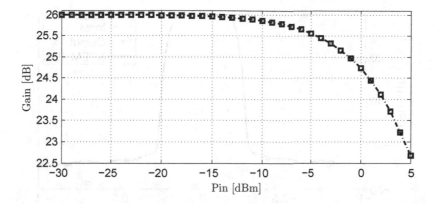

Fig. 3. Gain versus average input power of simulated PA.

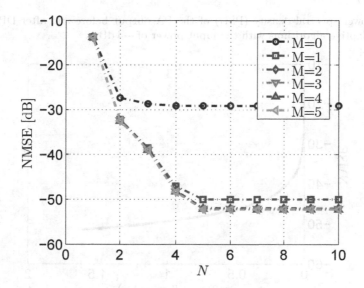

Fig. 4. NMSE versus N and M.

Figure 5 shows effectiveness in canceling the spectral regrowth of the proposed approach. for the input power of -4 dBm. It can be seen that there is a significant spectral regrowth reduction after DPD. The adaptive RPEM algorithm converges after 10-K samples, as shown in Fig. 6.

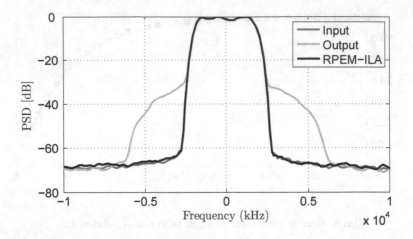

Fig. 5. Power spectral density (PSD) of the PA output before and after DPD using various adaptive algorithms with the input power of −4 dBm.

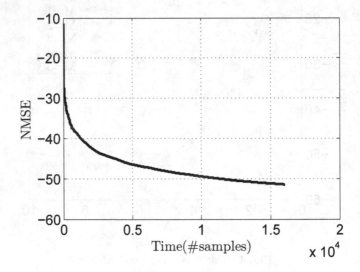

Fig. 6. Learning curves for adaptive RPEM algorithms.

Figures 7(a) and (b) respectively show the ACPR and EVM performance of the proposed DPD for various input power levels. From these figures, one can observe that the proposed DPD technique shows a significant performance improvement in terms of ACPR and EVM. It obtains the ACPR values almost equal to those of input. Furthermore, after applying RPEM-ILA, the EVM values are significantly reduced and less than 0.26%, which shows excellent performance in in-band distortion mitigation. This is because the RPEM algorithm makes

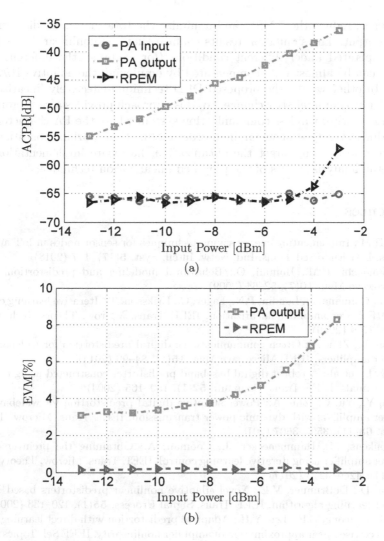

Fig. 7. ACPR and EVM performance after the proposed DPD using adaptive RPEM algorithm. (a) ACPR. (b) EVM.

the PD coefficient estimates more consistent and precise in steady state. The simulation results have clarified the improvements of the proposed technique compared with LMS [15] and RLS-based [10,12] ILA methods.

4   Conclusions

In this paper, an adaptive ILA linearization using the RPEM algorithm has been proposed. Thanks to the time-varying forgetting factor, the PD coefficient estimates are consistent and accurate in steady state, leading to speed up the convergence, reduce the NMSE, and minimize the total nonlinear distortion at the PA output. The simulation results confirm that the nonlinear distortion of the PA operated under different conditions (for example, the different input powers), can be almost fully compensated by employing the adaptive ILA with RPEM. In other words, the proposed DPD technique effectively linearizes the PA even if its characteristics change. So, this approach provides a very promising solution for future wireless communication system where the PA characteristics change due to the type of signal, high-order modulation, working condition, etc. The future works will target the optimization, hardware implementation and more detail analysis results of the proposed linearization technique.

References

1. Vo, H.M.: Implementing energy saving techniques for sensor nodes in IoT applications. EAI Endorsed Trans. Ind. Netw. Intell. Syst. **5**(17), 1–7 (2018)
2. Ghannouchi, F.M., Hammi, O.: Behavioral modeling and predistortion. IEEE Microwave Mag. **10**(7), 52–64 (2009)
3. Chani-Cahuana, J., Landin, P.N., Fager, C., Eriksson, T.: Iterative learning control for RF power amplifier linearization. IEEE Trans. Microw. Theory Tech. **64**(9), 2778–2789 (2016)
4. Guan, L., Zhu, A.: Green communications: digital predistortion for wideband RF power amplifiers. IEEE Microwave Mag. **15**(7), 84–99 (2014)
5. Ding, L., et al.: A robust digital baseband predistorter constructed using memory polynomials. IEEE Trans. Commun. **52**(1), 159–165 (2004)
6. Guo, Y., Yu, C., Zhu, A.: Power adaptive digital predistortion for wideband RF power amplifiers with dynamic power transmission. IEEE Trans. Microw. Theory Tech. **63**(11), 3595–3607 (2015)
7. Schoukens, M., Hammenecker, J., Cooman, A.: Obtaining the preinverse of a power amplifier using iterative learning control. IEEE Trans. Microw. Theory Tech. **65**(11), 4266–4273 (2017)
8. Zhou, D., DeBrunner, V.E.: Novel adaptive nonlinear predistorters based on the direct learning algorithm. IEEE Trans. Signal Process. **55**(1), 120–133 (2007)
9. Choi, S., Jeong, E.R., Lee, Y.H.: Adaptive predistortion with direct learning based on piecewise linear approximation of amplifier nonlinearity. IEEE Sel. Topics Signal Process. **3**(3), 397–404 (2009)
10. Suryasarman, P.M., Springer, A.: A comparative analysis of adaptive digital predistortion algorithms for multiple antenna transmitters. IEEE Trans. Circuits Syst. I **62**(5), 1412–1420 (2015)
11. Eun, C., Powers, E.J.: A new Volterra predistorter based on the indirect learning architecture. IEEE Trans. Signal Process. **45**(1), 223–227 (1997)
12. Morgan, D.R., Ma, Z., Ding, L.: Reducing measurement noise effects in digital predistortion of RF power amplifiers, vol. 4, pp. 2436–2439, May 2003
13. Paaso, H., Mammela, A.: Comparison of direct learning and indirect learning predistortion architectures. In: Proceedings of IEEE International Symposium on Wireless Communication Systems, pp. 309–313, October 2008
14. Hussein, M.A., Bohara, V.A., Venard, O.: On the system level convergence of ILA and DLA for digital predistortion. In: Proceedings of 2012 International Symposium on Wireless Communication Systems (ISWCS), pp. 870–874, August 2012

15. Mohr, B., Li, W., Heinen, S.: Analysis of digital predistortion architectures for direct digital-to-RF transmitter systems. In: Proceedings of 2012 IEEE 55th International Midwest Symposium on Circuits and Systems (MWSCAS), pp. 650–653, August 2012

16. Söderström, T., Stoica, P. (eds.): System Identification. Prentice-Hall Inc., Upper Saddle River (1988)

17. Abdelrahman, A.E., Hammi, O., Kwan, A.K., Zerguine, A., Ghannouchi, F.M.: A novel weighted memory polynomial for behavioral modeling and digital predistortion of nonlinear wireless transmitters. IEEE Trans. Ind. Electron. **63**(3), 1745–1753 (2016)

18. Mkadem, F.: Behavioral modeling and digital predistortion of wide- and multi-band transmitter systems. Ph.D. dissertation (2014)

19. Feng, X., Wang, Y., Feuvrie, B., Descamps, A.-S., Ding, Y., Yu, Z.: Analysis on LUT based digital predistortion using direct learning architecture for linearizing power amplifiers. EURASIP Wirel. Commun. Netw. **2016**(1), 132 (2016)

20. Kwon, J., Eun, C.: Digital feedforward compensation scheme for the non-linear power amplifier with memory. IJISTA **9**, 326–334 (2010)

21. Eun, C., Powers, E.J.: A predistorter design for a memory-less nonlinearity preceded by a dynamic linear system. In: Proceedings of GLOBECOM 1995, vol. 1, pp. 152–156, November 1995

22. Saleh, A.A.M.: Frequency-independent and frequency-dependent nonlinear models of TWT amplifiers. IEEE Trans. Commun. **COM–29**(11), 1715–1720 (1981)

LBP-Based Edge Information
for Color Texture Classification

Duc Phan Van Hoai and Vinh Truong Hoang[(⊠)]

Faculty of Information Technology, Ho Chi Minh City Open University,
Ho Chi Minh City, Vietnam
{1551010028duc,vinh.th}@ou.edu.vn

Abstract. In this paper, we propose to extract two types of feature from Neighbor-Center Difference Image (NCDI). NCDI is a variant of Local Binary Pattern (LBP) and originally used as input for Convolutional Neural Network (CNN). NCDI is a high dimensional feature and thus histograms are extracted from these NCDI features to mainly store useful information for statistical analysis. Two types of histograms are extracted from NCDI and then concatenated to further capture useful information. Experimental results on several benchmark color texture datasets show that the proposed approaches outperform the original LBP with a large margin (in accuracy) on several benchmark color texture datasets.

Keywords: LBP · Neighbor-Center Different Image ·
Color texture classification

1 Introduction

Texture analysis is one of the most active area of research in computer vision with a wide range of real-life applications, including industrial inspection, medical magnetic resonance imaging, materials science. In reality, the texture of the same material or object varies in illumination, orientation, scale, and rotation. Therefore, it needs a robust descriptor to characterize and discriminate different classes. Various approaches have been proposed to overcome these drawbacks in illumination, orientation, and other visual appearance problems. There exist many works to propose a new efficiency and discriminant image descriptors. Most approaches are based on local and global techniques.

One simple yet efficient local descriptor is Local Binary Pattern (LBP) introduced by Ojala et al. [1]. LBP is a computational efficiency operator with high discriminative power and robustness against illumination. However, it may not work properly for noisy images due to its threshold function [2]. Various variants of LBP and its extension have been proposed to minimize its limitation [3]. Lu et al. [4] have proposed Neighbor-Center Difference Vector (NCDV) which is extracted by subtracting the center pixel value from its neighboring pixel values. They extract NCDV features of different sizes from several non-overlapped

© ICST Institute for Computer Sciences, Social Informatics and Telecommunications Engineering 2019
Published by Springer Nature Switzerland AG 2019. All Rights Reserved
T. Q. Duong et al. (Eds.): INISCOM 2019, LNICST 293, pp. 232–239, 2019.
https://doi.org/10.1007/978-3-030-30149-1_19

blocks of training samples. For NCDV features extracted from each block, they train one projection to map it into a binary feature vector. Then, they cluster these binary codes into a codebook and encode these binary codes within the same face as a histogram feature vector. Finally, an age ranker is trained on these histograms. Their approach has shown to provide very good results on several datasets.

Recently, Wu and Lin [5] have proposed a new descriptor based on NCDV to capture edge information, namely Neighbor-Center Difference Image (NCDI). Normally, Convolutional Neural Network (CNN) model takes RGB images as input. However, Wu and Lin [5] have proposed to feed CNN model with hand-crafted feature NCDI which collects NCDV from all patches to reconstruct the image. This approach has shown to improve the accuracy in the facial expression recognition task. Edge information is useful to a wide range of computer vision tasks but NCDI is a high dimensional feature. Therefore, we propose to extract two types of histogram feature from NCDI to mainly store useful information and use it for texture classification.

The rest of this paper is organized as follows. In Sect. 2, LBP, NCDI, and K-Nearest Neighbors are briefly reviewed. Section 3 introduces the feature extracting methods. Next, four benchmark color texture datasets and experimental results are introduced in Sect. 4. Finally, the conclusion is discussed in Sect. 5.

2 Related Work

2.1 Local Binary Pattern (LBP)

LBP is a powerful local descriptor to deal with texture and related to classification tasks. LBP operator takes values of points on a circular neighborhood, thresholds the pixel values of the neighborhood at the value of the central pixel value. The binary results are then used to form an integer LBP code. The formula to compute the $\text{LBP}_{P,R}$ code from P circular neighbors of radius R is defined as:

$$\text{LBP}_{P,R} = \sum_{i=0}^{P-1} \theta \left(g_i - g_c \right) \times 2^i \tag{1}$$

where g_c is the value of central pixel and g_i is the value of ith neighborhood pixel. The threshold function $\theta \left(. \right)$ is defined as:

$$\theta \left(t \right) = \begin{cases} 1 & \text{if } t \geq 0 \\ 0 & \text{otherwise} \end{cases} \tag{2}$$

LBP is a computational efficiency descriptor with high discriminative power and robustness against illumination. However, LBP has several disadvantages, it loses intensity information due to the threshold function and may not work properly for noisy images [6]. Many variants of LBP have been proposed to minimize LBP's limitation [3].

2.2 Neighbor-Center Difference Image (NCDI)

LBP may lose intensity information due to the threshold function $\theta(.)$. In order to tackle this issue, several approaches have been proposed, one of these approaches is NCDI which is proposed by Wu and Lin [5]. NCDI is extracted on a grayscale image by iterating each pixel (x, y) and subtracting its value g_c from P neighboring pixel values $\{g_i\}_{i=1}^{P}$.

$$\text{NCDI}\,(x, y)_i = g_i - g_c \tag{3}$$

Finally, $\text{NCDI}_{i=1}^{P}$ were concatenated to create a multi-channel image. The P-channel of NCDI are extracted from a grayscale image will have edge information in P directions.

2.3 K-Nearest Neighbors Classifier

The K-Nearest Neighbors classifier (K-NN) is among the simplest classifiers of all machine learning algorithms and it is widely used in texture classification. To classify a testing image, firstly, the distance in the feature space between the testing image and each training images is computed. Then, the testing image is assigned to the class that has the highest number of images among K nearest neighbors. K is a user-defined constant. If K = 1, the testing image is assigned as the class of the nearest neighbor in the feature space. The value of K is commonly set to 1 and the distance metric is usually L1 or Euclidean. The L1 distance d_{L1} between two feature vectors a, b is computed as follow:

$$d_{L1} = \sum_i |a_i - b_i| \tag{4}$$

Where a_i, b_i is the ith value in the feature vector a, b respectively.

3 Proposed Approach

The edge information is useful for several computer vision tasks. The 8-channel NCDI feature has shown to provide better results for facial expression recognition [5]. However, The 8-channel NCDI feature is a very high dimensional feature. Therefore, we investigate to extract two types of NCDI histogram feature to reduce the dimension and apply it for texture classification.

- The first histogram feature is obtained from each NCDI by counting the frequency of each value from -255 to 255. These histograms have information about the intensity of difference between pixel values.
- To further capture the edge information from NCDI, the second histogram feature is extracted from each NCDI by counting the frequency of 256 LBP values. These histograms have information about the correlation of difference between pixel values.

A LBP value at each point only stores information of which neighbor pixel value is larger than the center pixel value. In case of the NCDI, at each pixel, it stores how much each neighbor pixel value larger than the center pixel value. Therefore, the histogram of each NCDIs has more information about the intensity of difference between pixel values. However, the histogram of each NCDI channels does not have correlation information of neighboring values, thus we propose to combine NCDI Histogram and LBP-NCDI histogram to incorporate the intensity of difference and the correlation of difference information.

4   Experiments

4.1   Dataset Description

The proposed approaches are evaluated on four benchmark color texture datasets, including New BarkTex [7], Outex-TC-00013 [1], USPTex [8] and STex. Training and testing set of each dataset is divided by the holdout method (as shown in Table 1).

Table 1. Summary of image datasets used in the experiment.

Dataset name	Image size	# class	# training	# test	Total
New BarkTex	64×64	6	816	816	1632
Outex-TC-00013	128×128	68	680	680	1360
USPTex	128×128	191	1146	1146	2292
STex	128×128	476	3808	3808	7616

4.2   Experimental Setup

In order to evaluate the proposed approaches, experiments are conducted on the same training and testing set of four benchmark color texture datasets by using the nearest neighbor (1-NN) classifier associated with the L1 distance. It is worth to note that the sophisticated classifier might provide a better classification performance (i.e SVM classifier), but at the cost of computing and tuning parameters.

Firstly, the baseline result (LBP RGB) is obtained by extracting three $LBP_{8,1}$ histograms from the three channels of RGB images.

Secondly, two types of proposed histogram feature are evaluated separately. To begin, 8-channel NCDI is extracted from each channel of RGB image. Then, $LBP_{8,1}$ histograms of NCDI and histograms of NCDI are extracted from these NCDIs. Next, 1-NN classifier is used to obtain the accuracy of each type of feature.

Finally, two proposed types of histogram feature are concatenated produce the result of the proposed approach.

All experiments are implemented in Matlab-2015b and conducted on a PC with a configuration of a CPU 4 cores 2.2 GHz, 8 GBs of RAM.

4.3 Results

Table 2 shows that the proposed approaches take a longer time to extract features. However, the proposed approach to combine two types of NCDI histograms is still very fast. It takes only 0.068 s to extract features from the three channels of a 128×128 RGB image.

Table 2. The computation time (in seconds) to extract features from a 128×128 RGB image of the proposed approaches compare with the original LBP.

Methods	Computation time
LBP RGB	0.008
LBP NCDI	0.047
NCDI Histogram	0.024
NCDI Histogram & LBP NCDI	0.068

Table 3. Classification accuracy (in %) of the original LBP approach and the proposed approaches on four texture datasets New BarkTex, Outex-TC-00013, USPTex, and STex. LBP RGB stands for LBP histogram feature extracted on three channel of RGB image. LBP NCDI is the approach that uses the LBP histogram feature extracted on NCDIs. NCDI Histogram is the histogram feature extracted by counting the frequency of each value in NCDIs. NCDI Histogram & LBP NCDI is the proposed approach that concatenates two types of histogram features.

Methods	New BarkTex	Outex-TC-00013	USPTex	STex
LBP RGB	76.6	86.0	85.3	85.5
LBP NCDI	74.4	88.9	82.3	87.8
NCDI Histogram	69.4	86.3	80.8	76.9
NCDI Histogram & LBP NCDI	78.3	90.2	88.7	89.1

Table 3 clearly shows that the combination of two proposed feature types outperforms the result of LBP histogram from RGB image. Comparing with features from LBP histograms of RGB image, the proposed approaches achieved significant gain of more than 1.7%, 4%, 3.4% and 3.5% in accuracy on New BarkTex, Outex-TC-00013, USPTex, and STEX dataset respectively.

Table 4. Classification accuracy (in %) of the proposed method compared with other approaches on four texture datasets New BarkTex, Outex-TC-00013, USPTex, and STex.

Methods	New BarkTex	Outex-TC-00013	USPTex	STex
Color ranges LBP [9]	71.0	86.2	79.1	-
Wavelet coefficients [10]	-	89.7	-	77.6
Color contrast occurrence matrix [11]	-	82.6	-	76.7
Soft color descriptors [12]	-	81.4	58.0	55.3
LBP and local color contrast [13]	71.0	85.3	82.9	-
CLBP [14]	72.8	84.4	72.3	-
Mix color order LBP histogram [15]	77.7	87.1	84.2	-
LTP [6]	76.1	90.6	88.4	87.0
LPQ [16]	66.2	81.4	86.6	87.6
TPLBP [17]	61.3	75.0	80.0	71.7
LBP Median [18]	72.3	83.0	84.1	81.9
NCDI Histogram & LBP NCDI	78.3	90.2	88.7	89.1

Table 4 shows that our proposed approach to concatenate two types of feature obtains better results than several other approaches. According to the experiments on STex dataset, the proposed approach outperforms other approaches by a margin of more than 1.5%. The classification results obtained on the USPTex dataset by the proposed method provides slightly better result than other approaches. In the case of Outex-TC-00013 dataset, the Mix color order LBP histogram approach give slightly better accuracy than ours. However, the proposed method outperforms that approach by improving 2.2%, 0.3%, and 2.1% on New BarkTex, USPTex and STex datasets, respectively.

5 Conclusion

In this paper, two types of histogram features are proposed to extract edge information from NCDI. These two types of histograms are then concatenated to further capture useful information from NCDIs. Experimental results on four benchmark color texture datasets show that the concatenation of features from NCDI outperforms the original LBP with a large margin in accuracy. Moreover, the proposed approach has achieved better results compare with several other approaches on four texture datasets.

The future of this work is to reduce the dimension of proposed features by feature selection method and apply it to other related computer vision tasks.

References

1. Ojala, T., Maenpaa, T., Pietikainen, M., Viertola, J., Kyllonen, J., Huovinen, S.: Outex - new framework for empirical evaluation of texture analysis algorithms. In: Object Recognition Supported by User Interaction for Service Robots, vol. 1, Quebec City, Quebec, Canada: IEEE Computer Society, pp. 701–706 (2002)
2. Tan, X., Triggs, B.: Enhanced local texture feature sets for face recognition under difficult lighting conditions. In: Zhou, S.K., Zhao, W., Tang, X., Gong, S. (eds.) AMFG 2007. LNCS, vol. 4778, pp. 168–182. Springer, Heidelberg (2007). https://doi.org/10.1007/978-3-540-75690-3_13
3. Liu, L., Fieguth, P., Guo, Y., Wang, X., Pietikäinen, M.: Local binary features for texture classification: taxonomy and experimental study. Pattern Recogn. **62**, 135–160 (2017)
4. Lu, J., Liong, V.E., Zhou, J.: Cost-sensitive local binary feature learning for facial age estimation. IEEE Trans. Image Process. **24**(12), 5356–5368 (2015)
5. Wu, B.-F., Lin, C.-H.: Adaptive feature mapping for customizing deep learning based facial expression recognition model. IEEE Access **6**, 12451–12461 (2018)
6. Tan, X., Triggs, B.: Enhanced local texture feature sets for face recognition under difficult lighting conditions. IEEE Trans. Image Process. **19**(6), 1635–1650 (2010)
7. Porebski, A., Vandenbroucke, N., Macaire, L., Hamad, D.: A new benchmark image test suite for evaluating color texture classification schemes. Multimedia Tools Appl. **70**, 543–556 (2014)
8. Backes, A.R., Casanova, D., Bruno, O.M.: Color texture analysis based on fractal descriptors. Pattern Recogn. **45**(5), 1984–1992 (2012)
9. Ledoux, A., Losson, O., Macaire, L.: Color local binary patterns: compact descriptors for texture classification. J. Electron. Imaging **25**(6), 061404 (2016)
10. El Maliani, A.D., El Hassouni, M., Berthoumieu, Y., Aboutajdine, D.: Color texture classification method based on a statistical multi-model and geodesic distance. J. Vis. Commun. Image Represent. **25**(7), 1717–1725 (2014)
11. Rios, A.M., Richard, N., Fernandez-Maloigne, C.: Alternative to colour feature classification using colour contrast ocurrence matrix, vol. 9534, June 2015
12. Bello, R., Bianconi, F., Fernández, A., González, E., Di Maria, F.: Experimental comparison of color spaces for material classification. J. Electron. Imaging **25**, 061406 (2016)
13. Cusano, C., Napoletano, P., Schettini, R.: Combining local binary patterns and local color contrast for texture classification under varying illumination. J. Opt. Soc. Am. A **31**(7), 1453 (2014)
14. Guo, Z., Zhang, L., Zhang, D.: A completed modeling of local binary pattern operator for texture classification. IEEE Trans. Image Process. **19**(6), 1657–1663 (2010)
15. Ledoux, A., Losson, O., Macaire, L.: Color local binary patterns: compact descriptors for texture classification. J. Electron. Imaging **25**, 061404 (2016)
16. Ojansivu, V., Rahtu, E., Heikkila, J.: Rotation invariant local phase quantization for blur insensitive texture analysis. In: 2008 19th International Conference on Pattern Recognition, Tampa, FL, USA, pp. 1–4. IEEE December 2008

17. Lior, W., Tal, H., Yaniv, T.: Descriptor based methods in the wild. In: Real-Life Images Workshop at the European Conference on Computer Vision (ECCV), October 2008

18. Hafiane, A., Palaniappan, K., Seetharaman, G.: Joint adaptive median binary patterns for texture classification. Pattern Recogn. **48**(8), 2609–2620 (2015)

A Vietnamese Sentiment Analysis System Based on Multiple Classifiers with Enhancing Lexicon Features

Bich-Tuyen Nguyen-Thi and Huu-Thanh Duong[(✉)]

Faculty of Information Technology, Ho Chi Minh City Open University,
97 Vo van Tan, Ward 6, District 3, Ho Chi Minh City, Vietnam
{1551010145tuyen, thanh.dh}@ou.edu.vn

Abstract. Today, a customer is easy to express his opinions about a bought products thanks to accelerated development of social networks and ecommerce websites. These opinions are very useful indicators to evaluate the degree of the customers' real satisfaction. From that, the traders will emerge the strategies and predict the trends to get the directions for their products and businesses in the future. In this paper, we have built a dataset and executed many experiments based on the multiple classifiers with complementing lexicon features to increase the accuracy of sentiment polarities. The experimental section shows good results and proves our approach is reasonable.

Keywords: Machine learning · Text mining · Natural language processing · Sentiment analysis · User behavior

1 Introduction

Sentiment analysis task is a hot trend in natural language processing applied widely in many domains, especially in ecommerce system, this identifies the sentiment polarity between positive and negative comments. Through sentiment, managers of the brand and product can quickly get an overview of customer attitudes towards their brand at a certain time or within a certain period of time. Moreover, sentiment analysis also points to brand or product strengths and weaknesses in the eyes of the customers, which aspects are being appreciated and what are the focal points of negative discussion. In addition to monitor sentiment changes over a long period of time will tell brand health, which helps the brand managers and marketers to re-evaluate the performance and giving directions for the future campaigns. This not only improves the effectiveness of business, but also helps to be more proactive in preventing and handling crisis.

In this paper, we have proposed the sentiment analysis solution based on multiple classifiers with enhancing lexicon features. Our main contribution is to build the Vietnamese dataset from reputation ecommerce websites for sentiment analysis, apply preprocessing techniques for the dataset such as word segmentation, lowercase transformation, punctuation removal and feed more features to the feature vectors such as the adjective phrases, negation and emotional icons replacement. The experiments of

© ICST Institute for Computer Sciences, Social Informatics and Telecommunications Engineering 2019
Published by Springer Nature Switzerland AG 2019. All Rights Reserved
T. Q. Duong et al. (Eds.): INISCOM 2019, LNICST 293, pp. 240–249, 2019.
https://doi.org/10.1007/978-3-030-30149-1_20

various aspects are executed to evaluate the well-known classifiers and various features to find the suitable solution for real-time applications.

In the rest of this paper is organized as follows: Sect. 2 presents the related works, Sect. 3 shows background and our approach. Next, Sect. 4 shows the experimental results. Finally, the conclusions and future works are presented in Sect. 5.

2 Related Works

Sentiment Analysis is a potential field attracted many research groups having studied with various experiences and approaches.

Medhat et al. [1] showed the main approaches for sentiment analysis problem including machine learning, lexicon-based and hybrid approaches. Actually, sentiment analysis is a kind of text classification problem identifying the user's comments as positive or negative polarities, so it can totally apply the machine learning algorithms with the linguistic features to sentiment polarities. About lexicon-based approach relies on the emotional lexicons and is divided into dictionary-based or corpus-based approaches built by statistical or semantic methods to find the emotional polarities. Besides, the hybrid approach combines these two approaches to utilize the advantages and limit the disadvantages of them in the hope of increasing more accuracy.

Although the lexicon-based approach also has drawbacks such as reliability of the lexicons, dependencies on their contexts or languages. But the emotional lexicons are important factors to recognize the customers' emotions in their comments, so it has many lexicon-based studies in particular languages, especially adjective, adjective phrases, verb and verb phrases. The adjective and verb phrases are the essential clues for sentiment analysis problem to enhance the accuracy of the adjective and verb phrases, also build the sentiment wordnet. Trinh et al. [2] proposed the lexicon-based approach for sentiment analysis with facebook data in Vietnamese and built the Vietnamese emotional lexicon dictionary, including noun, verb, adjective, adverb based on the English emotional analysis applied to the Vietnamese language and used support vector machine classifier to identify the emotions. Tran et al. [3] proposed a fuzzy language computation based on Vietnamese linguistic characteristics to provide an effective method for computing the sentiment polarity of verb phrases.

Deep learning is also an trending solution mentioning a lots in the recent years. Vo et al. [4] integrated the advantages of CNN (Convolutional Neural Network) and LSTM (Long ShortTerm Memory) for sentiment analysis with their Vietnamese proposed corpus as comments/reviews in ecommerce websites. Araque et al. [5] enhanced deep learning sentiment analysis with ensemble techniques, including ensemble classifiers and ensemble features.

Our approach has based on the multiple classifiers and complemented the lexicon features. According to the experiments of Duong and Truong Hoang [6], we choose logistic regression, SVM (Suport Vector Machine), random forest, OVO, OVR which are classfiers obtains the best score in current.

Next, it presents the theory background of our approach and the experiments to evaluate the well-known classifiers and also the dataset.

3 Background

3.1 Feature Extraction

Each dimension of comment vectors is $tf \times idf$ weight of a term, where tf is the number of the terms appearing in a document, df is the number of the documents containing a term, idf is the inversion of df weight. This weight is widely used in natural language processing because tf shows the importance of the term, but if that term also appears many times in other documents, it may be a less meaning word, so incorporating with idf to punish that one. Deng et al. [7] executed the exhaustive experiments showing $tf \times idf$ has still gotten the high score in text classification, our approach also chooses this one for dimensions of the feature vectors. This weight is calculated as follows:

$$(tf \times idf)_{w_i} = freq(w_i)log\frac{N}{1+df} \tag{1}$$

Subject to N is the number of the comments, $freq(w_i)$ is the frequency of w term in the i-th comment. Besides, it also incorporates unigram and bigram for feature vectors.

3.2 The Vietnamese Phrases

Woking in Vietnamese processing will face the first challenge as word segmentation. It's different from English which the words are the tokens divided by the space characters, Vietnamese words may one token or two tokens such as tốt (good), xuất sắc (excellent), tuyệt vời (wonderful), hoàn hảo (perfect). The second one is Part of Speech (POS) tagging which is used to assign parts of speech to each words such as noun, verb, adjective, adverb helps to increase more semantic to texts. They are important problems in natural language processing, this study has used pyvi[1] library for Vietnamese word segmentation (F1 score 0.979) and POS (F1 score 0.925), and relies on them to get the adjective phrases to increase the semantic dimensions for feature vectors. The adjective phrases are the essential indicators in sentiment analysis to determine the degrees of customers' satisfaction. An adjective phrase in Vietnamese has the structure:

$$\text{P1} < center\ adj > \text{P2}$$

The previous (**P1**) and post (**P2**) sub-sections may be lack, but the center adjective is required. **P1** is often the adverbs of complementing for the center adjectives. For example, "*rất tốt*" (very good): "*tốt*" (good) is an adjective and "*rất*" (very) is the adverb of complementing for "tốt". **P2** is often adverbs of degree. For example "đẹp quá" (so beautiful): "*đẹp*" (beautiful) is an adjective and "*quá*" (so) is the adverb of degree. It also may be noun, adjective, verb making more clear the features of the center adjective. For example, "*anh ấy khó thuyết phục*" (it is difficult to convince him): "khó" (difficult) is the adjective and "thuyết phục" (convince) is the verb.

---

[1] https://pypi.org/project/pyvi/.

The adjective phrases are especially important to distinguish strong satisfied and only satisfied polarities. The special point is to distinguish two of them as the adverb complementing for the adjective to increase the emotion of the adjective word. For example, a user's said: *"tôi rất hài lòng về sản phẩm"* (I'm very pleased about the product) is more satisfied than *"tôi hài lòng về sản phẩm"* (I'm pleased about the product) or *"tôi tạm hài lòng về sản phẩm"* (I'm a little bit pleased about the product), *"sản phẩm đẹp quá"* (the product is so beautiful) is more satisfied than *"sản phẩm đẹp"* (the product is beautiful). Moreover, comments are the short texts, so the study focuses on two forms of the adjective phrases as **P1 <center adj>** and **<center adj> P2**.

3.3   SVM

SVM (Support Vector Machine) is a strong classifier using both regression and classification problem. The main idea is to find a hyperland (calling M_0) to divide the dataset into various groups in the multi-dimensions space. Firstly, SVM is used for the binary problem and the linearly separable dataset, it means M_0 divides the dataset into two groups and gets the same distance with two support vector hyperlands of those two groups (see Fig. 1 is the dashed lines). The support vectors of groups are the data points having the nearest distance to M_0. Clearly, these data points are more important than other ones in finding M_0-. The cost function of M_0 forms as follows:

$$y = w^T x + b \qquad (2)$$

Where $y_n(w^T x_n + b) \geq 1, \forall n = 1, 2, \ldots, N$ (N is the number of data points) and b is the bias. It needs to find w and b have satisfied as below

$$(w, b) = argmin_{w,b} \frac{\|w\|_2^2}{2} \qquad (3)$$

This is an optimal problem with constraints which can solve by Lagrange function. It means to need find the roots of the following equation, where α_n are Lagrange multipliers $\forall n = 1, 2, \ldots, N$

$$\mathcal{L}(w, b, \alpha) = \frac{\|w\|_2^2}{2} - \sum_{n=1}^{N} \alpha_n \left(y_n(w^T x_n + b) - 1 \right) \qquad (4)$$

After solving this, the category of a data point is calculated by $f(x) = sign(w^T x + b)$.

If the dataset has any noise data points (only nearly separable linear dataset), the set of the data points will be divided into safe, unsafe and wrong data point areas (see Fig. 1). It combines with slack variables (ξ) for this division, slack variable of the i-th data point (d_i) is calculated: $\xi_i = |w^T x_i + b - y_i|$, it means

- $\xi_i = 0 \rightarrow d_i$ is the safe data point (belongs to right category).
- $0 < \xi_i < 1 \rightarrow d_i$ is unsafe data point (still belongs to right category, but in area between M_0 and support vectors hyperland of that category).
- $\xi_i > 1 \rightarrow d_i$ is the wrong data point (belongs to wrong category).

It needs to find w and b of the following equation satisfying $y_i(w^T x_i + b) \geq 1 - \xi_i$

$$(w, b, \xi) = argmin_{w,b,\xi} \frac{\|w\|_2^2}{2} + C \sum_{i=1}^{N} \xi_i \tag{5}$$

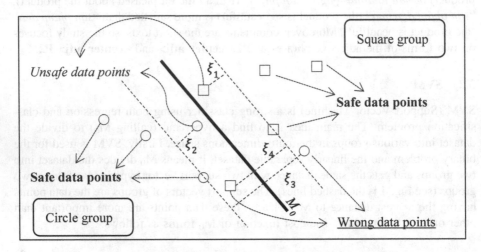

Fig. 1. Illustration for the nearly separable linear dataset.

For the non-linear separable dataset needs to find the transformation to transform the dataset from the non-linear separable space to the linear separable space.

3.4 Logistic Regression

Logistic regression is a statistical approach to determine the relationship between the dependent variable y and a set of independent variables x. The prediction of a data point is a probability of each category by logistic function and uses a threshold $\in [0, 1]$ to determine it belongs to that category or not, the form of logistic function is as follows:

$$y = logistic(w^T x + b) \tag{6}$$

It looks like the linear models which needs to estimate the coefficients w^T and b from training phase. However, the logistic function is a non-linear function, the most often using is sigmod function as follows:

$$f(x) = \frac{1}{1 + e^{-x}} \tag{7}$$

Although the name contains "regression" word, but it's used more in classification problems. This approach is high score and good performance, easy to implement, so it's widely used in various fields of machine learning.

3.5 Random Forest

Random forest is one of the most used algorithms today applied to both regression and classification problem developed Breiman in 2001. The main idea is to combine the decision trees into a single model, each decision tree is built from a random subset of the features in the training dataset using classification and regression trees (CART) technique. For the original random forest, the label of an unlabeled data point is determined via majority voting. Afterward, there are many studies to boost its performance, our context uses average of probabilistic prediction via scikit-learn[2].

The decision tree is built by splitting the training dataset recursively from the root node, CARD uses Gini Index to measure the impurity of data for the split feature in a decision node, this one is calculated as:

$$G = 1 - \sum_{i=1}^{N} (p_i)^2 \tag{8}$$

Where N is the number of categories, p_i which is probability of a data point in a subset belongs to the i-th category. The split process continues recursively until the decision tree reaches the max deep or can't split subsets anymore.

Random forest is a simple algorithm, obtains great result and avoid the big problem in machine learning as overfitting. However, its main limitation is slow and not effective in real-time prediction when the number of subtrees is large.

3.6 OVO and OVR

OVO (One-vs-One) classifier has got each pairwises of the categories and applied the binary classifiers, the final category of the document is decided by the majority voting of them. So, if having c categories, it decomposes $c(c - 1)/2$ iterations for pairwises of the categories to execute the binary classifier. This is a simple approach and obtains the good experiments, but gets high computational cost and takes a long time for training.

Another approach has the less computational cost than OVO, namely OVR (One-vs-Rest) classifier, this approach has executed c binary classifiers of c categories, each i-th one defined whether the documents belong c_i category or a probability of the document belongs to that category. The final category based on the probability.

In our experiments, we use linear SVM for each iterations of OVO and OVR.

4 Approach and Experiments

Our dataset has collected 7200 comments on the ecommerce websites such as tiki.vn, thegioididong.com, fptshop.com.vn grouped into StrongSatisfied (2380 comments), Satisfied (2440 comments), UnSatisfied (2380 comments) sentiment polarities manually. This dataset shows more challenges than the original sentiment analysis problem

---

[2] https://scikit-learn.org/stable/.

which only contains two sentiment polarities as Positive and Negative. For evaluation, the dataset is preprocessed and complemented the useful lexicons, the comments is vectorized with each dimension of feature vectors is $tf \times idf$ weight of terms or phrases, they are feeded to the multiple classifiers for sentiment polarities.

Based on a comparative evaluation of preprocessing techniques of Symeonidis et al. [8] by measuring their accuracy in twitter sentiment classification, we also applied some preprocessing techniques including lowercase transformation, number removal, punctuation removal to improve the accuracy. Next, we have complemented the adjective phrases by using the label of the part of speech to the comments and the adjective structure to indicate the adjective phrases. Then, it replaces negation lexicons and emotional icons, a list of negation lexcions is manually determined such as "không" (not), "chưa" (not yet), "chẳng" (not) and two lists of positive lexicons and negative lexicions based on Vietnamese SentiWordNet built by Vu et al. [9] and our collected dataset. For those ones, each negation lexicon which follows as a positive lexicon is replaced by "not_positive" lexicon or a negative lexicon is replaced by "not_negative" lexicon. Similarly, it also prepares a list of emotional positive and negative icons, positive icons are replaced by "positive" lexicion and negative icons are replaced by "negative" lexicon.

In order to prove our approaches, we executes various experiments with the well-known classifiers. Since the dataset hasn't been much big enough yet, so we have used k-fold cross-validation method to evaluate, this divides the dataset into k subsets and executes k iterations. For each iteration, one of the subsets is used for the testing data, the remaining ones are used for the training data. It uses k as 5 for the experiments and all of the experiments are executed on Macbook Pro (2017) 2.8 GHz Intel Core i7, RAM 16 GB 2133 MHz. The first experiment executes with the original features, Table 1 shows the average of F1 scores, average of training and testing time (in seconds) when executing the multiple classifiers without preprocessing and feeding the proposed feature indicators. Where F1 score is the weighted average of precision and recall, this reaches the best score at 1 and worst score at 0.

Table 1. Executing the multiple classifiers with $tf \times idf$ weight of features.

Classifiers	Parameters	Average of F1 score	Training time (s)	Testing time (s)
Logistic regression	*multi_class = ovr; solver = lbfgs*	0.801	**0.210**	**0.001**
	multi_class = multinomial; solver = lbfgs	0.809	0.523	**0.001**
Random forest	*subtrees = 10*	0.734	0.346	0.005
	subtrees = 50	0.787	1.683	0.022
	subtrees = 80	0.792	2.684	0.035
	subtrees = 100	0.802	3.343	0.043
SVM	*kernel = linear; C = 1e5*	0.784	4.432	0.913
	kernel = rbf; gamma = auto	0.766	4.752	0.826
OVR	*linear SVM*	0.804	10.088	2.300
OVO	*linear SVM*	**0.817**	4.299	2.964

Table 2. Executing the multiple classifiers with preprocessing the dataset and implementing the lexicon features.

Classifiers	Parameters	Average of F1 score	Training time (s)	Testing time (s)
Logistic regression	*multi_class = ovr; solver = lbfgs*	0.829	1.289	0.003
	multi_class = multinomial; solver = lbfgs	0.837	2.032	**0.002**
Random forest	*subtrees = 10*	0.765	**1.070**	0.007
	subtrees = 50	0.807	5.377	0.034
	subtrees = 80	0.812	8.468	0.060
	subtrees = 100	0.818	10.694	0.078
SVM	*kernel = linear; C = 1e5*	0.839	9.750	2.103
	kernel = rbf; gamma = auto	0.847	10.067	2.097
OVR	*linear SVM*	0.842	20.246	4.936
OVO	*linear SVM*	**0.850**	8.887	6.324

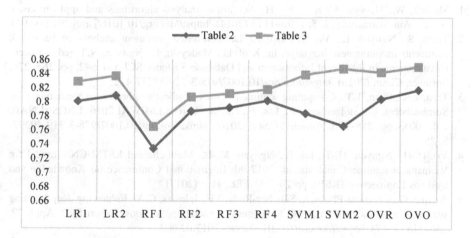

Fig. 2. The F1 scores stats of the experimental results

Table 2 shows the results in the same strategies as Table 1, but applying the proprocessing techniques and complementing the useful lexicons. Observing the F1 score stats of Fig. 2 and the time stats of Table 2, the results are improved and OVO classifier still obtains the highest score for now, but training and testing time are also rather high, this is a big disadvantage for the bigger dataset and difficult to apply the real-time prediction. In other hands, logistic regression classifier obtains a little bit lower than OVO classifier, but training and testing time are much faster, so it's a nice option to deploy in the real-time applications.

5 Conclusions and the Future Works

In this paper, we have built the sentiment dataset and used the multiple classifiers approach with enhancing lexicon features to the comment's vectors to improve the accuracy of sentiment polarities and executed various experiments to prove our approach and suggest the suitable solution for the real-time prediction. In the experiment, it has obtained the good results.

In future works, we will build a Vietnamese sentiment dictionary, investigate misspelling words, wrong grammars of the sentences, synonymy words, antonymous words, slang, also more for emotional icons/symbols, negation. Besides, the dataset will continue to broaden in more fields.

References

1. Medhat, W., Hassan, A., Korashy, H.: Sentiment analysis algorithms and applications: a survey. Ain Shams Eng. J. **5**, 1093–1113 (2014). https://doi.org/10.1016/j.asej.2014.04.011
2. Trinh, S., Nguyen, L., Vo, M., Do, P.: Lexicon-based sentiment analysis of Facebook comments in vietnamese language. In: Król, D., Madeyski, L., Nguyen, N.T. (eds.) Recent Developments in Intelligent Information and Database Systems. SCI, vol. 642, pp. 263–276. Springer, Cham (2016). https://doi.org/10.1007/978-3-319-31277-4_23
3. Tran, T.K., Phan, T.T.: Computing sentiment scores of adjective phrases for vietnamese. In: Sombattheera, C., Stolzenburg, F., Lin, F., Nayak, A. (eds.) MIWAI 2016. LNCS (LNAI), vol. 10053, pp. 288–296. Springer, Cham (2016). https://doi.org/10.1007/978-3-319-49397-8_25
4. Vo, Q.-H., Nguyen, H.-T., Le, B., Nguyen, M.-L.: Multi-channel LSTM-CNN model for Vietnamese sentiment analysis. In: 2017 9th International Conference on Knowledge and Systems Engineering (KSE), pp 24–29. IEEE, Hue (2017)
5. Araque, O., Corcuera-Platas, I., Sánchez-Rada, J.F., Iglesias, C.A.: Enhancing deep learning sentiment analysis with ensemble techniques in social applications. Expert Syst. Appl. **77**, 236–246 (2017). https://doi.org/10.1016/j.eswa.2017.02.002
6. Duong, H.-T., Truong Hoang, V.: A survey on the multiple classifier for new benchmark dataset of vietnamese news classification. In: 2019 11th International Conference on Knowledge and Smart Technology (KST). IEEE, Phuket, pp 23–28 (2019)
7. Deng, X., Li, Y., Weng, J., Zhang, J.: Feature selection for text classification: a review. Multimed. Tools Appl. **78**, 3797–3816 (2019). https://doi.org/10.1007/s11042-018-6083-5
8. Symeonidis, S., Effrosynidis, D., Arampatzis, A.: A comparative evaluation of pre-processing techniques and their interactions for Twitter sentiment analysis. Expert Syst. Appl. **110**, 298–310 (2018). https://doi.org/10.1016/j.eswa.2018.06.022
9. Vu, X.-S., Song, H.-J., Park, S.-B.: Building a vietnamese SentiWordNet using vietnamese electronic dictionary and string kernel. In: Kim, Y.S., Kang, B.H., Richards, D. (eds.) PKAW 2014. LNCS (LNAI), vol. 8863, pp. 223–235. Springer, Cham (2014). https://doi.org/10.1007/978-3-319-13332-4_18

10. Mirończuk, M.M., Protasiewicz, J.: A recent overview of the state-of-the-art elements of text classification. J. Expert. Syst. Appl. **106**, 36–54 (2018)
11. Hussein, D.M.E.-D.M.: A survey on sentiment analysis challenges. J. King Saud Univ.-Eng. Sci. **30**, 330–338 (2018). https://doi.org/10.1016/j.jksues.2016.04.002
12. Devika, M.D., Sunitha, C., Ganesh, A.: Sentiment analysis: a comparative study on different approaches. Procedia Comput. Sci. **87**, 44–49 (2016). https://doi.org/10.1016/j.procs.2016.05.124
13. Ban, D.V., Thung, H.V.: Ngữ pháp tiếng Việt, "Vietnamese Grammar". Vietnam Education Publisher, Hanoi (1998)

Security and Privacy

A Security Proof of the GLP Signature Scheme

Thanh Xuan Khuc[1]([⊠]), Minh Kim Bui[2], and Hien Chu[3]

[1] Institute of Cryptography Science and Technology,
21 Truc Khe, Lang Ha, Dong Da, Hanoi, Vietnam
khucxuanthanh@gmail.com
[2] Ho Chi Minh University of Science, Vietnam National University,
223 Nguyen Van Cu, District 5, Ho Chi Minh City, Vietnam
kmath93@gmail.com
[3] Ho Chi Minh City University of Education,
280 An Duong Vuong, District 5, Ho Chi Minh City, Vietnam
hienchu.1610@gmail.com

Abstract. In 2012, Tim Güneysu, et al. proposed the GLP signature scheme, a practical and efficient post-quantum signature scheme. It is built on the modification of Vadim Lyubashevsky's idea of constructing previous signature schemes. It has a significantly smaller signature and key size than prior signature scheme. The design of the GLP is a foundation to construct newer signature schemes such as Bai-Galbraith, Dilithium. However, Tim Güneysu has only given the description of the GLP signature scheme that has not yet given a detailed security proof for this scheme. Therefore, in this paper, we will present a full security proof for the GLP signature scheme. Specifically, we show that the GLP signature scheme is EU-CMA secure in the random oracle model.

Keywords: Latticed-based signature · R-SIS problem · The GLP signature scheme · Post-quantum cryptography

1 Introduction

The security of currently popular signature schemes is based on hard problems in number theory, such as integer factorization or discrete logarithm problem. More than 20 years ago, Peter Shor proposed an efficient algorithm to solve these hard arithmetic problems on a quantum computer [Sho99]. Therefore, as soon as quantum computers achieve sufficient computational power, widely-used signature schemes will be insecure. This urges researchers to search for new cryptography primitives resistant to quantum computing-based attacks. Due to the recent innovative development of quantum computers, post-quantum cryptography becomes more and more crucial. Among the candidates for post-quantum cryptography, lattice-based cryptography is the most potential candidates.

T. Q. Duong et al. (Eds.): INISCOM 2019, LNICST 293, pp. 253–268, 2019.
https://doi.org/10.1007/978-3-030-30149-1_21

The last few years have witnessed surprising development of lattice-based cryptography, especially constructions of new signatures schemes. At present, lattice-based signatures have become more practical and competitive with classical signatures related to efficiency and security level. Among the methods of constructing signatures, Fiat-Shamir technique is attracting the attention of the cryptography community for its effectiveness and practicality. The first Fiat-Shamir lattice-based signature was introduced by Lyubashevsky [Lyu08] in 2008. This scheme based on the hardness of the shortest vector problem. Later, several improved schemes was proposed in [Lyu12, DDLL13, BG14, DLL+17]. One of them, the Dilithium signature scheme, has been submitted to NIST to standardize as a post-quantum signature.

The GLP signature scheme [GLP12][1] is constructed by adapting the idea from [Lyu12] and [Lyu08]. Concretely, GLP follows the idea that ephemeral values in signing algorithm are chosen randomly from a uniform distribution over a set the same as in [Lyu08], instead of from a Gaussian distribution in [Lyu12]. Moreover, the way how to select a valid signature of GLP is the same as the proposed way in [Lyu08] (in [Lyu08] the signatures is simply checked to be in some interval or not, while [Lyu12] uses rejection sampling to output a valid signature). However, due to the usage of optimized parameters to reduce the signature size and key size, the security proof of GLP was founded on the proof in [Lyu12]. Key size and signature size of GLP is significantly smaller than those of other signature schemes in [Lyu12, Lyu08]. Besides, the construction idea of GLP is a foundation to construct later improved scheme such as [BG14, DLL+17].

Our contribution: In [GLP12], the authors described the GLP signature scheme but did not prove its security in details (concretely, [GLP12] provided a guideline to prove based on [Lyu12]). Therefore, our main contribution is to present a full security proof for the GLP signature scheme based on previous technique in [Lyu08, Lyu12]. We show that the GLP signature scheme is EU-CMA secure in the random oracle model under assuming the hardness of the worst-case lattice problems.

Organization of the paper: In Sect. 2, we recall the definitions of SIS problem and its hardness on ideal lattices. The description of GLP is recalled in Sect. 3. In Sect. 4, we present a security proof of the GLP signature scheme in the random oracle model. Finally, the conclusion of this paper is described in Sect. 5.

2   Preliminaries

Throughout the paper, we will assume that $n = 2^\alpha$ where α is a positive integer, $p \equiv 1 \bmod 2n$ and $R^{p^n} = \mathbb{Z}_p[x]/\langle x^n + 1 \rangle$. Each element of R^{p^n} can be presented as a polynomial of which the degree is at most $n - 1$ and the coefficients are in $\left[-\frac{p-1}{2}, \frac{p-1}{2}\right]$. We let $R_k^{p^n}$ be a subset of the ring R^{p^n} in which the elements of $R_k^{p^n}$ can be presented as polynomials with degree at most $n - 1$ and the

---

[1] A later scheme version is given in [GLP15]. But, it still does not contain a full security proof.

coefficients in $[-k, k]$, where $|k| < \frac{p-1}{2}$. For any set S, let $s \leftarrow_\$ S$ mean that s is chosen uniformly at random from S. We denote $U(0,1)$ to be the uniform distribution on $(0,1)$. Let $\mathbf{s} = s_0 + s_1 x + \ldots + s_{n-1} x^{n-1}$ be a polynomial, we denote $\|\mathbf{s}\|_\infty = \max_{i=0}^{n-1} |s_i|$.

Definition 1 (R-SIS$_{p,n,\gamma,\beta}$ problem, [Lyu12]). *Given the polynomials $\mathbf{a}_1, \mathbf{a}_2, \ldots, \mathbf{a}_\gamma$ are chosen uniformly random from R^{p^n}, find the polynomials $\mathbf{s}_1, \mathbf{s}_2, \ldots, \mathbf{s}_\gamma \neq 0$ in $R_\beta^{p^n}$ such that $\mathbf{a}_1 \mathbf{s}_1 + \mathbf{a}_2 \mathbf{s}_2 + \ldots + \mathbf{a}_\gamma \mathbf{s}_\gamma = 0$.*

Definition 2 (R-SIS$_{p,n,\gamma,k}$ distribution, [Lyu12]). *The $R\text{-}SIS_{p,n,\gamma,k}$ distribution is sampled by choosing $\mathbf{a}_1, \mathbf{a}_2, \ldots, \mathbf{a}_\gamma$ uniformly random from R^{p^n}, the polynomials $\mathbf{s}_1, \mathbf{s}_2, \ldots, \mathbf{s}_\gamma$ uniformly random from $R_k^{p^n}$ and outputting $(\mathbf{a}_1, \mathbf{a}_2, \ldots, \mathbf{a}_\gamma, \mathbf{t} = \mathbf{a}_1 \mathbf{s}_1 + \mathbf{a}_2 \mathbf{s}_2 + \ldots + \mathbf{a}_\gamma \mathbf{s}_\gamma)$.*

Definition 3 (R-SIS$_{p,n,\gamma,k}$ decision problem, [Lyu12]). *Given $(\mathbf{a}_1, \mathbf{a}_2, \ldots, \mathbf{a}_\gamma, \mathbf{t})$, distinguish whether $(\mathbf{a}_1, \mathbf{a}_2, \ldots, \mathbf{a}_\gamma, \mathbf{t})$ are generated from the $R\text{-}SIS_{p,n,\gamma,k}$ distribution or chosen uniformly random from $(\underbrace{R^{p^n} \times \ldots \times R^{p^n}}_{\gamma}, R^{p^n})$.*

The R-SIS$_{p,n,\gamma,\beta}$ problem is consider as a hard problem. Solving the problem R-SIS$_{p,n,\gamma,\beta}$ have the hardness as solving worst-case lattice problems in ideal lattices [LM06].

Definition 4 (DCK$_{p,n}$ problem, [GLP12]). *Distinguish between the uniform distribution over $R^{p^n} \times R^{p^n}$ and the distribution $(\mathbf{a}, \mathbf{a}\mathbf{s}_1 + \mathbf{s}_2)$, where $\mathbf{s}_i$ are uniformly random in $R_1^{p^n}$ and $\mathbf{a}$ is uniformly random in R^{p^n}.*

As in [GLP12], the DCK$_{p,n}$ problem is also considered as a hard problem. The following lemma indicates a reduction from the DCK$_{p,n}$ problem to the R-SIS$_{p,n,2,3\alpha+1}$ decision problem. In other words, if one can solve the R-SIS$_{p,n,2,3\alpha+1}$ decision problem, then one can solve the DCK$_{p,n}$ problem.

Lemma 1 [Lyu12]. *Let α be a non-negative integer such that $GCD(2\alpha+1, p) = 1$, then there exists a polynomial-time reduction from the DCK$_{p,n}$ problem to the R-SIS$_{p,n,2,3\alpha+1}$ decision problem.*

Proof. See [Lyu12], Lemma 3.6, p. 7. □

The following lemma shows that, with suitable parameters, if one can solve the R-SIS$_{p,n,2,\beta}$ problem then one can solve the DCK$_{p,n}$ problem.

Lemma 2. [2] *If $8\beta n \leq p$ then there is a polynomial-time randomized reduction from the DCK$_{p,n}$ problem to the R-SIS$_{p,n,2,\beta}$ problem.*

---

[2] This lemma is stated based on Lemma 3.7 in [Lyu12]. Namely, we give a reduction for the hard problems on the ideal lattices instead of the lattice in $\mathbb{R}^n$ as in Lemma 3.7.

Proof. Given an instance $(\mathbf{a}, \mathbf{t})$ of the $\text{DCK}_{p,n}$ problem. Using the $\text{R-SIS}_{p,n,2,\beta}$ oracle on $(\mathbf{a}, 1)$ to find $\mathbf{s}'_1, \mathbf{s}'_2 \in \mathcal{R}_\beta^{p^n}$ such that $\mathbf{a}\,\mathbf{s}'_1 + \mathbf{s}'_2 = 0$. If $\mathbf{t} = \mathbf{a}\mathbf{s}_1 + \mathbf{s}_2$ then

$$\mathbf{s}'_1\,\mathbf{t} = \mathbf{s}'_1\,(\mathbf{a}\mathbf{s}_1 + \mathbf{s}_2) = \mathbf{a}\,\mathbf{s}'_1\,\mathbf{s}_1 + \mathbf{s}'_1\,\mathbf{s}_2 = -\mathbf{s}'_2\,\mathbf{s}_1 + \mathbf{s}'_1\,\mathbf{s}_2.$$

Since $\|\mathbf{s}_1\|_\infty, \|\mathbf{s}_2\|_\infty \le 1$, $\|\mathbf{s}'_1\|_\infty, \|\mathbf{s}'_2\|_\infty \le \beta$, and [[Lyu08], Lemma 2.8], we have

$$\|\mathbf{s}'_2\,\mathbf{s}_1\|_\infty \le n\|\mathbf{s}'_2\|_\infty\|\mathbf{s}_1\|_\infty \le n\beta$$
$$\|\mathbf{s}'_1\,\mathbf{s}_2\|_\infty \le n\|\mathbf{s}'_1\|_\infty\|\mathbf{s}_2\|_\infty \le n\beta$$

Hence $\|\mathbf{s}'_1\,\mathbf{t}\|_\infty = \|-\mathbf{s}'_2\,\mathbf{s}_1 + \mathbf{s}'_1\,\mathbf{s}_2\|_\infty \le 2n\beta \le \frac{p}{4}$. On the other hand, if $\mathbf{t}$ is uniformly random R^{p^n} then $\|\mathbf{s}'_1\,\mathbf{t}\|_\infty$ will also be uniformly random in $\mathbb{Z}_p$. Hence, in order to solve the $\text{DCK}_{p,n}$ problem, distinguisher simply looks at the value $\|\mathbf{s}'_1\,\mathbf{t}\|_\infty$. If $\|\mathbf{s}'_1\,\mathbf{t}\|_\infty \le \frac{p}{4}$ then he says that $(\mathbf{a}, \mathbf{t})$ is an instance of the $\text{DCK}_{p,n}$ problem. Otherwise, he says that $(\mathbf{a}, \mathbf{t})$ is chosen uniformly from $R^{p^n} \times R^{p^n}$. In the case $(\mathbf{a}, \mathbf{t})$ is an instance of the $\text{DCK}_{p,n}$ problem, the distinguisher will be correct. However, in the case of the uniform distribution $(\mathbf{a}, \mathbf{t})$ is uniformly on $R^{p^n} \times R^{p^n}$, he will make an error with probability $\frac{1}{2}$ (because $\mathbf{t}$ is chosen uniformly from R^{p^n}, the probability that $\|\mathbf{s}'_1\,\mathbf{t}\|_\infty \le \frac{p}{4}$ is $\frac{1}{2}$). $\qquad\square$

A signature scheme includes three algorithms: Keygen, Sign and Verify. Keygen takes as input a security parameter and outputs a public key pk and secret key sk. Sign takes as input a message μ and a secret key sk, outputs a signature Σ. Verify takes as input a message μ, signature Σ and public key pk, output valid (1) or invalid (0). We require that, with the non-negligible probability, $\text{Verify}(\mu, \Sigma, pk) = 1$.

The standard security notion for signatures is existential unforgeability under chosen message attacks (EU-CMA). Consider the following game of challenger $\mathcal{C}$ and forger $\mathcal{F}$. Firstly, the challenger generates a key pair (sk, pk) and sends pk to $\mathcal{F}$. The forger takes as input public key pk. Besides, the forger can make the polynomial queries for signatures on messages $\mu_1, \ldots, \mu_Q$ of its choice. For the i-th query, the challenger answers (μ_i, Σ_i). The output of the forger is (μ^*, Σ^*). It wins the game if $\text{Verify}(\mu^*, \Sigma^*, pk) = 1$, and $\mu^* \ne \mu_i$ for any $i = 1, \ldots, Q$. The signature scheme is called EU-CMA secure if there is no polynomial-time $\mathcal{F}$ whose success probability in the above game is non-negligible.

3   The GLP Signature Scheme

The signature scheme has three main algorithms: key generation, signing and verifying. In particular, the signing and verifying algorithm can access to the random oracle H is defined in [GLP12] as follows: $H : \{0,1\}^* \to \{\mathbf{v} : \mathbf{v} \in R_1^{p^n}, \sum_{i=0}^{n-1} |v_i| = 32\}$, where $\mathbf{v} = v_0 + v_1 x + \ldots + v_{n-1}x^{n-1} \in R_1^{p^n}$.

Key Generation Algorithm

Input: Parameters (n, p).
Output: Secret key $(\mathbf{s}_1, \mathbf{s}_2)$ and public key $(\mathbf{a}, \mathbf{t})$.

1. Choose secret key $\mathbf{s}_1, \mathbf{s}_2 \leftarrow_\$ R_1^{p^n}$.
2. Choose $\mathbf{a} \leftarrow_\$ R^{p^n}$.
3. Compute $\mathbf{t} \leftarrow \mathbf{a}\mathbf{s}_1 + \mathbf{s}_2$.
4. Return secret key $(\mathbf{s}_1, \mathbf{s}_2)$ and public key $(\mathbf{a}, \mathbf{t})$.

Signing Algorithm

Input: Parameters (n, p, k), message μ, secret key $(\mathbf{s}_1, \mathbf{s}_2)$ and $\mathbf{a}$.
Output: A signature for message μ.

1. Choose $\mathbf{y}_1, \mathbf{y}_2 \leftarrow_\$ R_k^{p^n}$.
2. Compute $\mathbf{c} \leftarrow H(\mathbf{a}\mathbf{y}_1 + \mathbf{y}_2, \mu)$.
3. Compute $\mathbf{z}_1 \leftarrow \mathbf{s}_1\mathbf{c} + \mathbf{y}_1, \mathbf{z}_2 \leftarrow \mathbf{s}_2\mathbf{c} + \mathbf{y}_2$.
4. If $\mathbf{z}_1, \mathbf{z}_2 \in R_{k-32}^{p^n}$ then
5.   Output $(\mathbf{z}_1, \mathbf{z}_2, \mathbf{c})$
6. Else return to Step 1.

Verifying Algorithm

Input: The parameters (n, p, k), public key $(\mathbf{a}, \mathbf{t})$, message μ and signature $(\mathbf{z}_1, \mathbf{z}_2, \mathbf{c})$.
Output: The signature is valid or invalid.

1. If $\mathbf{z}_1, \mathbf{z}_2 \in R_{k-32}^{p^n}$ and $\mathbf{c} = H(\mathbf{a}\mathbf{z}_1 + \mathbf{z}_2 - \mathbf{t}\mathbf{c}, \mu)$ then
2.   The signature is valid.
3. Else
4.   The signature is invalid.

The secret keys are random polynomials $\mathbf{s}_1, \mathbf{s}_2 \leftarrow_\$ R_1^{p^n}$. The public key is $(\mathbf{a}, \mathbf{t})$, where $\mathbf{a} \leftarrow_\$ R^{p^n}$ and $\mathbf{t} = \mathbf{a}\mathbf{s}_1 + \mathbf{s}_2 \in R^{p^n}$.

To sign message μ, firstly, the signer chooses random polynomials $\mathbf{y}_1, \mathbf{y}_2 \in R_k^{p^n}$. Then the signer compute $\mathbf{c} = H(\mathbf{a}\mathbf{y}_1 + \mathbf{y}_2, \mu)$ and $\mathbf{z}_1 = \mathbf{s}_1\mathbf{c} + \mathbf{y}_1, \mathbf{z}_2 = \mathbf{s}_2\mathbf{c} + \mathbf{y}_2$. Before outputting the signature for the message μ, the signer will check if $\mathbf{z}_1, \mathbf{z}_2$ has belong to $R_{k-32}^{p^n}$ or not. If $\mathbf{z}_1, \mathbf{z}_2 \in R_{k-32}^{p^n}$ then signer will output $(\mathbf{z}_1, \mathbf{z}_2, \mathbf{c})$ as the signature for μ. Otherwise, the signer regenerate $\mathbf{y}_1, \mathbf{y}_2$ and recalculate the signature. The signer will perform the signing algorithm until it can give $\mathbf{z}_1, \mathbf{z}_2 \in R_{k-32}^{p^n}$. With the parameters chosen as in Table 1, Lemma 3 will show that the average signer needs to make approximately 7 times to generate a valid signature.

To verify the signature $(\mathbf{z}_1, \mathbf{z}_2, \mathbf{c})$, the verifier simply checks that $\mathbf{z}_1, \mathbf{z}_2 \in R_{k-32}^{p^n}$ and that $\mathbf{c} = H(\mathbf{a}\mathbf{z}_1 + \mathbf{z}_2 - \mathbf{t}\mathbf{c}, \mu)$. If $\mathbf{z}_1, \mathbf{z}_2$ are generated from the signing algorithm then we have $\mathbf{z}_1, \mathbf{z}_2 \in R_{k-32}^{p^n}$ and

$$H(\mathbf{a}\mathbf{z}_1 + \mathbf{z}_2 - \mathbf{t}\mathbf{c}, \mu) = H(\mathbf{a}(\mathbf{s}_1\mathbf{c} + \mathbf{y}_1) + (\mathbf{s}_2\mathbf{c} + \mathbf{y}_2) - (\mathbf{a}\mathbf{s}_1 + \mathbf{s}_2)\mathbf{c}, \mu)$$
$$= H(\mathbf{a}\mathbf{y}_1 + \mathbf{y}_2, \mu) = \mathbf{c}$$

Table 1. The GLP signature scheme parameters

		I	II
1.	n	512	1024
2.	p	8383489	16760833
3.	k	2^{14}	2^{15}
4.	The average number of executions to generate a valid signature	7	7
5.	Signature size (bit) $\approx 2n \log{(2(k-32)+1)} + n$	15869	33789
6.	Secret key size (bit) $\approx 2n \log(3)$	1623	3246
7.	Public key size $\approx n \log p$	11775	24574

The Table 1 [GLP12] shows the parameters n, p, k, the average number of executions to generate a valid signature, signature size and keys size used in the GLP signature scheme.

The parameter k controls the trade-off between the security and the run time of the scheme. The smaller k gets, the more secure the scheme becomes and the shorter the signatures get but the time to sign will increase. The authors of the implementation of [GLP12] suggest $k = 2^{14}, n = 512$ and $p = 8383489$ for ≈ 80 bits of security and $k = 2^{15}, n = 1024$ and $p = 16760833$ for >256 bits of security.

The size of the signature is the number of bits used to represent $(\mathbf{z}_1, \mathbf{z}_2, \mathbf{c})$. Since $\mathbf{z}_1, \mathbf{z}_2 \in R_{k-32}^{p^n}$, $\mathbf{z}_1, \mathbf{z}_2$ can be represented by $2n \log{(2(k-32)+1)}$. And $\mathbf{c}$ can be represented by n bits. Therefore, the signature size can be represented by $2n \log{(2(k-32)+1)} + n$ bits.

Since $\mathbf{s}_1, \mathbf{s}_2 \in R_1^{p^n}$, the size of secret key can be represented by $2n \log(3)$. The public key consists of two polynomials $(\mathbf{a}, \mathbf{t})$, where $\mathbf{a}$ is shared by all users (therefore, can be viewed as part of the signature scheme) and $\mathbf{t}$ is the individual component of each user. Therefore, the public key size with each user is just a component $\mathbf{t} \in R^{p^n}$ and can be represented by $n \log p$.

Lemma 3. [3] *For any* $\mathbf{w} \in R^{p^n}$ *such that* $\|\mathbf{w}\|_\infty \le 32$,

$$\Pr\left[\mathbf{w} + \mathbf{y} \in R_{k-32}^{p^n} : \mathbf{y} \leftarrow_{\$} R_k^{p^n}\right] = \left(1 - \frac{64}{2k+1}\right)^n.$$

Proof. Let some $\mathbf{w} \in R^{p^n}$ such that $\|\mathbf{w}\|_\infty \le 32$ and consider $\mathbf{w}$ as a vector of dimension n with coefficients w_j (for $1 \le j \le n$) having absolute value at most 32. Then the sum $\mathbf{w} + \mathbf{y}$ will belong to $R_{k-32}^{p^n}$ if for every coefficient w_i the corresponding coefficient of $\mathbf{y}$ (denoted y_j) is in the range

$$[-(k-32) - w_j, k - 32 - w_j] \tag{1}$$

Since every coefficient y_j is generated randomly in the range $[-k, k]$ and $|w_j| \le 32$, the range (1) is contained in the range of possible coefficient y_j of $\mathbf{y}$. The

---

[3] This lemma is stated based on Lemma 6.1 in [Lyu08].

probability that y_j is in the range (1) is $\frac{2(k-32)+1}{2k+1} = 1 - \frac{64}{2k+1}$. Therefore,

$$\Pr\left[\mathbf{w} + \mathbf{y} \in R_{k-32}^{p^n} : \mathbf{y} \xleftarrow{\$} R_k^{p^n}\right] = \left(1 - \frac{64}{2k+1}\right)^n.$$

$\square$

Since $\|\mathbf{s}_1\|_\infty \leq 1, \|\mathbf{s}_2\|_\infty \leq 1$ and $\|\mathbf{c}\|_1 \leq 32$, we have $\|\mathbf{s}_1\mathbf{c}\|_\infty \leq 32$ and $\|\mathbf{s}_2\mathbf{c}\|_\infty \leq 32$. Hence, we have

$$\Pr\left[\mathbf{s}_1\mathbf{c} + \mathbf{y}_1 \in R_{k-32}^{p^n} : \mathbf{y}_1 \xleftarrow{\$} R_k^{p^n}\right] = \left(1 - \frac{64}{2k+1}\right)^n$$

and

$$\Pr\left[\mathbf{s}_2\mathbf{c} + \mathbf{y}_2 \in R_{k-32}^{p^n} : \mathbf{y}_2 \xleftarrow{\$} R_k^{p^n}\right] = \left(1 - \frac{64}{2k+1}\right)^n$$

Moreover, $\mathbf{z}_1 = \mathbf{s}_1\mathbf{c} + \mathbf{y}_1$ and $\mathbf{z}_2 = \mathbf{s}_2\mathbf{c} + \mathbf{y}_2$, the probability that $\mathbf{z}_1, \mathbf{z}_2$ belong to $R_{k-32}^{p^n}$ is

$$\Pr\left[\mathbf{z}_1 \in R_{k-32}^{p^n} \wedge \mathbf{z}_1 \in R_{k-32}^{p^n} : \mathbf{y}_1, \mathbf{y}_2 \xleftarrow{\$} R_k^{p^n}\right] = \left(1 - \frac{64}{2k+1}\right)^{2n}.$$

We can see that if k is too small then the probability that $\mathbf{z}_1, \mathbf{z}_2 \in R_{k-32}^{p^n}$ is very low. The below table show the relationship between n, k and the average number of executions to generate a valid signature (Table 2).

Table 2. The relationship between n, k and the average number of executions to generate a valid signature

n	k	The average number of executions to generate a valid signature
512	2^{14}	7
512	2^{15}	3
512	2^{16}	2
1024	2^{14}	55
1024	2^{15}	7
1024	2^{16}	3
1024	2^{17}	2

4   Security Proof of the GLP Signature Scheme

In this section, we show that the GLP signature scheme is EU-CMA secure in the random oracle model, assuming the hardness of the $DCK_{p,n}$ problem and the

R-SIS$_{q,n,2,\beta}$ problem. In [GLP12], it's a bit confusing when saying that the security of the GLP signature scheme is based only on the hardness of the DCK$_{p,n}$ problem. Because [GLP12] assumes that there is a polynomial-time reduction from the DCK$_{p,n}$ problem to the R-SIS$_{q,n,2,\beta}$ problem as Lemma 2. However, we can easily see that if the parameters n, p, k chosen as in [GLP12] (Table 1) then the condition $8\beta n \leq p$ in Lemma 2 is not satisfied. In other words, there is no polynomial-time reduction from the DCK$_{p,n}$ problem to the R-SIS$_{q,n,2,\beta}$ problem with the parameters chosen as in [GLP12] by using Lemma 2.

Firstly, we see that if a adversary can derive the secret key $(\mathbf{s}_1, \mathbf{s}_2)$ from the public key $(\mathbf{a}, \mathbf{t})$ then he can be used to solve DCK$_{p,n}$ problem. Therefore, the security of the GLP signature scheme must be based on DCK$_{p,n}$ problem.

The following theorem shows the security of the GLP signature scheme also need to be based on R-SIS$_{q,n,2,\beta}$ problem. Assume, for contradiction, that there exists a polynomial-time forger $\mathcal{F}$ breaking the EU-CMA security of the signature scheme with non-negligible advantage. Then, we can use $\mathcal{F}$ to solve the R-SIS$_{q,n,2,\beta}$ problem.

Theorem 1. [4] *If there is a polynomial-time adversary who can produce a valid signature with success probability δ after making s queries to the signing oracle and h queries to the random oracle H, then there exists a polynomial-time algorithm to solve R-SIS$_{p,n,2,\beta}$ problem, where $\beta = (2(k - 32) + 64k')$, $k' = 3 \left\lceil \frac{2^{\frac{100}{n} - 1}\sqrt{p} - 1}{3} \right\rceil + 1$, with probability at least $\approx \frac{\delta^2}{2(h+s)}$.*

Proof. This theorem is proven through Lemmas 5 and 6 using the hybrid arguments. To be more precise, Lemma 5 shows that the actual signing algorithm and the actual key generation algorithm can be replaced by the key generation algorithm Hybrid 3 and the signing algorithm Hybrid 3 (those are obtained from the actual key generation algorithm and the actual signing algorithm. The advantage of the adversary in distinguishing the actual algorithms and the Hybrid 3 algorithms is negligible). Therefore, if the adversary is able to produce a valid signature with probability δ for the actual key generation algorithm and actual signing algorithm, then he can also produce a valid signature with probability δ for the key generation algorithm Hybrid 3 and signing algorithm Hybrid 3.

In Lemma 6, we show that if an adversary is able to produce a valid signature with probability δ for the key generation algorithm Hybrid 3 and signing algorithm Hybrid 3, then we can use that signature to solve the R-SIS$_{p,n,2,\beta}$ problem, where $\beta = (2(k - 32) + 64k')$, $k' = 3 \left\lceil \frac{2^{\frac{100}{n} - 1}\sqrt{p} - 1}{3} \right\rceil + 1$, with success probability at least

$$\left(\frac{1}{2} - 2^{-200}\right)\left(\delta - 2^{-200}\right)\left(\frac{\delta - 2^{-200}}{h + s} - 2^{-200}\right) \approx \frac{\delta^2}{2(h+s)}.$$

□

---

[4] This theorem is stated based on Theorem 5.1 in [Lyu12]. Namely, we provide an additional algorithms Hybrid 3 to prove the security of the GLP signature scheme.

The Hybrid algorithms are described as follows.

Key Generation Algorithm Hybrid 1 and Hybrid 2

Input: Parameters (n, p).
Output: Secret key $(\mathbf{s}_1, \mathbf{s}_2)$ and public key $(\mathbf{a}, \mathbf{t})$.

1. Choose secret key $\mathbf{s}_1, \mathbf{s}_2 \leftarrow_\$ R_1^{p^n}$.
2. Choose $\mathbf{a} \leftarrow_\$ R^{p^n}$.
3. Compute $\mathbf{t} \leftarrow \mathbf{a}\mathbf{s}_1 + \mathbf{s}_2$.
4. Return secret key $(\mathbf{s}_1, \mathbf{s}_2)$ and public key $(\mathbf{a}, \mathbf{t})$.

Signing Algorithm Hybrid 1

Input: Parameters (n, p, k), message μ, secret key $(\mathbf{s}_1, \mathbf{s}_2)$ and $\mathbf{a}$.
Output: A signature for μ.

1. Choose $\mathbf{y}_1, \mathbf{y}_2 \leftarrow_\$ R_k^{p^n}$.
2. Choose $\mathbf{c} \leftarrow_\$ \{\mathbf{v} \in R_1^{p^n} : \sum_{i=1}^{n} |v_i| = 32\}$
3. Compute $\mathbf{z}_1 \leftarrow \mathbf{s}_1\mathbf{c} + \mathbf{y}_1, \mathbf{z}_2 \leftarrow \mathbf{s}_2\mathbf{c} + \mathbf{y}_2$.
4. If $\mathbf{z}_1, \mathbf{z}_2 \in R_{k-32}^{p^n}$ then
5.   Output a signature $(\mathbf{z}_1, \mathbf{z}_2, \mathbf{c})$
6.   Program[5] $\mathbf{c} = H(\mathbf{a}\mathbf{z}_1 + \mathbf{z}_2 - \mathbf{t}\mathbf{c}, \mu)$
7. Else return to Step 1.

Signing Algorithm Hybrid 2 and Hybrid 3

Input: Parameters (n, p, k), message μ, secret key $(\mathbf{s}_1, \mathbf{s}_2)$ and $\mathbf{a}$
Output: A signature for μ.

1. Choose $\mathbf{c} \leftarrow_\$ \{\mathbf{v} \in R_1^{p^n} : \sum_{i=1}^{n} |v_i| = 32\}$
2. Choose $\mathbf{z}_1 \leftarrow_\$ R_{k-32}^{p^n}, \mathbf{z}_2 \leftarrow_\$ R_{k-32}^{p^n}$.
3. Choose $u \leftarrow_\$ U(0, 1)$
4. Set $M \leftarrow \left(1 - \frac{64}{2k+1}\right)^{2n}$
5. If $u \leq M$ then
6.   Output a signature $(\mathbf{z}_1, \mathbf{z}_2, \mathbf{c})$
7.   Program $\mathbf{c} = H(\mathbf{a}\mathbf{z}_1 + \mathbf{z}_2 - \mathbf{t}\mathbf{c}, \mu)$
8. Else return to Step 1.

---

[5] When it is queried, the oracle H is programmed to return a random $\mathbf{c} \in \{\mathbf{v} \in R_1^{p^n} : \sum_{i=1}^{n} |v_i| = 32\}$ without checking whether that value has been used before.

Key Generation Algorithm Hybrid 3

Input: Parameters (n, p, k'), where $k' = 3\alpha + 1, \alpha = \left\lceil \frac{2^{\frac{100}{n}-1}\sqrt{p}-1}{3} \right\rceil$.

Output: Secret key (s_1, s_2) and public key (a, t).

1. Choose secret key $s_1, s_2 \xleftarrow{\$} R_{k'}^{p^n}$.
2. Choose $a \xleftarrow{\$} R^{p^n}$.
3. Compute $t \leftarrow as_1 + s_2$.
4. Return secret ket (s_1, s_2) and public key (a, t).

The following lemma shows that, if (s_1, s_2) is randomly chosen from $R_{k'}^{p^n} \times R_{k'}^{p^n}$ in which k' is reasonably chosen, then with high probability, there exists $(s_1', s_2') \in R_{k'}^{p^n} \times R_{k'}^{p^n}$ different from (s_1, s_2) satisfying $as_1 + s_2 = as_1' + s_2'$.

Lemma 4. [6] *For any $a \in R^{p^n}$ and (s_1, s_2) randomly chosen from $R_{k'}^{p^n} \times R_{k'}^{p^n}$, in which $k' \geq 2^{\frac{100}{n}-1}\sqrt{p}$. Then, with probability at least $1 - 2^{-200}$, there exists (s_1', s_2') different from (s_1, s_2) satisfying $as_1 + s_2 = as_1' + s_2'$.*

Proof. For any $a \in R^{p^n}$, consider the map f_a defined as follows:

$$f_a : R_{k'}^{p^n} \times R_{k'}^{p^n} \to R^{p^n}$$
$$(s_1, s_2) \mapsto as_1 + s_2$$

We see that the cardinality of R^{p^n} is $|R^{p^n}| = p^n$ and the cardinality of $R_{k'}^{p^n} \times R_{k'}^{p^n}$ is $(2k'+1)^{2n}$. Therefore, the probability of choosing $(s_1, s_2) \in R_{k'}^{p^n} \times R_{k'}^{p^n}$ such that there are no collisions is at least

$$\frac{p^n}{(2k'+1)^{2n}} \leq \frac{p^n}{\left(2^{\frac{100}{n}}\sqrt{p}+1\right)^{2n}} < \frac{p^n}{2^{200}p^n} = \frac{1}{2^{200}}$$

In other words, with probability at least $1 - 2^{-200}$, there exists (s_1', s_2') different from (s_1, s_2) such that $as_1 + s_2 = as_1' + s_2'$. $\square$

Lemma 5. *Let $\mathcal{D}$ be a distinguisher who can querry to the random oracle H. Moreover, $\mathcal{D}$ can query the actual key generation algorithm, the actual signing algorithm, the key generation algorithm Hybrid 3 and the signing algorithm Hybrid 3. If $\mathcal{D}$, after making h queries to the random oracle H and s queries to the signing algorithm (either actual or Hybrid 3), then the advantage of $\mathcal{D}$ in distinguishing the actual key generation algorithm and the actual signing algorithm with the key generation algorithm Hybrid 3 and the signing algorithm Hybrid 3 is less than $s(s - 1 + 2h) 2^{-(n+1)}$.*

Proof. We will use the hybrid arguments to prove the Lemma. Specifically, we will use three Hybrid key generation and signing algorithms (the verification

---

[6] This lemma is stated based on Lemma 5.2 in [Lyu12].

algorithms are the same to the actual one, so we do not mention them here). The key generation algorithm Hybrid 1 and Hybrid 2 are similar to the actual key generation algorithm while the key generation Hybrid 3 is different from the actual one in that $(\mathbf{s}_1, \mathbf{s}_2)$ is uniformly chosen from $R_{k'}^{p^n} \times R_{k'}^{p^n}$ instead of $R_1^{p^n} \times R_1^{p^n}$. The signing algorithm Hybrid 1 is different from the actual one in that $\mathbf{c}$ is randomly chosen from $\{\mathbf{v} \in R_1^{p^n} : \sum_{i=0}^{n-1} |v_i| = 32\}$ instead of being computed by $\mathbf{c} = H(\mathbf{a}\mathbf{z}_1 + \mathbf{z}_2 - \mathbf{t}\mathbf{c}, \mu)$. The signing algorithm Hybrid 2 is different from the Hybrid 1 in that in the Hybrid 2, $\mathbf{z}_1, \mathbf{z}_2$ are uniformly chosen from $R_{k-32}^{p^n}$ instead of being computed as $\mathbf{z}_1 = \mathbf{s}_1\mathbf{c} + \mathbf{y}_1, \mathbf{z}_2 = \mathbf{s}_2\mathbf{c} + \mathbf{y}_2$. The signing algorithm Hybrid 3 is similar to the Hybrid 2 one. We will show that with a suitable choice of parameters, the advantage of the distinguisher $\mathcal{D}$ in distinguishing the Hybrid algorithms is negligible.

First of all, we show that the distinguisher $\mathcal{D}$ has advantage less than $s(s - 1 + 2h)2^{-(n+1)}$ in distinguishing the actual signing algorithm with the Hybrid 1 one. The only difference between those two algorithms is that in the Hybrid 1 algorithm, the output of the random oracle H is randomly chosen from $\{\mathbf{v} \in R_1^{p^n} : \sum_{i=0}^{n-1} |v_i| = 32\}$ and then is programmed as the answer of $H(\mathbf{a}\mathbf{z}_1 + \mathbf{z}_2 - \mathbf{t}\mathbf{c}, \mu) = H(\mathbf{a}\mathbf{y}_1 + \mathbf{y}_2, \mu)$ without checking whether the hash value $(\mathbf{a}\mathbf{y}_1 + \mathbf{y}_2, \mu)$ has been queried to H or not. Therefore, in the signing algorithm Hybrid 1, there may be the case that with the same input, two queries to H may produce two different outputs, while with the actual algorithm one gets the same output. And this is the only point that the distinguisher can distinguish between the actual signing algorithm with the Hybrid 1 algorithm. In other words, the advantage of the distinguisher in distinguishing these two algorithms is the probability that the signing algorithm Hybrid 1 produces two outputs with the same query to H.

Since $\mathcal{D}$ makes h queries to H and s queries to the signing algorithm, there are at most $s + h$ values $(\mathbf{a}\mathbf{y}_1 + \mathbf{y}_2, \mu)$ established. Now, we show that for every call to the signing algorithm Hybrid 1, the probability of generating $\mathbf{y}_1, \mathbf{y}_2$ such that $\mathbf{a}_1\mathbf{y}_1 + \mathbf{y}_2$ is equal to the previous queried value is less than 2^{-n}. Given $\mathbf{t}$ arbitrarily in R^{p^n}, we have

$$\Pr\left[\mathbf{a}\mathbf{y}_1 + \mathbf{y}_2 = \mathbf{t}; \mathbf{y}_i \leftarrow_\$ R_k^{p^n}\right] = \Pr\left[\mathbf{y}_2 = (\mathbf{t} - \mathbf{a}\mathbf{y}_1); \mathbf{y}_i \leftarrow_\$ R_k^{p^n}\right]$$

$$\leq \max_{\mathbf{t}' \in R^{p^n}} \Pr\left[\mathbf{y}_2 = \mathbf{t}'; \mathbf{y}_2 \leftarrow_\$ R_k^{p^n}\right]$$

$$\leq \left(\frac{2k+1}{p}\right)^n$$

By hypothesis $k \ll p$, then

$$\Pr\left[\mathbf{a}\mathbf{y}_1 + \mathbf{y}_2 = \mathbf{t}; \mathbf{y}_1 \leftarrow_\$ R_k^{p^n}, \mathbf{y}_2 \leftarrow_\$ R_k^{p^n}\right] < 2^{-n}.$$

We see that the previous queried value can be generated in the random oracle H or in signing algorithm Hybrid 1. Hence, the probability of getting a collision after making s queries is at

$$\left(\binom{s}{2} + sh \right) 2^{-n} = s(s - 1 + 2h)2^{-(n+1)}$$

Now, we need to show that outputs of the signing algorithms Hybrid 1 and Hybrid 2 are indistinguishable. Indeed, the element $\mathbf{c}$ in both algorithms are computed in a similar way. The main difference is the way $\mathbf{z}_1, \mathbf{z}_2$ are computed. In the Hybrid 1 algorithm, $\mathbf{z}_1 = \mathbf{s}_1 \mathbf{c} + \mathbf{y}_1, \mathbf{z}_2 = \mathbf{s}_2 \mathbf{c} + \mathbf{y}_2$ and $\mathbf{z}_1, \mathbf{z}_2$ are output only if $\mathbf{z}_1, \mathbf{z}_2 \in R_{k-32}^{p^n}$. By Lemma 3, the probability that the Hybrid 1 algorithm produces a valid signature is $\left(1 - \frac{64}{2k+1}\right)^{2n}$, whereas the Hybrid 2 chooses $\mathbf{z}_1, \mathbf{z}_2$ uniformly from $R_{k-32}^{p^n}$ and the probability of producing a valid signature is

$$\Pr\left[u \leq \left(1 - \frac{64}{2k+1}\right)^{2n} : u \leftarrow_{\$} U(0,1) \right] = \left(1 - \frac{64}{2k+1}\right)^{2n}.$$

Therefore, $\mathcal{D}$ cannot distinguish whether $(\mathbf{z}_1, \mathbf{z}_2, \mathbf{c})$ is outputted by the Hybrid 1 algorithm or the Hybrid 2 algorithm.

Next, we show that $\mathcal{D}$ cannot distinguish the public keys generated by the key generation algorithm Hybrid 3 and the Hybrid 2 algorithm. Indeed, by the hardness of $\mathrm{DCK}_{p,n}$, the distinguisher $\mathcal{D}$ cannot distinguish between the outputs of the Hybrid 2 algorithm and the uniform distribution on $R^{p^n} \times R^{p^n}$. By Lemma 1, there is a reduction from solving $\mathrm{DCK}_{p,n}$ to solving the R-SIS$_{p,n,2,(3\alpha+1)}$ decision problem. Hence $\mathcal{D}$ cannot distinguish between the outputs the Hybrid 3 algorithm with the uniform distribution on $R^{p^n} \times R^{p^n}$ (if it does then it is possible to solve the R-SIS$_{p,n,2,(3\alpha+1)}$ decision problem, from which one can solve the $\mathrm{DCK}_{p,n}$ problem). Therefore, $\mathcal{D}$ cannot distinguish between the outputs by the Hybrid 3 algorithm and the Hybrid 2 algorithm.

As a conclusion, we see that the actual key generation algorithm and the actual signing algorithm can be replaced by the corresponding Hybrid 3 algorithms. The advantage of the distinguisher in distinguishing between the actual algorithms and the Hybrid 3 algorithms is less than $s(s - 1 + 2h)2^{-(n+1)}$. $\qquad\square$

Lemma 6. *Assume that there exists a polynomial-time forger $\mathcal{F}$ who can produce a valid signature with success probability δ after making at most s queries to the signing algorithm Hybrid 3 and at most h queries to the random oracle H. Then there exists a polynomial-time algorithm to efficiently solve the R-SIS$_{p,n,2,\beta}$ problem, in which $\beta = (2(k - 32) + 64k')$, $k' = 3\left\lceil \frac{2^{\frac{100}{n}-1}\sqrt{p}-1}{3} \right\rceil + 1$ with success probability at least*

$$\left(\frac{1}{2} - 2^{-200} \right) (\delta - 2^{-200}) \left(\frac{\delta - 2^{-200}}{h + s} - 2^{-200} \right).$$

Proof. Denote by $D_H := \{\mathbf{v} \in R_1^{p^n} : \sum_{i=0}^{n-1} |v_i| = 32\}$ the range of the random oracle H. Given $\mathbf{a} \in R^{p^n}$, choose $\mathbf{s}_1, \mathbf{s}_2 \leftarrow_{\$} R_{k'}^{p^n}$ to be the secret key and compute

$\mathbf{t} = \mathbf{a}_1\mathbf{s}_1 + \mathbf{s}_2$. The public key is $(\mathbf{a}, \mathbf{t})$. Set $t = h + s$ to be the upper bound of the total number of times that H is queried or programmed by $\mathcal{F}$. One query to H can be made by the forger or H can be programmed by the signing algorithm Hybrid 3 when the forger requests a signature of some message. Next we choose a random coins ϕ for the forger, a random coins ψ for the signer (using Hybrid 3) and choose $\mathbf{r}_1, \ldots, \mathbf{r}_t \leftarrow_\$ D_H$ to be the answers of the random oracle.

Now, we consider a sub-algorithm $\mathcal{A}$ with input $(\mathbf{a}, \mathbf{t}, \phi, \psi, \mathbf{r}_1, \ldots, \mathbf{r}_t)$. The algorithm $\mathcal{A}$ will run $\mathcal{F}$ by supplying to $\mathcal{F}$ the public key $(\mathbf{a}, \mathbf{t})$ and the random coins ϕ as input. Whenever $\mathcal{F}$ requests a signature of a message, $\mathcal{A}$ runs the signing algorithm Hybrid 3 and uses the random coins ψ to generate a signature. During the signing process, the random oracle H is programmed and answers the first value in $(\mathbf{r}_1, \ldots, \mathbf{r}_t)$ which has not been used before. Here $\mathcal{A}$ will save a table consisting of all queries to H, and when a same query is requested again, the random oracle will output the same answer. The forger can also make queries to the random oracle. In this case, the answer is done in the same way. Whenever $\mathcal{F}$ stops and produces a forgery (with probability δ), the algorithm $\mathcal{A}$ will take the output of $\mathcal{F}$ as its output.

With probability δ, the forger $\mathcal{F}$ will produce a message μ and a signature $(\mathbf{z}_1, \mathbf{z}_2, \mathbf{c})$ satisfying $\mathbf{z}_1, \mathbf{z}_2 \in R_{k-32}^{p^n}$ and $\mathbf{c} = H((\mathbf{a}\mathbf{z}_1 + \mathbf{z}_2 - \mathbf{t}\mathbf{c}), \mu)$. Note that if the random oracle was not queried or programmed with input $\mathbf{w} = (\mathbf{a}\mathbf{z}_1 + \mathbf{z}_2 - \mathbf{t}\mathbf{c})$ then the probability that $\mathcal{F}$ produces $\mathbf{c}$ with $\mathbf{c} = H(\mathbf{w}, \mu)$ is $1/|D_H|$. Hence, with probability $1 - 1/|D_H|$, $\mathbf{c}$ has to be one of the values $\mathbf{r}_i$ with $1 \leq i \leq t$. Thus the probability that $\mathcal{F}$ succeeds in a forgery and $\mathbf{c}$ is one of the values $\mathbf{r}_i$ (with $1 \leq i \leq t$) is at least $\delta - 1/|D_H|$. Indeed, let A be the event that $\mathcal{F}$ succeeds in a forgery and B be the event that $\mathbf{c}$ is one of the values $\mathbf{r}_i$. Then we have

$$\Pr[A \cap B] = \Pr[A] + \Pr[B] - \Pr[A \cup B]$$
$$\geq \delta + 1 - \frac{1}{|D_H|} - 1 = \delta - \frac{1}{|D_H|}.$$

Assume that j is the index such that $\mathbf{c} = \mathbf{r}_j$, with $1 \leq j \leq t$. Then there are two possibilities as follows:

- Either $\mathbf{r}_j$ is an answer of a query of $\mathcal{F}$ to the random oracle,
- or $\mathbf{r}_j$ is programmed in the signing process.

In the second case, assume that, when signs a message μ', the signer programs the random oracle as $H((\mathbf{a}\mathbf{z}_1' + \mathbf{z}_2' - \mathbf{t}\mathbf{c}), \mu') = \mathbf{c}$. If the forger produces a valid signature $(\mathbf{z}_1, \mathbf{z}_2, \mathbf{c})$ for μ then $\mu \neq \mu'$ or $(\mathbf{z}_1, \mathbf{z}_2) \neq (\mathbf{z}_1', \mathbf{z}_2')$ (or both are different). Because if $\mu = \mu'$ and $(\mathbf{z}_1, \mathbf{z}_2) = (\mathbf{z}_1', \mathbf{z}_2')$, the adversary just outputs a message with a signature that he already saw. If $\mu \neq \mu'$ then we have a collision for H, since

$$H((\mathbf{a}\mathbf{z}_1' + \mathbf{z}_2' - \mathbf{t}\mathbf{c}), \mu') = \mathbf{c} = H((\mathbf{a}\mathbf{z}_1 + \mathbf{z}_2 - \mathbf{t}\mathbf{c}), \mu).$$

If $\mu = \mu'$ then $(\mathbf{z}_1, \mathbf{z}_2) \neq (\mathbf{z}_1', \mathbf{z}_2')$ and

$$H((\mathbf{a}\mathbf{z}_1' + \mathbf{z}_2' - \mathbf{t}\mathbf{c}), \mu) = \mathbf{c} = H((\mathbf{a}\mathbf{z}_1 + \mathbf{z}_2 - \mathbf{t}\mathbf{c}), \mu).$$

Thus if $\mathbf{a}\mathbf{z}_1' + \mathbf{z}_2' - \mathbf{t}\mathbf{c} \neq \mathbf{a}_1\mathbf{z}_1 + \mathbf{z}_2 - \mathbf{t}\mathbf{c}$ then we find a collision for H. On the other hand, if $\mathbf{a}\mathbf{z}_1' + \mathbf{z}_2' - \mathbf{t}\mathbf{c} = \mathbf{a}_1\mathbf{z}_1 + \mathbf{z}_2 - \mathbf{t}\mathbf{c}$ then

$$\mathbf{a}(\mathbf{z}_1 - \mathbf{z}_1') + (\mathbf{z}_2 - \mathbf{z}_2') = 0.$$

Since $\|\mathbf{z}_1\|_\infty, \|\mathbf{z}_2'\|_\infty \leq k - 32$, one has $\|\mathbf{z}_1 - \mathbf{z}_1'\|_\infty, \|\mathbf{z}_2 - \mathbf{z}_2'\|_\infty \leq 2(k - 32)$. Therefore we can solve the R-SIS$_{p,n,2,\beta}$ problem where $\beta = 2(k - 32)$ with success probability at least $\delta - 1/|D_H|$.

Now we come back to the first case, i.e., $\mathbf{r}_j$ is an answer when $\mathcal{F}$ makes a query to the random oracle. In this case, first we save the signature $(\mathbf{z}_1, \mathbf{z}_2, \mathbf{r}_j)$ of the forger $\mathcal{F}$ for μ. Then, we generate new random elements $\mathbf{r}_j', \ldots, \mathbf{r}_t' \leftarrow_\$ D_H$. Next we run again the algorithm $\mathcal{A}$ with input $(\mathbf{a}, \mathbf{t}, \phi, \psi, \mathbf{r}_1, \ldots, \mathbf{r}_{j-1}, \mathbf{r}_j', \ldots, \mathbf{r}_t')$. By the General Forking Lemma [[BN06], Lemma 1], the probability that $\mathbf{r}_j' \neq \mathbf{r}_j$ and the forger uses $\mathbf{r}_j'$ as an answer for a query to the random oracle is at least

$$\left(\delta - \frac{1}{|D_H|}\right)\left(\frac{\delta - 1/|D_H|}{t} - \frac{1}{|D_H|}\right),$$

and so with the above probability, $\mathcal{F}$ produces a signature $\mathbf{z}_1', \mathbf{z}_2', \mathbf{r}_j'$ for μ and

$$\mathbf{a}\mathbf{z}_1 + \mathbf{a}\mathbf{z}_2 - \mathbf{t}\mathbf{c} = \mathbf{a}\mathbf{z}_1' + \mathbf{a}\mathbf{z}_2' - \mathbf{t}\mathbf{c}' \tag{2}$$

where $\mathbf{c} = \mathbf{r}_j$ and $\mathbf{c}' = \mathbf{r}_j'$. Plug $\mathbf{t} = \mathbf{a}_1\mathbf{s}_1 + \mathbf{s}_2$ into (2), we obtain

$$\mathbf{a}(\mathbf{z}_1 - \mathbf{c}\mathbf{s}_1 - \mathbf{z}_1' + \mathbf{c}'\mathbf{s}_1) + (\mathbf{z}_2 - \mathbf{c}\mathbf{s}_2 - \mathbf{z}_2' + \mathbf{c}'\mathbf{s}_2) = 0. \tag{3}$$

Since $\|\mathbf{z}_1\|_\infty, \|\mathbf{z}_1'\|_\infty, \|\mathbf{z}_2\|_\infty, \|\mathbf{z}_2'\|_\infty \leq k - 32$ and $\|\mathbf{s}_1\mathbf{c}\|_\infty, \|\mathbf{s}_1\mathbf{c}'\|_\infty, \|\mathbf{s}_2\mathbf{c}\|_\infty, \|\mathbf{s}_2\mathbf{c}'\|_\infty \leq 32k'$, one gets

$$\|\mathbf{z}_1 - \mathbf{c}\mathbf{s}_1 - \mathbf{z}_1' + \mathbf{c}'\mathbf{s}_1\|_\infty, \|\mathbf{z}_2 - \mathbf{c}\mathbf{s}_2 - \mathbf{z}_2' + \mathbf{c}'\mathbf{s}_2\|_\infty \leq (2(k - 32) + 64k').$$

Set $\mathbf{z}_1 - \mathbf{c}\mathbf{s}_1 - \mathbf{z}_1' + \mathbf{c}'\mathbf{s}_1 = \mathbf{u}_1$ and $\mathbf{z}_2 - \mathbf{c}\mathbf{s}_2 - \mathbf{z}_2' + \mathbf{c}'\mathbf{s}_2 = \mathbf{u}_2$. If $\mathbf{u}_1, \mathbf{u}_2 \neq 0$ then we can solve the R-SIS$_{p,n,2,\beta}$ problem with $\beta = (2(k - 32) + 64k')$. Therefore, it suffices to show that $\mathbf{u}_1, \mathbf{u}_2 \neq 0$ with probability at least $\frac{1}{2} - 2^{-200}$. By Lemma 4, with probability at least $1 - 2^{-200}$, there exists $(\mathbf{s}_1', \mathbf{s}_2') \neq (\mathbf{s}_1, \mathbf{s}_2) \in R_{k'}^{p^n}$ such that $\mathbf{a}\mathbf{s}_1' + \mathbf{s}_2' = \mathbf{a}\mathbf{s}_1 + \mathbf{s}_2$. If $0 = \mathbf{u}_1 = \mathbf{z}_1 - \mathbf{c}\mathbf{s}_1 - \mathbf{z}_1' + \mathbf{c}'\mathbf{s}_1$ then $\mathbf{z}_1 - \mathbf{c}\mathbf{s}_1' - \mathbf{z}_1' + \mathbf{c}'\mathbf{s}_1' \neq 0$. Indeed, assume that

$$\mathbf{z}_1 - \mathbf{c}\mathbf{s}_1 - \mathbf{z}_1' + \mathbf{c}'\mathbf{s}_1 = 0 \text{ and } \mathbf{z}_1 - \mathbf{c}\mathbf{s}_1' - \mathbf{z}_1' + \mathbf{c}'\mathbf{s}_1' = 0$$

Then $\mathbf{z}_1 - \mathbf{c}\mathbf{s}_1 - \mathbf{z}_1' + \mathbf{c}'\mathbf{s}_1 = \mathbf{z}_1 - \mathbf{c}\mathbf{s}_1' - \mathbf{z}_1' + \mathbf{c}'\mathbf{s}_1'$. Hence

$$(\mathbf{c} - \mathbf{c}')(\mathbf{s}_1 - \mathbf{s}_1') = 0. \tag{4}$$

Because $\|\mathbf{c}\|_1, \|\mathbf{c}'\|_1 \leq 32$ and $\|\mathbf{s}_1\|_\infty, \|\mathbf{s}_1'\|_\infty \leq k'$, we have $\|(\mathbf{c} - \mathbf{c}')(\mathbf{s}_1 - \mathbf{s}_1')\|_\infty \leq 256k'$. By the choice of parameters, $256k' < p$. Thus if $(\mathbf{c} - \mathbf{c}')(\mathbf{s}_1 - \mathbf{s}_1') = 0$ over $R^{p^n} = \mathbb{Z}_p[x]/\langle x^n + 1\rangle$ then $(\mathbf{c} - \mathbf{c}')(\mathbf{s}_1 - \mathbf{s}_1') = 0$ over $\mathbb{Z}[x]/\langle x^n + 1\rangle$. Since n is a power of 2, $x^n + 1$ is irreducible in $\mathbb{Z}[x]$. Hence $\mathbb{Z}[x]/\langle x^n + 1\rangle$ is an integral domain. Then, $(\mathbf{c} - \mathbf{c}')(\mathbf{s}_1 - \mathbf{s}_1') = 0$ which implies $\mathbf{c} - \mathbf{c}' = 0$ or

$s_1 - s'_1 = 0$. So $c \neq c'$ and hence $s_1 = s'_1$ (which contradicts to $s_1 \neq s'_1$ above). Similarly, if $0 = u_2 = z_2 - cs_2 - z'_2 + c's_2$ then $z_2 - cs'_2 - z'_2 + c's'_2 \neq 0$. On the other hand, at the beginning, we do not know whether $z_1 - cs_1 - z'_1 + c's_1 = 0$ and $z_2 - cs_2 - z'_2 + c's_2 = 0$ or $z_1 - cs_1 - z'_1 + c's_1 \neq 0$ and $z_2 - cs_2 - z'_2 + c's_2 \neq 0$. Therefore, the probability that $z_1 - cs_1 - z'_1 + c's_1 \neq 0$ and $z_2 - cs_2 - z'_2 + c's_2 \neq 0$ is exactly the probability that the secret key is (s'_1, s'_2). Denote by E the event that there exists a related key (s'_1, s'_2) and by F the event that the secret key is (s'_1, s'_2). Then the probability that $z_1 - cs_1 - z'_1 + c's_1 \neq 0$ and $z_2 - cs_2 - z'_2 + c's_2 \neq 0$ is

$$\Pr[E \cap F] = \Pr[E] + \Pr[F] - \Pr[E \cup F]$$

$$\geq 1 - 2^{-200} + \frac{1}{2} - 1 = \frac{1}{2} - 2^{-200}.$$

Therefore, the probability that $z_1 - cs_1 - z'_1 + c's_1 \neq 0$ and $z_2 - cs_2 - z'_2 + c's_2 \neq 0$ is at least $\frac{1}{2} - 2^{-200}$. Moreover, since $\mathcal{A}$ does not use the secret keys as input and does not use those secret keys for signing, the forger cannot know which secret key used in signing is (s_1, s_2) or (s'_1, s'_2).

Hence in case r_j is an answer for a query of $\mathcal{F}$ to the random oracle, we can solve the R-SIS$_{p,n,2,\beta}$ problem where $\beta = (2(k-32) + 64k')$ with probability at least

$$\left(\frac{1}{2} - 2^{-200}\right)\left(\delta - \frac{1}{|D_H|}\right)\left(\frac{\delta - 1/|D_H|}{h+s} - \frac{1}{|D_H|}\right).$$

Since $\beta = 2(k-32)$ in the first case is less than $\beta = (2(k-32) + 64k')$ in the later case and the success probability of solving the R-SIS$_{p,n,2,\beta}$ problem in the first case is greater than the success probability of solving the R-SIS$_{p,n,2,\beta}$ problem in the second case, the success probability will be the small one. Therefore, if there exists a polynomial-time forger $\mathcal{F}$ with success probability δ after making s queries to the signing algorithm Hybrid 3 and h queries to the random oracle H, then there exists a polynomial-time algorithm to solve the R-SIS$_{p,n,2,\beta}$ problem, where $\beta = (2(k-32) + 64k')$, with success probability at least

$$\left(\frac{1}{2} - 2^{-200}\right)(\delta - 2^{-200})\left(\frac{\delta - 2^{-200}}{h+s} - 2^{-200}\right).$$

Since $|D_H| = 2^{32}\binom{n}{32}$. For $n = 512$, we have $|D_H| \approx 2^{200}$ and for $n = 1024$, we have $|D_H| \approx 2^{233}$. □

5   Conclusion

In this paper, we present a full security proof for the GLP signature scheme. Concretely, we show that the GLP signature scheme is EU-CMA secure in the random oracle model, assuming the hardness of the DCK$_{p,n}$ problem and the R-SIS$_{q,n,2,\beta}$ problem. Based on the statements and arguments in [Lyu12, Lyu08], we gave the Lemmas 2, 4, Lemmas 5, 6 and Theorem 1. The optimal version of the GLP digital signature scheme can be proved in the same way as this paper.

Acknowledgements. The authors are grateful to Duong Hoang Dung and Trieu Quang Phong for helpful comments and discussions on drafts of this paper.

References

[BG14]   Bai, S., Galbraith, S.D.: An improved compression technique for signatures based on learning with errors. In: Benaloh, J. (ed.) CT-RSA 2014. LNCS, vol. 8366, pp. 28–47. Springer, Cham (2014). https://doi.org/10.1007/978-3-319-04852-9_2

[BN06]   Bellare, M., Neven, G.: New multi-signatures and a general forking lemma. Full version of this paper (2006). http://www.cs.ucsd.edu/users/mihir

[DDLL13] Ducas, L., Durmus, A., Lepoint, T., Lyubashevsky, V.: Lattice signatures and bimodal gaussians. In: Canetti, R., Garay, J.A. (eds.) CRYPTO 2013. LNCS, vol. 8042, pp. 40–56. Springer, Heidelberg (2013). https://doi.org/10.1007/978-3-642-40041-4_3

[DLL+17] Ducas, L., Lepoint, T., Lyubashevsky, V., Schwabe, P., Seiler, G., Stehlé, D.: Crystals-dilithium: digital signatures from module lattices. Technical report, Cryptology ePrint Archive, Report 2017/633 (2017)

[GLP12] Güneysu, T., Lyubashevsky, V., Pöppelmann, T.: Practical lattice-based cryptography: a signature scheme for embedded systems. In: Prouff, E., Schaumont, P. (eds.) CHES 2012. LNCS, vol. 7428, pp. 530–547. Springer, Heidelberg (2012). https://doi.org/10.1007/978-3-642-33027-8_31

[GLP15] Güneysu, T., Lyubashevsky, V., Pöppelmann, T.: Lattice-based signatures: optimization and implementation on reconfigurable hardware. IEEE Trans. Comput. **64**(7), 1954–1967 (2015)

[LM06]   Lyubashevsky, V., Micciancio, D.: Generalized compact Knapsacks are collision resistant. In: Bugliesi, M., Preneel, B., Sassone, V., Wegener, I. (eds.) ICALP 2006. LNCS, vol. 4052, pp. 144–155. Springer, Heidelberg (2006). https://doi.org/10.1007/11787006_13

[Lyu08] Lyubashevsky, V.: Towards Practical Lattice-Based Cryptography. University of California, San Diego (2008)

[Lyu12] Lyubashevsky, V.: Lattice signatures without trapdoors. In: Pointcheval, D., Johansson, T. (eds.) EUROCRYPT 2012. LNCS, vol. 7237, pp. 738–755. Springer, Heidelberg (2012). https://doi.org/10.1007/978-3-642-29011-4_43

[Sho99] Shor, P.W.: Polynomial-time algorithms for prime factorization and discrete logarithms on a quantum computer. SIAM Rev. **41**(2), 303–332 (1999)

Toward a Trust-Based Authentication Framework of Northbound Interface in Software Defined Networking

Phan The Duy[✉], Do Thi Thu Hien, Nguyen Van Vuong,
Nguyen Ngoc Hai Au, and Van-Hau Pham

Information Security Laboratory, University of Information Technology,
VNU-HCM, Ho Chi Minh City, Vietnam
{duypt,hiendtt,haupv}@uit.edu.vn,
{14521108,14520041}@gm.uit.edu.vn

Abstract. Software Defined Networking (SDN) – a new rising terminology of network is recently gained more and more interest in both academic and industrial field. Not only decoupling of its control plane and data plane, SDN also provides the whole view of entire network for better and more flexible network management. Despite the benefits of the global view of the whole network, SDN with a single point of failure at the controller encounters some drawbacks and additional challenge for security. A malicious OpenFlow application (OF app) can access to SDN controller to perform illegal activities due to the lack of the authentication protocol in Northbound interface to ensure that only trusted, and authorized applications access critical network resources. The information about the whole network, such as topology data, flow information or statistics can be retrieved. Even worse the entire network can be controlled from the compromised controller. In this paper, we introduce Trust Trident - a framework of securing trustworthy authentication between applications and controller, with the controller-independent capability. It gives network administrator a fully and fine-grained observation of OF apps communicating with the controller. Threats in Northbound interface and counter measurements by our plugin are classified and evaluated according to the threat categories from the STRIDE methodology.

Keywords: Northbound interface · Trust authentication · SDN

1 Introduction

Conventional network architecture has remained mostly unchanged over the last decades from its launching stage and proved to be difficult to monitor in large scale size network. In this circumstance, SDN has recently emerged to become one of the promising technologies for the 5G era and the future Internet. The architecture of SDN, shown in Fig. 1, separates the network control and forwarding functions from network device, which allows the network to be programmed by the application and monitored from a centralized controller. However, like any other centralized architecture, the advent of controller also brings up issues of security and availability for the network.

T. Q. Duong et al. (Eds.): INISCOM 2019, LNICST 293, pp. 269–282, 2019.
https://doi.org/10.1007/978-3-030-30149-1_22

According to Kreutz et al. [1], SDN itself may be a target of some threat attack vectors. Controller, controller-data interface and controller-application interface are identified as three of critical positions which can be easy to exploit. In the survey of SDN security [2], Scott-Hayward et al. described a categorization of the potential attacks to which the architecture is vulnerable. It is composed of unauthorized access, data leakage, data modification, malicious/compromised applications, denial of service, misconfiguration issues, and system level SDN security. To provide protection in SDN architecture and implementation, it is critical to entail the three basic properties of security concept such as confidentiality, integrity and availability of information. This can be achieved by applying authorization, authentication and encryption for a secure operation of the network.

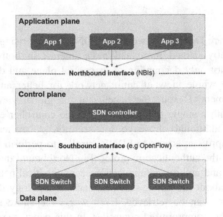

Fig. 1. Architecture of SDN

Regarding to controller position, when the controller is compromised, the information about the entire network, like topology data, flow information, incoming connections or statistics can be disclosed to unintended parties. Even in a bad situation, a whole network can be manipulated in an unauthorized way from the compromised one. In addition, whereas the control-data interface also known as Southbound interface (SBI) is dominant with OpenFlow standard, there is no standard at Northbound interface (NBI) for OF apps locating in the application plane connecting to controller. Many controllers choose RESTful as an implementation for NBI to decouple OF app (which can be run on a different host) and controller. These applications can request information from the controller and external sources, then send instructions to the controller in order to control the network, such as adding or removing flow rules. Due to the lack of standardization, the NBI is less explored than the SBI which has many suggestions for prevention and mitigation of its own vulnerabilities.

When it comes to NBI's vulnerabilities, they all stem from malicious or compromised OF applications. Malicious OF apps can cause data leakage if no control policy is applied in NBI. Whether the application is malicious from the time of its installation, or compromised by an attacker at a future date, has the same effect in these situations.

Once such an application has access to the NBI, the attacker could exploit vulnerabilities of lack of authentication and fine-grained control, then seize full control over
the controller to disrupt service or listen in on communication. It can be feasible
through hijacking, man-in-the-middle attacks in the communication between controller
and network applications where attackers can use malicious apps to compromise
controller for illegal actions, according to research of SDN security challenges and
countermeasures [3].

Moreover, in the AIM-SDN study [4], Dixit et al. proposed a threat model aiming
to identify several vulnerabilities of SDN which heavily relies on its datastores to
program and control the network. They conducted several attacks which compromise
availability, integrity, and confidentiality of the network by exploiting potential vulnerabilities. In their scope, these attacks stem from a semantic gap problem between
different abstractions as part of the SDN network and the datastore design implemented
in SDN controllers. Northbound interface API is also described as vital element to be
easily vulnerable to many specific attacks on involved SDN entities according to CIA
triad of security concept (confidentiality, integrity and availability) for controller. For
example, the consistency and accuracy of the information that is stored in the datastores
and passed to the network can be manipulated using Northbound or Southbound
channels with SDN controller. If the configuration is installed in the network from NBI
at the time not primarily intended by the administrator, unauthorized and undesired
traffic may be allowed in the network. In addition, unchecked storage and improper
management of the stored information could lead to memory overflows and impact the
controllers' availability. With their experiments, they realize that an OF app with
RESTful privileges could install configuration (flow rules) in the SDN controllers
without any responsibility of ensuring validity of rules. There is no limit or threshold of
flow rules that an OF app can install, also controllers will always accept a new flow
configuration.

Meanwhile, DELTA framework, a SDN-focused security assessment tool for
security flaws testing, introduced by Lee et al. [5], having capability of reinstantiating
published SDN attacks in diverse test environments. It successfully reproduced 20
known attack scenarios across diverse SDN controller environments and discovered
seven unknown SDN application mislead attacks. There are many attacks related to
NBI such as Service-Unregistration Attack which enables applications can freely
register to parse control messages arriving from the switch. The malicious or compromised applications can dynamically change the services of other applications
without constraint, and potentially with malicious intent. Additionally, legitimate
applications could be no longer in an ACTIVE state after undergoing Application
Eviction Attacks.

As mentioned above, those situations related to vulnerabilities in NBI for communication of OF apps and controller are proven that it is crucial to build an appropriate OF app monitoring mechanism with fine-grained control policy to secure SDN
network. Hence, this paper proposes a framework of trust authentication between OF
apps and controller to prevent malicious OF app from accessing network resource by
fine-grained control policy. By certifying whether OF apps are trusted in their intended
activities, our approach can secure a controller from losing of control, information
disclosure and a resource exhaustion attack originating from malicious OF app through

NBI. Our framework can operate independently with controller since its plugin architecture and OF app isolation from critical entities in SDN without any significant impact on transparency and functions of application. Trustworthiness of network application is also observed and regulated for applying an appropriate treatment policy into OF app sending commands to controller via NBI. Dissecting the methods of exploiting vulnerabilities in OF apps is out of scope of this research, but the consequences for SDN controller and counter measures are relevant in the coverage of NBI. A mechanism of flow rule conflict detection and prevention is not also discussed here due to our focus on trustworthiness of OF apps via its privileges and API calls.

The remainder of this paper is structured as follows. Section 2 gives an overview of Northbound interface's security issues and the related works. The proposed mechanism of our approach is introduced in Sect. 3. In Sect. 4, the experiments are conducted to analyze the proper working and efficiency of our model. Finally, we draw conclusions and propose some future work in Sect. 5.

2 Background and Related Work

2.1 SDN NBI Security and Trust Management

NBI is a component located in application-control interface, allows OF app interact with controller - the brain of SDN to request and send commands via NBI's APIs. If an application is insecure, or has vulnerabilities, a malicious actor uses such an otherwise legitimate OF app to send harmful instructions to the controller in many ways.

To illustrate security level, threat modeling is used as a procedure for evaluating and optimizing application, network, system by identifying objectives, vulnerabilities and then making counter measures to prevent or mitigate the impacts of threats to the system. According to research [6], there are diverse methods to assess security of a system, such as PASTA, STRIDE, Trike, UMLSec, CORAS,... However, STRIDE model proposed by Microsoft, which is standing for six categories of threats, is considered as the most appropriate methodology for classifying security issues without implementation. Therefore, STRIDE is selected for indicating vulnerabilities in SDN NBI in this research. Table 1 shows the threats categories of STRIDE methodology.

Regarding to Spoofing category, credentials of authentication can be guessed or disclosed by listening in and capturing; then rogue actor can get permission to perform any actions in the controller. Moreover, due to the lack of authentication mechanism in NBI that mandates the OF apps running commands in the network, some harmful or rogue OF apps can take advantage of this privilege and disrupt network operation thereby violating the confidentiality, integrity and availability of the network. Besides, data is transferred between controller and OF app can be altered and tampered with owing to having no encryption technique, according to Tampering vector. Without cryptography-based data transmission, network state information can be disclosed to unintended parties, according to Information Disclosure. In Repudiation category, it is easy for a malicious actor to perform operations anonymously if there is no secure logging policy for sending instructions to controller over NBI. In addition, in the context of Denial of Service, NBI could be unavailable for legitimate OF apps to properly work if compromised ones torrentially dispatch large amount of traffic or resource-intensive requests to the controller. Finally, relating to Elevation of Privileges type, OF apps should only have the least amount of privileges required for their operations. An OF app with rich-permission can perform actions with serious impacts to the whole network. So, the method of authorization plays an important role in fine-grained control of applications over NBI.

Table 1. Threats categories of STRIDE methodology

Threat category	Security object	Description of reversion
Spoofing	Authentication	Lack of identification in a distinctive way
Tampering	Integrity	Susceptibility to be altered, modified, tampered in whole or in part
Repudiation	Non-repudiation	Not tracking its actions, its events and their role actors
Information disclosure	Confidentiality	Revealing, disclosing its features and communications
Denial of Service	Availability	Resources are partially or totally inaccessible
Elevation of privileges	Authorization	Susceptibility to be accessed by any element without restriction

Additionally, every OF application should be monitored for each behavior with a trust value showing that it is trusted based on its functions and pre-defined permission to communicate with the controller in a trustworthy manner. Thus, attributes of OF apps is required for real-time tracking and adjusting the trust of OF apps working through NBI. All of their historical actions are always under observation and permanently kept for further report about characteristics of behaviors in the network over time. Access control and reliability can be achieved by checking whether an application is trusted or not, in order to enable OF app request, command and consume network resources in the controller.

2.2 Related Work

Despite the gaps in security, though, SDN continues to be an emerging alternative solution to the problems of modern networks. Therefore, the urgency of enhancing security in SDN and protection for the controller are more and more concerned. In this context, several solutions of a secured Northbound interface used for the communication between applications and SDN controller have been proposed. Our work is inspired by prior work in SDN security and vulnerability analysis techniques, particularly regarding NBI security as the follows.

Firstly, a comprehensive analysis of the 22 separate vectors for potential abuse or direct attack that arise from diverse location in SDN is introduced by Yoon et al. [7]. With this research, unauthorized network-view manipulation (Internal Data Storage Modification), unauthorized application management through controller are indicated as some of the NBI vulnerabilities of SDN.

Aliyu et al. [8] proposed a trust management framework for the access of OF apps to the controller. Each OF app's permission is defined in the terminology of four tasks *Read, Write, Notify* and *SystemCall* and this framework takes the responsibility of ensuring no OF app can encroach its granted permission. Moreover, the trust established between applications and controller are also periodically monitored and updated based on their operations.

In [9], Porras et al. developed a kernel solution called FortNOX to overcome the problem of malicious OF app's intrusion to the controller and unauthorized flow rule installation. SE-Floodlight [10] is another solution extended from FortNOX. Both of them categorize applications into three groups and apply authentication based on these roles. However, with the responsibility of processing a large volume of requests, an additional feature of access control and conflict verification can lead the controller to be overloaded and suffered an increasement in response time.

Consider trust as an important factor, Isong et al. [11] introduced a model aiming to protect the controller by requiring a specific OF app to meet at least a trust level to communicate with the controller and consume network resources. This approach creates a trust matrix of applications corresponding to their identity used for effective resources management in SDN.

Besides, ControllerSEPA [12] utilizes a repacking service to manage the transferred data from the controller to OF app, which prevents controllers from being affected by malicious applications. It also provides management services of authentication, authorization, accounting (AAA) based on the features and granted permission of a specific OF app, while these applications communicate with controller via a TLS-enabled northbound interface. Developed as a plug-in model, its operations take place outside of the controller, without significant effect on the overall performance of the SDN control plane will occur.

In comparison with the aforementioned studies, our framework has a proactive approach like a plugin model in ControllerSEPA's idea to enable authentication and access control tasks is performed independently with controller to meet SDN network brain's performance. In addition, trust degree of OF app is also paid attention to monitor for manipulation and adjustment its actions according to the matching between actual behaviors and intended function description with granted permission of network application in SDN. In our trust management principle, a trustworthiness in trust relationship is represented by a trust value which is collected, stored or updated in our framework for further analysis at later time. Then, controller can certify how much trusted an application is to be eligible to execute different actions related to network resource consumption. Moreover, a treatment policy for OF app with assigned trust level is designed to give a relevant counteraction for network administrator, such as blocking or limiting the involvement of OF app with controller through NBI in case of exceeding declaration of actions.

3 Trust Trident – A Trust-Based Authentication Framework

This section presents our proposed framework for trust-based authentication, named Trust Trident, which plays a role as a plug-in intercepting communication between OF apps in application plane and the SDN controller, in order to ensure the security of this centralized component. The name Trust Trident firstly indicates the capability of this framework which provides three services of authentication, authorization, and accounting for security. The architecture of this framework is given in Fig. 2, consisting of four main modules: Authentication module, Authorization module, Accounting module and Repacking service. Additionally, TrustStore and database are used to enable our framework to store trusted certificates and privileges of OF apps when interacting with controller.

3.1 OF App in Context of Trust Trident

In the scope of our research, considered OF apps are applications which communicate with controller for requesting information via its REST API. The plug-in keeps in its knowledge some identity information of a specific application to be able to distinguish one to one.

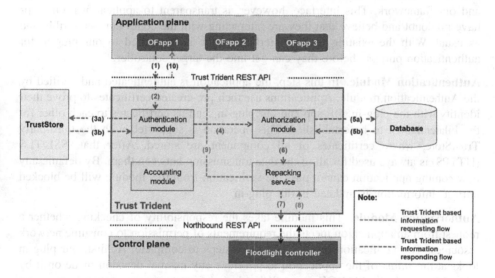

Fig. 2. Trust Trident framework

Certificate. Each application joining in Trust Trident-enabled SDN network must have a certificate for authentication purpose. These certs are pre-created when an application registers itself to the plug-in to gain permission for later requests.

Trust_value. This value represents the measure of trust of our plug-in on a specific application, in the range of 0 to 1000. In our context, the higher *trust_value* is, the more reliable the application is.

Status. An OF app can have its status in *enabled or disabled.* Besides depending on *trust_value* of application, the plug-in can also considers allowing or denying applications from accessing SDN controller based on the current status of it. The administrator can use this property to manually block a potential malicious OF app due to its suspected behaviors.

Permission on a Resource. Inherited from [8], our plug-in defines privileges of an OF app for consuming specific resources in terms of four groups: *READ* for information queries, *WRITE* for modifying or configuring on the network, *NOTIFY* for getting informed about events, *SYSTEM CALL* for requests on system resources.

3.2 Requesting via Trust-Based Authentication Framework

To illustrate, our proposed framework will be dissected in the context of processing a request sent from an OF app to query information from the SDN controller.

Trust Trident REST API. This component, which has similar features like the common Northbound REST API, plays as a communication interface between OF app and our framework. This interface, however, is transparent to applications, OF apps have no doubt and believe that they are interacting with the controller via northbound as usual. With the existing of this interface, traffic is forwarded to our plug-in for authentication purpose before they can get into the target controller.

Authentication Module. In this step, the application is authenticated and verified by the Authentication module. Applications use their pre-created certificates to prove their identity with the plug-in. OF app and our plug-in send its certificates to each other for the bilateral trust to be established. This operation is supported with the third-party TrustStore, where certificates of all component are stored. After that, SSL/TLS (HTTPS) is always used for all of the data transmission between them. By default, any new coming application cannot pass the verification from this module will be blocked until its information is updated in our plug-in.

Authorization Module. This module takes the responsibility of checking whether a request from an application meets the requirements of permission to consume network resources. The verification process takes two steps to complete. At first, our plug-in looks at the *status* of the application to prioritize any manual decision made on it by administrator. A disabled OF app will be denied from accessing network resources regardless of their privileges. In case of active applications, this module makes a mapping of requested information or operation to their corresponding required permissions based on the data stored in the database, then verify if the application has properly granted one to make this request.

This module also keeps an eye on the trust_value of the application in the view of our plug-in. It re-evaluates this value based on the validation of received requests. The regular request will remain the trust in the high value, but over-privilege ones can cause a drop in trust_value as a punishment, which is 5 points per unprivileged request. Until a pre-defined threshold is met, this module warns the administrator about applications that need more attention.

Repacking Service. Upon the authentication and permission verification process, legal requests need to be sent to controller unchanged to be transparent to OF apps when receiving response data. Our plug-in has this duty performed by Repacking service. Being a traffic forwarder, this component operates like an application in the application plane sending requests to the controller via its Northbound interface, which is REST API in this case. Controller sends a response containing network resources data to applications through Repacking service then.

Accounting Module. Accounting module keeps record of every request sending to the controller, including the application name, the permission of OF app and the *trust_-value* of it at the time of that event. This information can be useful in the system audit trail process. Our framework also provides a web interface for administrator to have an overview of underlying operations and receive warning messages of suspicious activities.

4 Experiment and Evaluation

4.1 Network Emulator

To design an experiment to evaluate our proposed mechanism, Mininet [13] is used to demonstrate a simple SDN network managed by a controller. We create a linear-type network with two switches and two hosts connect to each of them. Mininet is run on a Linux virtual machine with Ubuntu 14.04 LTS.

As mentioned above, Floodlight [14] is our choice of SDN controller to develop our solution as its plug-in. Trust Trident is operated on the same machine as Floodlight controller in order to simplify the interception of communication traffic between this core component and OF apps.

Our tested OF apps locate in another machine of Ubuntu, then they connect to controller to request for specific information or features on the managed SDN network.

4.2 Permission and Trust Configuration

For experiment purpose, we pre-define some configurations of mappings between API requests and permissions as well as the trust policy used in our plug-in.

Request and Permission Mappings. In the scope of this paper, we demonstrate and evaluate our proposed plug-in in the case of analyzing some common API requests used by many OF apps. These APIs are categorized into four above-mentioned permissions based on their requested information or operation on the SDN network.

Trust Policy. These configurations are applied based on the *trust_value* of a specific OF app to protect the control plane of the SDN network.

Deactivate Policy. This is the policy for applications with the *trust_value* less than *deactivate trust_value*, which is 50 in our case. Matched applications will be changed to disabled status and no request can be made to the controller during this time.

Low Trust Policy. Applications whose *trust_value* are lower than 75 – the *warning trust_value* are targets of this policy. Legitimate requests from these OF apps are still processed by the controller without any interruptions, but some warning messages will be notified to the network administrator to have more attention on the behaviors of these applications.

High Trust Policy. Applications with decent behaviors will be presented in the high value of *trust_value*, which is classified to the best condition of a given application in the communication with SDN controller. These applications are trusted by plug-in and their operations are considered as harmless to the SDN network.

Default Policy. The policy is applied when no other policy matching its current status.

4.3 Our Scenarios

Four scenarios of OF apps with different pre-granted permissions and requested operations on SDN network will be tested to observe the proper behavior of our plug-in. Our aim is to verify whether the suggested plug-in can catch and analyze these requests for the authentication purpose to prevent controller from communicating with malicious applications. STRIDE methodology is used to evaluate the security object of NBI after running Trust Trident. The information of tested applications is given in Table 2.

Table 2. OF applications used for evaluation

Application name	Granted permission	Requested information/operation	Required permission
Information of SDN	No permission	SDN controller version checking	READ
		SDN topology information	READ
		SDN uptime checking	NOTIFY
SDN Firewall management	READ, WRITE	SDN Firewall status checking	READ
		Firewall rule listing	READ
		Firewall rule configuring	WRITE
SDN Firewall management version 2.0	READ, WRITE	SDN Firewall status checking	READ
		Firewall rule listing	READ
		Firewall rule configuring	WRITE
		SDN uptime checking	NOTIFY

Scenario 1. Authentication of New OF App. A new OF app – called *Information of SDN*, makes its first connection to the SDN network and requests the following information: the version of Floodlight running on the SDN controller, the uptime of SDN and its topology information. This application communicates with the controller in the first time, therefore it has not yet authenticated with the controller as well as our plug-in and no specific permission is assigned to it.

Scenario 2. A Well-Behaved OF App. This is an example of an authenticated application named *SDN Firewall management* with given privileges for specific requests of Firewall status checking, Firewall rule listing and configuring on SDN controller. This application has relevant permissions allowing it to perform these requests without any warnings or impediments from our plug-in. All information transmitted between our plugin and OF app are protected from tampering (targeting on the integrity of data) and information disclosure (targeting on the confidentiality of data) by HTTPS connection.

Scenario 3. Over-Privileged Requests. In case of upgraded applications, some of the new features are added into the current version of applications. *SDN Firewall management* application version 2.0 with some updates compared to the version at scenario 2 can be an example, which introduces a new capability of system uptime monitoring. However, the authentication and authorization of the plug-in have not yet updated this fresh feature and no permission related to it is granted to application.

Scenario 4. Manual Punishment for Consecutive Bad Behaviors. In this test, we observe the decision made by our plug-in in the case of applications trying to request unallowed information or operations for many consecutive times, which alerts the administrator many times about this strange behavior. *SDN Firewall management* is still used as an example in this scenario.

4.4 Evaluation Results

The overall result of our tested scenario can be viewed in Table 3, along with the information about the requests from applications and the corresponding privileges needed for them to be allowed. Comparing these permissions to the granted one of application, we expect some proper operations of our plug-in, also keep an eye on its actual handling tasks to make an evaluation of this proposed solution.

Scenario 1. Authentication of New OF App. Due to the lack of authentication information of this new applications in the knowledge of the plug-in, our solution is not able to identify if this application is a safe one to communicate with controller. Thus, all of its requested information and operation are denied and returned errors as responses. When an OF app desires to use NBI in order to access network resources, it must register with our plugin by declaring which API it wants to access. Subsequently, administrator could consider whether an OF app is granted specific permissions with default *trust_value* or not.

Table 3. Evaluation results

Scenario	Granted permission	Request	Requested permission	Expected operation	Confirmed behavior
(1) Authentication of new OF app	No permission	Controller version checking	READ	Failed to authentication and communicate with controller	Failed to communicate with controller
		Topology information	READ		
		Uptime checking	NOTIFY		
(2) A well-behaved OF app	READ, WRITE	SDN Firewall status checking	READ	All requests are allowed and forwarded to controller	All requests are allowed and forwarded to controller, *trust_value* remains unchanged
		Firewall rule listing	READ		
		Firewall rule configuring	WRITE		
(3) Over-privileged requests	READ, WRITE	SDN Firewall status checking	READ	Requests are allowed and forwarded to controller	Requests are allowed and forwarded to controller
		Firewall rule listing	READ		
		Firewall rule configuring	WRITE		
		SDN uptime checking	NOTIFY	Denied due to over-privilege	Response with error message, *trust_value* is dropped
(4) Manual punishment for consecutive bad behaviors	READ, WRITE	SDN Firewall status checking	READ	Denied due to disabled status of a bad-behavior application	Response with error message, *trust_value* is increased to re-gain the allowed status
		Firewall rule listing	READ		
		Firewall rule configuring	WRITE		

Scenario 2. A Well-Behaved OF App. Requests from example application are legitimate based on its granted permission. As a result, the SDN controller and then our plug-in response it with the corresponding information of Firewall status or the list of Firewall rules, through HTTPS connection. In this context, the integrity and confidentiality are ensured for both parties during bilateral communication. The *trust_value* of this OF app is remained with the current value.

Scenario 3. Over-Privileged Requests. According to the saved authentication information in the plug-in, features of firewall management have been allowed for this application, therefore the result is returned to it without any error. However, the new capability of requesting system uptime is denied, at the same time the *trust_value* of this application is decreased as a punishment for over-privileged behaviors. In the case of the dropped *trust_value* matched the *Deactivate policy*, this application is disabled for further requests.

Scenario 4. Manual Punishment for Consecutive Bad Behaviors. The tested application has sent multiple over-privileged requests, which results in the drop of *trust_value* to 65, therefore administrator has been notified with the number of about 10 warning messages and decides to disallow this application. The punished application, still with a reasonable *trust_value*, is in disabled status so it cannot request even allowed information or operations. However, these abused activities are still recorded in the plugin while the trust of this application can be re-gained manually by the administrator after receiving enough number of legitimate requests in the blocking time.

With mentioned experiments, our framework can secure NBI between controller and OF apps, compliant with STRIDE methodology. Table 4 summarizes the capability of Trust Trident in assuring security properties against different threat categories.

Table 4. Trust Trident evaluation for security assurance in NBI with STRIDE

Threat category	Security object	Solution	Scenario
Spoofing	Authentication	Certificate	1, 2
Tampering	Integrity	HTTPS	2
Repudiation	Non-repudiation	Logging, Accounting	4
Information disclosure	Confidentiality	HTTPS	2
Denial of service	Availability	Out of scope	Out of scope
Elevation of privileges	Authorization	Fine-grained control	3, 4

5 Conclusion

In this paper, we propose a trust-based authentication framework in the NBI, which manages the secure communication between OF apps and SDN controller, to protect and prevent controller - the vital component from being compromised and taken advantages to control the whole SDN network. Our plug-in approach provides the capability of intercepting requests from applications to further analyzing without any interruption or significant effects on operations of the controller. It is attributed to a trust management and fine-grained control activity using privileges. The mechanism of trust aims at giving applications the required privileges to perform in the network and not to exceed declared limit of operation. The experiment results show that this solution is feasible to be used as an effective protection method for controller in many scenarios. In the future, auto-detecting and preventing malicious OF app without the administrator as a supervisor, also integrating mechanism of flow rule verification into this framework can be a potential research problem in enhancing the security of the NBI as well as SDN network.

Acknowledgement. This work is funded by University of Information Technology, VNU-HCM under grant number of D1-2019-09.

References

1. Kreutz, D., Ramos, F.M.V., Veríssimo, P.E., Rothenberg, C.E., Azodolmolky, S., Uhlig, S.: Software-defined networking: a comprehensive survey. In: Proceedings of the IEEE (2014)
2. Scott-Hayward, S., Natarajan, S., Sezer, S.: A survey of security in software defined networks. IEEE Commun. Surv. Tutor. **18**(1), 623–654 (2015)
3. Li, W., Meng, W., Kwok, L.F.: A survey on OpenFlow-based software defined networks: security challenges and countermeasures. J. Netw. Comput. Appl. **68**, 126–139 (2016)
4. Dixit, V.H., Doupé, A., Shoshitaishvili, Y., Zhao, Z., Ahn, G.-J.: AIM-SDN: attacking information mismanagement in SDN-datastores. In: Proceedings of the 2018 ACM SIGSAC Conference on Computer and Communications Security (CCS 2018), pp. 664–676. ACM, New York
5. Lee, S., Yoon, C., Lee, C., Shin, S., Yegneswaran, V., Porras, P.: DELTA: a security assessment framework for software-defined networks. In: Network & Distributed System Security Symposium (2017)
6. Chikhale, A., Khondoker, R.: Security analysis of SDN cloud applications. In: Khondoker, R. (ed.) SDN and NFV Security. LNNS, vol. 30, pp. 19–38. Springer, Cham (2018). https://doi.org/10.1007/978-3-319-71761-6_2
7. Yoon, C., et al.: Flow wars: systemizing the attack surface and defenses in software-defined networks. IEEE/ACM Trans. Netw. **25**(6), 3514–3530 (2017)
8. Aliyu, L., Bull, P., Abdallah, A.: A trust management framework for network applications within an SDN environment. In: Proceedings of 31st International Conference on Advanced Information Networking and Applications Workshops (WAINA), Taipei, Taiwan (2017)
9. Porras, P., Shin, S., Yegneswaran, V., Fong, M., Tyson, M., Gu, G.: A security enforcement kernel for OpenFlow networks. In: Proceedings of the 1st Workshop on Hot Topics in Software Defined Networks, Helsinki, Finland (2012)
10. Cheung, S., Fong, M., Porras, P., Skinner, K., Yegneswaran, V.: Securing the software-defined network control layer. In: Proceedings of the 2015 Network and Distributed System Security Symposium (NDSS), San Diego, California (2015)
11. Isong, B., Kgogo, T., Lugayizi, F., Kankuzi, B.: Trust establishment framework between SDN controller and applications. In: Proceedings of 18th IEEE/ACIS International Conference on Software Engineering, Artificial Intelligence, Networking and Parallel/Distributed Computing (SNPD), Kanazawa, Japan (2017)
12. Tseng, Y., Zhang, Z., Naït-Abdesselam, F.: ControllerSEPA: a security-enhancing SDN controller plug-in for OpenFlow applications. In: Proceeding of 17th International Conference on Parallel and Distributed Computing, Applications and Technologies (PDCAT), Guangzhou, China (2016)
13. Mininet - An instant virtual network on your laptop (or other PC). http://mininet.org/
14. Floodight Controller - Project Floodlight. https://floodlight.atlassian.net/wiki/spaces/floodlightcontroller/pages/1343514/Tutorials

Detecting Signalling DoS Attacks on LTE Networks

Ginés Escudero-Andreu[1]([✉]) [iD], Konstantinos Kyriakopoulos[1,2] [iD],
James A. Flint[1] [iD], and Sangarapillai Lambotharan[1] [iD]

[1] Wolfson School of Engineering, Loughborough University,
Loughborough LE11 3TU, UK
{g.escudero-andreu,k.kyriakopoulos,j.a.flint,s.lambotharan}@lboro.ac.uk
[2] Institute for Digital Technologies, Loughborough University London,
London E15 2GZ, UK

Abstract. As mobile communications increase their presence in our life, service availability becomes a crucial player for the next generation of cellular networks. However, both 4G and 5G systems lack of full protection against Denial-of-Service (DoS) attacks, due to the need of designing radio-access protocols focused on providing seamless connectivity. This paper presents a new method to detect a DoS attack over the Radio Resource Control (RRC) layer, offering three original metrics to identify such attack in a live Intrusion Detection System (IDS). The proposed metrics evaluate the connection release rate, the average session establishment and the session success rate to identify the attack. The presented results provide an average detection rate above 96%, with an average false positive rate below 3.8%.

Keywords: LTE security · DoS attacks · RRC signaling attack · Dempster-Shafer

1 Introduction

The 4th Generation (4G) of mobile cellular networks has been designed to provide high-speed access to broadband mobile services even in the worse scenarios, such as high mobility scenarios, overcrowded cells or rural areas; without detriment to the *Quality-of-Service* (QoS). Additionally, 4G systems are expected to provide safe communications among a huge number of cellular users, which is growing constantly every year.

The major contribution of the 4G is the portability of the entire network architecture into a flat, all IP infrastructure where all the services are provided over IP networking and circuit switching is no longer used. This improvement facilitates easy mobility among different radio-access technologies, such as

This work has been supported by the Gulf Science, Innovation and Knowledge Economy Programme of the UK Government under UK-Gulf Institutional Link grant IL 279339985.

© ICST Institute for Computer Sciences, Social Informatics and Telecommunications Engineering 2019
Published by Springer Nature Switzerland AG 2019. All Rights Reserved
T. Q. Duong et al. (Eds.): INISCOM 2019, LNICST 293, pp. 283–301, 2019.
https://doi.org/10.1007/978-3-030-30149-1_23

Worldwide Inter-operability for Microwave Access (WiMAX), *Wireless Fidelity* (WiFi), *Global System for Mobile communications* (GSM) or *Universal Mobile Telecommunications System* (UMTS), besides making easy backward compatibility with previous technologies.

However, compatibility with heterogeneous access technologies, which are provided by multiple *Mobile Network Operators* (MNOs), produces an increase in the number of Radio Access Network (RAN) hangovers. The main consequence of this effect is the mandatory negotiation of strict security policies between the MNOs with the purpose of defining trust policies and authorize users' migration, as well as defining secure countermeasures against identified vulnerabilities.

Since the first release of the 3rd *Generation Partnership Project* (3GPP) for the Long Term Evolution (LTE) standard, several publications have pointed out important shortcomings with regards to the security capabilities of this technology [1–4]. Most of the weaknesses were identified in the RAN, which is the most vulnerable part of the entire system due to its wireless nature and attacks can be easily performed remotely without having physical access to the infrastructure (Fig. 1).

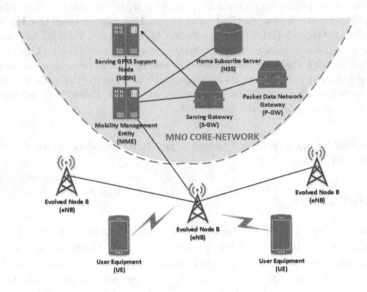

Fig. 1. LTE End-to-End network architecture

The security capabilities of LTE to protect the radio link are based in predefined secure domains with limited scopes and security contexts associated to each user, which are established during the user attachment procedure. The security context enables the encryption of the communications and is built based on a master key, K, previously shared between the mobile operator and the UE. Additional keys are dynamically computed on every session to guarantee their freshness, protecting the master key from being directly used.

The procedure in which the session keys are derived is called EPS-AKA [5], or Authentication and Key Agreement Protocol for *Evolved Packet System*, and it is triggered during the initial attachment of the mobile device. The EPS-AKA protocol plays an important role into the establishment of an initial security context for each user. At the same time, it exhibits most of the identified vulnerabilities of LTE, such as breach of privacy for the user's identity, weak mutual authentication between core-network elements or lack of perfect forward secrecy into the key hierarchy [6].

The AKA protocol has been redefined in recent specification drafts [7,8] for the 5th Generation (5G) of mobile communications, to enhance the privacy for the identity of the mobile subscriber. The *International Mobile Subscriber Identity* (IMSI) has been replaced by the *Subscriber Permanent Identifier* (SUPI), which should never be transferred through the radio channel.

Instead, an encrypted version of it is used as identifier, named as *Subscription Concealed Identifier* (SUCI). However, the 3GPP recognises several scenarios in which the privacy of the SUPI cannot be guaranteed [7]. Specifically, during the execution of any emergency services and/or whenever the Mobile Equipment (ME) is in use of the null-scheme, as the home network has not provisioned the Home Network Public Key in the *Universal Subscriber Identity Module* (USIM). These two exceptional cases make the next generation of cellular networks also vulnerable against the aforementioned DoS attack over the RRC layer.

This paper focuses in the detection of a particular vulnerability of the EPS-AKA protocol which allows the attacker to perform a DoS attack against the core network. Section 2 describes the vulnerability itself, how LTE networks are secured and a description on how to perform a local DoS attack over the *Radio Resource Control* (RRC) layer. Section 3 presents the detection methodology, defines the metrics proposed to detect the attack and explains the application of Dempster-Shafer theory [9] as a data fusion technique. Section 4 presents the experimental environment and Sect. 5 presents and evaluates the results. Finally, conclusions and future work are stated in Sect. 6.

2   Background Work

Looking at the work carried out by the research community, two main research areas are clearly defined: the identification and impact assessment of the studied signalling DoS attack, and the enhancements on the AKA mechanism.

The first publication acknowledging the studied signalling DoS attack was presented in [2], where the authors described the entire process of launching the attack. First, the UE is lured to transmit the IMSI value instead of using the temporary identity. This action allows the attacker to gather the required list of legitimate IMSIs, and finally perform the signalling attack as explained in Sect. 2.2.

Other signalling DoS and intelligent jamming attacks have been identified on LTE networks [10,11], exploiting the initial allocation of radio bearers to exhaust the resources within the radio cell and disrupt the service. The authors in [10] successfully managed to replicate the attack in an OPNET simulator, and provide a detection mechanism without evaluating its performance thoroughly. However, such type of attacks go beyond the scope of this work, which focuses only on DoS attacks above the physical layer.

The other area of research has focused on improving the existing AKA mechanism proposed by the 3GPP. The disclosure of the IMSI during the initial UE attachment to the RAN is the main vulnerability exploited for performing DoS attacks [12] in LTE networks.

The research community has proposed interesting solutions to deal with this shortcoming, proposing a list of amendments on the original AKA protocol. The author in [13] enhanced the AKA mechanism to provide mutual authentication between RAN and UE. The protocol includes a new concept of USIM card [13], called *Enhanced Subscriber Identity Module* (ESIM), which is capable of computing pseudo-random values. The new feature enables the generation of challenge requests inside the UE, which are used to confirm the identity of the serving network. However, this proposal adds additional computational load to the original radio access mechanisms, and imposes further challenges when maintaining compatibility with legacy systems.

The authors in [12] modified the security architecture to act as a wireless public key infrastructure and used digital certificates to confirm identity. The certificates have to be provided in advance to all involved entities in the authentication process, to be able to gain access to the radio system. This requirement makes more difficult the actual implementation on a real LTE deployment.

A more robust solution is introduced in [14,15] that combines passwords with fingerprints and public keys to provide full mutual authentication. Unfortunately, the high computational cost to execute the Diffie-Hellman key agreement and mutual authentication, and the requirement of storing biometric parameters, make its implementation less viable in a commercial deployment.

Due to the aforementioned inconveniences, the authors of this paper believe that further research is required to improve the existing LTE standard without actually modifying the specification documents. A detection mechanism is proposed to provide security against signalling DoS attacks while still transferring the IMSI in clear-text whenever the use of temporary identities is not possible.

2.1 Authentication and Key Agreement (AKA) Preliminaries

User identification is the first step before gaining access to the network, as shown in the blue section of Fig. 2. UE establishes contact with the nearest *evolved-Node B* (eNB) triggering the registration process. During the first attempt, there is no *Globally Unique Temporary Identity* (GUTI) available for the UE to be identified by the *Mobile Management Entity* (MME). Therefore, the MME sends a User Identity Request message.

A reply message is made by the *Mobile Equipment* (ME) with its IMSI transferred in clear text, because no security context has been established before. This action, does not comply with the user identity confidentiality requirements [16], which exposes the user identity to eavesdropping attacks over the radio interface.

Once the MME receives the User Identity Response, a GUTI is allocated and paired with the corresponding IMSI. The Temporary Mobile Subscriber Identity (TMSI) may change due to different reasons, being no necessary to transfer the IMSI unless serving network can not retrieve a new TMSI from the GUTI.

Before the establishment of a security context, mutual authentication between the ME and the UE is achieved using the EPS-AKA protocol [5]. The process is triggered by the MME, after the user is successfully identified. Figure 2 shows the sequence diagram.

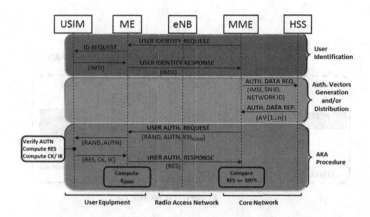

Fig. 2. Authentication sequence during EPS-AKA (Color figure online)

During the *Authentication Vector* (AV) generation phase, i.e. green section of Fig. 2, the MME checks the stored key material and its freshness. If there is any AV available, it will be used to start the authentication process. Otherwise, MME requests new AVs from the *Home Subscriber Server* (HSS). Whenever the HSS has no available AVs, 3GPP specifies [5] that multiple vectors should be computed and stored for future use, increasing the computational load in the core network. Each AV is composed according to the equation

$$AV := RAND \parallel AUTN \parallel XRES \parallel K_{ASME} \qquad (1)$$

where:

RAND is the challenge to prove the user authenticity.
AUTN is the parameter to prove freshness of authentication vector and serving network authenticity.
XRES is the expected response to the challenge.
K$_{ASME}$ is an identifier to derive the same key hierarchy in both end points.

The AKA process starts with a User Authentication Request message, composed of three parameters: RAND, AUTN and KSI_{ASME}, the Key Session Identifier used by the ME to generate the same key value for K_{ASME}.

Once the ME receives the message, it retrieves the KSI_{ASME} parameter and passes the other parameters to the USIM. The USIM verifies the freshness of the authentication vector, deriving the sequence number from the AUTN parameter. If the derived value matches with the expected sequence, a challenge response RES is computed and sent back to the UE. Then, two keys are derived from the master key K, one for integrity (IK) and another for confidentiality (CK).

A User Authentication Response is sent back to the MME, generating on it the same key pairs CK/IK and completing the AKA process. Now, both end points are able to generate the same key material following the scheme of Fig. 3.

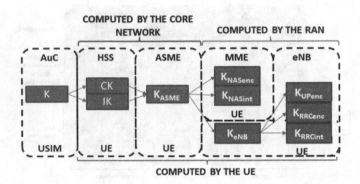

Fig. 3. Key hierarchy for E-UTRAN

Each time an AKA process is called, key material is re-generated based on the new value of K_{ASME}. Master key K is securely stored in the HSS and IMSI, without being transmitted or used directly. It is only used to derive the entire key hierarchy.

2.2 RRC Signalling Attack

The attack exploits a vulnerability in the specifications of the RRC layer [17], which is the third level of the *Open Systems Interconnection* (OSI) model [18], shown in Fig. 5. The attack was originally identified in [2], and subsequently acknowledged in several publications, such as [19–21].

In [2], the authors performed a simulation with the purpose of getting measurements to assess the impact of the attack inside the core network. The results conclude that there is no requirement of having high-computational hardware in order to perform this attack, since it can easily be launched by using a non-sophisticated equipment, such as a normal desktop PC. The attack is able to collapse the system in just 30 s by using a number of previously gathered legitimate IMSI values which are leveraged to send 500 service requests per second, following a Poisson distribution [2].

The aim of this attack is to consume available resources inside the core network by flooding the system with radio service requests which contain previously collected, legitimate IMSIs. Initially, the malicious user retrieves legitimate user identities or IMSI from the radio channel by luring the user to connect to a rogue MME and force the transmission of the IMSI with an *Identity Request* message. The approach to gather all IMSI is widely explained in [23].

Once the attacker has collected enough number of legitimate identities to perform the attack, it is ready to initiate it by sending *RRCConnectionRequest* spoofed messages including one of the eavesdropped IMSIs within each request. MME receives the service request and retrieves an authentication vector from the HSS. Because there is no means to identify the injected messages as illegitimate, the authentication process continues with both MME and HSS validating the user identity by checking the value of the IMSI.

MME sends a *RRCConnectionSetup* message and launches a timer while awaiting for a *RRCConnectionSetupComplete* message, which never arrives. Once the timer is expired, the authentication session is cancelled and all the occupied resources are released. HSS requires to consume hardware resources, such as RAM memory and CPU usage, in order to compute each authentication vector, as well as for storing the derived session keys until the authentication session is completed or fails. The attacker only has to simultaneously initiate a limited number of *RRCConnectionRequest* in order to exhaust the available resources of the HSS and collapse the cell service completely.

In conclusion, this attack has a twofold impact on the system performance. First, legitimate users are unable to connect to the network, since the available resources are depleted by the attacker. Secondly, by serving multiple spoofed service requests, the core network is collapsed. The collapse occurs due to a heavy computing load registered on the HSS to generate and distribute the authentication vectors (Fig. 4).

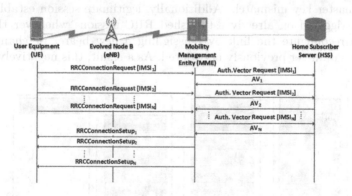

Fig. 4. Data-flow diagram of a DoS attack over the RRC layer.

3   Detection Methodology

3.1   Proposed Metrics

Three metrics are proposed in this work that have been specially designed to
expose the characteristics of the examined signalling DoS attack. To the best
of the authors' knowledge, this is the first proposed methodology for detecting
such attack and the main contribution of the work presented on this paper. All
the novel metrics used for detecting the signalling DoS attack are extracted from
the same layer, the RRC layer [16] and are presented below.

Connection Release Rate (CRR). Since the attacker will never be able
to successfully complete the authentication phase, the RRC connection will be
closed with a rrcConnectionRelease message sent by the base station. Because
the RRC connectivity remains active, this would never occur under normal circumstances, unless the node is experiencing handover. The CRR is given by

$$CRR(x) = \frac{\#CR_X}{\Delta T} \qquad (2)$$

where CR_x is the number of Connection Request messages on the current sliding
window, x, and ΔT is the duration of the window.

Average RRC Session Establishment (SMT). A UE is considered to have
established an RRC session once it is able to allocate a radio link to initiate
the RRC authentication phase. This metric evaluates the frequency of established RRC sessions within a certain period of time, x. During the attack, all
the attacking nodes will trigger a new RRC session establishment every time
the eNB rejects the previous request due to an incorrect challenge response or
due to a master key mismatch. Additionally, legitimate session establishments
might be triggered on already established RRC sessions whenever the MME
decides to reconfigure the link by reassigning a new bearer or changing any
additional parameter previously negotiated. As a result, this negatively impacts

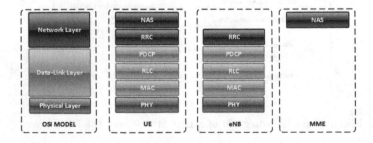

Fig. 5. Protocol stack for LTE control plane traffic [22] against OSI model [18]

the effectiveness of this metric to distinguish between rogue and legitimate node behaviour.

$$ASE(x) = \frac{\sum SE_X}{n^2} \tag{3}$$

where SE_x is the average Session Establishment duration for the sliding window x and n is the number of established sessions within the sliding window.

Session Success Rate (SR). This metric evaluates the number of successful RRC sessions established between each node and the serving eNB. Since the attacker nodes will not be able to complete any RRC session establishment, this metrics is expected to positively characterise the attack and reduce the uncertainty.

$$SSR(x) = \sum_{0}^{x} \frac{\#AA - \#AF}{\#CR} \tag{4}$$

where $\#AA$, $\#AF$, $\#CR$ are the number of Attach Accepted, Attach Failure and total number of Connection Request messages for the current window, x, respectively.

3.2 Dempster-Shafer Theory

The *Dempster-Shafer* (D-S) theory of evidence [9] is used to fuse the evidence collected from the proposed metrics. The D-S theory starts by defining a Frame of Discernment, which is a finite set of all possible, mutually exclusive propositions about a specific problem domain. In our case, a frame in the LTE network might either be normal or malicious (attack) and this is represented as: $\Theta = \{N, A\}$. Θ is also exhaustive, which means that one proposition from the set has to be true.

Given a Frame of Discernment, any hypothesis or proposition, A, is an element of the power set $P(\Theta)$ with $|P(\Theta)| = 2^\Theta$. Note that the power set contains all possible subsets of Θ, including the empty set ($\varnothing$) and the universal set, i.e. the Frame of Discernment, Θ, itself. Thus, the formula to represent the power set is as follows: $P(\Theta) = \{N, A, \{N, A\}, \varnothing\}$. In the case where a hypothesis has only one element, i.e. no subsets, it is referred to as a *singleton*. In our case, hypotheses A and N are singletons. However, $\{A, N\}$ is not singleton as it has A and N as subsets.

In Bayesian probability theory, all hypotheses are singletons. However, this is not necessary in the theory of evidence. As a consequence, in the case of assigning belief towards a non singleton hypothesis, $H \subseteq \Theta$, there is no explicit commitment of belief towards a subset of H, $A \subseteq H$. Thus, the theory of evidence gives the freedom to explicitly assign belief to uncertainty. For example, by assigning a belief value to $\{A, N\}$, there is no explicit information on whether A is more probable than N [24]. As a result, the additivity rule is no longer required. In other words, in the theory of evidence framework, the belief in a hypothesis and its complement can be less than one, i.e. $Bel(H) + Bel(\neg H) \leq 1$ [24].

Each proposition from the Frame of Discernment, $H \subseteq \Theta$, is assigned a belief value within the range $[0, 1]$ through a mass probability function, $m : 2^{\Theta} \rightarrow [0, 1]$.

The mass function, also known as basic probability assignment or basic belief assignment, follows the following three conditions [9]:

$$\sum_{H \subseteq \Theta} m(H) = 1 \tag{5}$$

$$m(H) \geq 0, \forall H \subseteq \Theta \tag{6}$$

$$m(\varnothing) = 0 \tag{7}$$

where:

$m(H)$ is the the amount of belief strictly supporting hypothesis H.

$m(\varnothing)$ is the probability of the empty set, which is equal to zero for a normalised mass function.

Once a mass value has been assigned for all the hypotheses, the rule of combination [9] defines the method to combine evidence from multiple observers about the same hypothesis. The fuse of evidences is only possible whenever all the observers are facing the same problem and set of hypotheses. In essence, this condition requires all the observers to share the same Frame of Discernment.

The rule of combination is applied to fuse the beliefs of two independent observers into a common belief, and it is defined by the following formula:

$$m_{comb}(H) = \frac{\sum_{X \cap Y = H} = m_1(X) * m_2(Y)}{1 - \sum_{X \cap Y = \varnothing} = m_1(X) * m_2(Y)} \forall H \neq \varnothing \tag{8}$$

where:

$m_{comb}(H)$ is the combined belief, by two independent observers, supporting the hypothesis H.

X, Y correspond to any supported hypothesis by observers 1 and 2 respectively.

$m_1(X)$ is the probability supporting hypothesis X as perceived by observer 1.

3.3   Detection Framework

All the necessary information required to detect the signalling DoS attack is extracted from captured traffic by constantly monitoring the network activity. Figure 6 presents the general view of the entire detection process, which is composed of four main phases.

The monitoring phase applies to the OSI model of the protocol stack and is represented in the left most side of Fig. 6. The selected LTE metrics, indicated

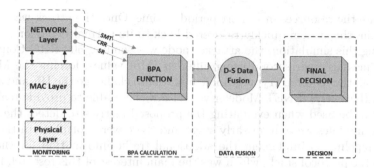

Fig. 6. General view of the detection process (Color figure online)

in red colour, are extracted from the Network layer. The output from this phase is a single buffer containing the individual samples of each monitored metric.

The *Basic Probability Assignment* (BPA) computation phase is responsible for calculating the beliefs for each of the hypotheses composing the power set in the D-S framework. To this end, using a sliding window of an experimentally selected [25], pre-defined size of 30 samples, the statistical BPA parameters are calculated during each time window, as described in [26]. These parameters, along with the actual metric samples extracted from each frame are introduced into the BPA function. The output of this phase is three distinct buffers containing a triplet set of BPA values, one for each hypothesis, per metric.

During the data fusion phase, the individual buffers obtained during the BPA computation phase are iteratively fused, using the D-S rule of combination (E.q. 8), treating each metric as an independent source of information and fusing them in pairs. The final output of this procedure is a triplet of beliefs for each frame, containing the beliefs for the three hypotheses.

Finally, in the decision phase, the resulting, per frame, fused beliefs for Attack, Normal and Uncertainty are received. The hypothesis with the highest probability is selected as the final decision. In the case where the Attack and Normal probabilities are equal, the decision of Normal is chosen. To the best of the authors' knowledge, the work presented on this paper is the first application of D-S theory for detecting DoS attacks on LTE networks.

4 Experimental Environment

4.1 Generating and Collecting the Data

This research project was initially focused on implementing the selected DoS attack within a simulated environment, aiming at verifying its feasibility beyond the theoretical approach, and evaluating its impact on the core-network components. Using OPNET Modeler Suite ver. 17.5, it was possible to confirm the side effects that running the *Radio Resource Control* (RRC) signaling attack could inflict on the *Home Subscriber Server* (HSS). Figure 7 shows the *Central Processing Unit* (CPU) utilisation registered in the HSS, where the attack was able to saturate the resources in a short period of time. Once the attack was stopped after 4 min, the system quickly recovered back to its normal behavior.

During this simulation, the attacker node was sending 500 RRC Connection Request messages per second using different IMSI values, forcing the MME to compute all the required cryptographic material to challenge the UE and authenticate it. However, OPNET Modeler ver. 17.5 was not able to provide real traffic captures to be used when evaluating the proposed metrics to detect the attack. The LTE modules were in an early stage and there were not plans in the development pipeline to implement the additional traffic information for the lower layers in the protocol stack, which was the main interest of this research project.

Due to the limitations with OPNET, a physical test-bed was designed using hardware-based emulating equipment, as discussed below, to run the required application-level services, and the core components composing a 4G deployment: Evolved Packet Core (EPC) entities, *Mobile Management Entity* (MME), *evolved-Node B* (eNB) and UE/s. The final architecture of the test-bed is displayed in Fig. 8 and includes the following components:

App Traffic Generator. The network traffic was managed in a Lenovo ThinkCentre M73 Tiny Desktop PC, equipped with an Intel Core i3-4130T Processor (2.9 GHz), 8 GB RAM and Windows 7 Pro 64 bits. This host was configured to run an FTP server with Microsoft IIS 6.0, allowing the UEs to perform file downloading via FTP to keep a continuous data flow throughout the duration of the emulation session.

eNode-B. The RAN infrastructure was emulated using an LTE Enterprise Femtocell board, manufactured by Mindspeed with model number M84300, including a Transcend T3310 chipset for implementing all the standard LTE modulation schemes. The femtocell station was configured to have 3 sectors, requiring only one sector to create a single cell for covering all the UEs required for conducting the experiments.

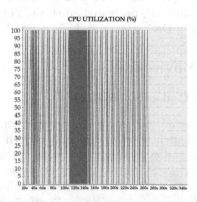

Fig. 7. CPU utilisation registered in the MME with 500 req/s

All the communications between the femtocell and the UE emulator were performed with wired connections and physical signal fading emulators, able to reproduce multiple radio path-loss schemes for signal attenuation. Free-space path-loss scheme was selected for this test-bed, making the UE emulator responsible for implementing the urban path-loss attenuation prediction on the registered radio measurements.

Evolved Packet Core (EPC). A single Aeroflex PXI 3000 modular platform was equipped with the Aeroflex LTE Base Station RF Measurements modules for providing the EPC capabilities to the test-bed. The equipment was configured to use FDD modulation for both downlink and uplink. All the UEs were registered in the HSS with the same master key, to reduce the computational complexity of each experiment.

UE Emulator. The UE emulation was managed in an Aeroflex E500 Network Tester, able to emulate the behaviour of up to 4000 UEs with the configuration used on this test-bed: 4 x Aeroflex TM500 modules interconnected to each other. The attacker's UEs were configured with an incorrect master key, forcing the unsuccessful authentication against the core network in the same manner the attack would occur in real life.

The data traffic load was generated on the UE side using a Spirent C50 Test-Center, model number C50-KIT-04-START, with 4-port 10/1 Gbps Ethernet SFP able to manage a volume of up to 40 Gbps. The TM500 was configured to emulate different groups of malicious and legitimate UE nodes, as described in Table 1.

4.2 Scenario Definition

The proposed LTE scenario complexity is a reduced representation of a commercial deployment. Specifically, the test-bed scenario is mainly composed of an LTE emulated eNB connected to an emulated LTE core network. During the RRC signalling attack implementation, there are two types of UE nodes; legitimate and rogue UEs. Both types of UEs are registered in the HSS as active subscribers. However, legitimate UEs use a valid master key (K) value, while rogue UEs have been configured with an incorrect K value, which forces a failure during the establishment of the RRC session.

In a real-life attack, the attacker would perform exactly the same actions, as the only information under its control would be the IMSI value of legitimate UEs, but it would not be able to compromise the private master key (K) associated to every mobile subscriber. Since the aim of a malicious user is to disrupt the service in the most efficient manner, the attacker's effect was replicated by multiple sets of rogue UEs acting as a single attacker node. Configuring multiple rogue UEs that attempt to establish RRC sessions in parallel, it is possible to reproduce the equivalent network traffic volume that would be produced by a single attacking UE targeting a commercial LTE cell.

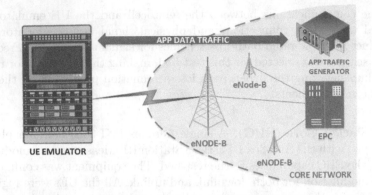

Fig. 8. LTE Test-bed architecture

Table 1. Emulation Scenarios for LTE experiments

Properties	Scenario 1	Scenario 2	Scenario 3
Legitimate UEs	200	50	200
Rogue UEs	200	450	200
Initial, attack phase	30 s	300 s	300 s
Final phase	95 s	221 s	431 s
Total duration	2 min 35 s	13 min 41 s	17 min 11 s

The ratio between legitimate and rogue nodes has been modified across the three scenarios, as described in Table 1 to evaluate the impact on core equipment when the rogue traffic load is equal (Scenarios 1 and 3) and higher (Scenario 2) than the traffic generated by the legitimate UEs. The simulation of the attack is composed by three phases: *initial phase*, where the legitimate nodes are activated; *attacking phase*, when the attacking nodes are activated in parallel to the existing legitimate nodes; and *final phase*, when the attacking nodes are deactivated to recover the normality on the network.

The duration of initial and attacking phases is set to 30 seconds for the Scenario 1, and 5 min for Scenarios 2 and 3; whereas the duration of the final phase varies on each scenario. Moreover, the time allocated to the initial and attacking phases includes the time required for initiating the legitimate and malicious UEs, having a gradual increase in the cellular network traffic received at the eNB.

5 Result Evaluation

The evaluation of the results have been conducted using the evaluation metrics of *Detection Rate* (DR), *False Positive Rate* (FPR) and *False Negative Rate* (FNR). The DR, also known as Recall, indicates the proportion of malicious frames detected in comparison with the total number of frames emitted by the attacker. This parameter offers a clear indication of how efficient the detection algorithm is. However, this value is not able to provide a fair assessment of the detection performance by itself and could lead to an error when analysed individually, since it does not take into account the negative effects of misclassifying malicious frames as normal (FN), or when a false alarm is raised (FP) as result of applying the detection algorithm.

$$DR = \frac{TP}{TP + FN} \tag{9}$$

$$FPR = \frac{FP}{TotalFrames} = \frac{FP}{TP + FP + TN + FN} \tag{10}$$

$$FNR = \frac{FN}{TP + FN} \tag{11}$$

This information has a direct impact onto the overall detection performance and must be taken into account when analysing the performance. Looking at the test evaluation in pattern recognition theory [27] and *Intrusion Detection System* (IDS) [28], it is possible to judge the performance in a more complete manner by evaluating the *Overall Successful Rate* (OSR), also known as accuracy, which takes into account the correctly classified frames against the total population. In addition, we calculated the *Precision* (P), which evaluates the number of frames correctly classified as malicious among the total number of frames classified as malicious by the algorithm and the F_1-Score, which is the harmonic mean of the DR and Precision, and evaluates the balance between these two parameters.

$$OSR = \frac{TP + TN}{TP + FP + TN + FN} \tag{12}$$

$$P = \frac{TP}{TP + FP} \tag{13}$$

$$F_1 - SCORE = \frac{2 * P * DR}{P + DR} \tag{14}$$

The above parameters are used to evaluate the results and obtain the research findings from the experiments, which are discussed in the section below and presented in Table 2.

5.1   Research Findings

The LTE experiments confirm the accuracy of the proposed metrics to detect the attack, no matter if they are used individually or as part of a set of metric combinations. Notably, if the individual performance of each metric is analysed, the CRR metric performs the strongest in detecting the attack, with a DR higher than 98% for all three scenarios presented in this paper. This is expected because the DoS attack creates an unusually large number of RRC Connection Requests, which might end up with a RRC Connection Release message when the attacker's requests are rejected.

Table 2. Results collected on LTE experiments (in percentage)

Scenario#	DR	OSR	FNR	FPR	Precision	F_1-Score	Metrics
1	98.095	87.227	1.904	11.526	84.774	90.949	CRR
1	97.143	74.143	2.857	23.988	72.598	83.096	CRR, SMT
1	96.190	97.508	3.809	0	100.00	98.058	CRR, SR
1	88.571	80.997	11.429	11.526	83.408	85.912	CRR, SMT, SR
1	84.762	66.044	15.238	23.988	69.804	76.559	SMT
1	83.810	89.408	16.190	0	100.00	91.192	SR
1	82.857	88.785	17.143	0	100.00	90.625	SMT, SR
2	100.00	100.00	0	0	100.00	100.00	CRR, SR
2	98.814	98.862	1.186	0	100.000	99.403	CRR
2	97.536	97.548	2.463	0.087	99.907	98.707	SMT
2	97.536	97.548	2.463	0.087	99.907	98.707	CRR, SMT
2	97.536	97.548	2.463	0.087	99.907	98.707	CRR, SMT, SR
2	96.989	97.023	3.010	0.087	99.906	98.426	SMT, SR
2	95.803	95.972	4.197	0	100.000	97.857	SR
3	99.915	98.011	0.084	1.917	97.765	98.828	CRR
3	99.915	98.011	0.084	1.917	97.765	98.828	CRR, SR
3	93.401	93.608	6.599	0.852	98.925	96.084	SMT
3	93.401	93.608	6.599	0.852	98.925	96.084	CRR, SMT
3	93.401	93.608	6.599	0.852	98.925	96.084	CRR, SMT, SR
3	79.272	81.747	20.728	0.852	98.736	87.940	SMT, SR
3	0.423	16.406	99.577	0	100.00	0.842	SR

However, to provide robustness to the detection algorithm, other metrics should be taken into account because normal network behaviour might also manifest with a high number of RRC Connection Requests. This could happen, for example, in the case of congested LTE cells without enough radio spectrum to cope with the demand. Under such circumstances, the individual analysis of the CRR would feed misleading information into the detection algorithm, necessitating to weigh in additional information captured by the SMT and SR metrics.

The best overall performance is obtained in Scenario 2, with the metric combination (CRR, SR), where every frame is correctly classified and all evaluation metrics reach their highest value. The DR, OSR, Precision and F_1-Score reach 100%. This result is specially important because Scenario 2 was specifically designed to facilitate the detection, with only 50 legitimate *User Equipments* (UEs) and 5 times more rogue UEs.

If the same metric combination is evaluated in the Scenario 1 and 3, where the legitimate to rogue UE ratio is equal to 1, the DR decreases down to 96.19% when the emulation duration is short, or 99.92% when the simulation duration is similar to Scenario 2. In both cases, the OSR and F_1-Score perform very well, remaining above 97.5% due to the accuracy on the classification of the collected network traffic.

The *Session Mean Time* (SMT) metric is the second best metric, bute requires a long duration (Scenario 3) for the emulation to guarantee the best performance. This metric is able to detect the attack with a DR of 97.54% and 93.4% for Scenarios 2 and 3, respectively. On the contrary, the OSR obtains a better result in Scenario 2, with a 97.55% against the 93.62% evaluated for Scenario 3 due to a 6.6% FNR. The reason for this minor difference is again the ratio between legitimate to rogue UEs, as only the legitimate UEs are able to successfully complete an RRC session and modify the average RRC session monitored by the SMT metric.

If the *Success Rate* (SR) or the CRR metrics are combined with the SMT metric, the individual detection performance suffers an important decrease of the DR and increase of the FNR. Scenario 1 registers the worse case, when the (SMT, SR) metric combination manages to obtain a decent 82.86% DR. This value implies a reduction of almost 15% on the DR if it is evaluated against its highest detection performance, when this metric is individually used in Scenario 2. However, even in this case, this metric combination is able to provide a high level of accuracy, with a 100% Precision and F_1-Score of 90.63%

The lowest result across all the conducted experiments is registered for the SR metric in Scenario 3, with an unacceptable 0.42% DR and FNR of 99.58%. The registered Precision is 100% due to the absence of FP cases, which boosts the OSR to a minimal 16.41%.

Although the SR is present in all low DR results across all scenarios, it adds value when combined with the CRR metric, as previously mentioned at the beginning of this section. The (CRR, SR) metric combination provides very good results for all three Scenarios. Besides the excellent performance in Scenario 2, it obtains a DR higher than 96% in both scenarios 1 and 3, which is accompanied with an OSR of 97.51% in Scenario 1, and 98.01% in Scenario 3.

Looking at the obtained results, it is possible to conclude that an attacker attempting to execute the signalling DoS attack should inject a similar number of RRC Connection Request messages equivalent to the actual number of legitimate requests registered on the network, reducing the attack duration to short periods of time. The optimisation of the attack implies a period of monitoring the channel, in an attempt to match the average legitimate RRC Connection Request messages registered within a certain time. Only by carefully tailoring the attack duration and injection rate, may the attacker be able to reduce the chances of being detected by the proposed algorithm.

6   Conclusions

This paper has presented three new metrics to detect DoS attacks over the RRC layer in LTE networks. The proposed solution has been experimentally validated using an emulated LTE test-bed, obtaining a sample data-set to evaluate the detection performance.

The detection mechanism is able to detect the attack with a DR higher than 98.81% and Precision higher than 88.77% using a single metric, the CRR. Furthermore, robustness has been added by combining the use of the CRR metric with two additional metrics, the SR and SMT, which are able to perform equally well in terms of DR or even improve the DR by 1.2% in specific cases.

The results have revealed how the attacker could optimise the attack, by adapting the average RRC connection Requests within the targeted network and reducing the attack's duration to short periods. Further research to improve this work should be focused on evaluating the algorithm in more congested networks, where the data fusion results have proven to be more successful to reduce the false positives.

References

1. Vintila, C., Patriciu, V.: Security analysis of LTE access network. In: Proceedings 10th International Conference Networks (ICN 2011), pp. 29–34 (2011)
2. Yu, D., Wen, W.: Non-access-stratum request attack in E-UTRAN. In: Computing, Communications and Applications, pp. 48–53 (2012)
3. Bilogrevic, I., Jadliwala, M., Hubaux, J.: Security issues in next generation mobile networks: LTE and femtocells. In: 2nd International Femtocell Workshop, pp. 1–3, Luton, UK (2010)
4. Purkhiabani, M., Salahi, A.: Enhanced authentication and key agreement procedure of next generation evolved mobile networks. In: 2011 IEEE 3rd International Conference on Communication Software and Networks, May 2011, pp. 557–563 (2011)
5. 3GPP TS 33.401: Technical Specification Group Services and System Aspects; 3GPP System Architecture Evolution (SAE); Security Architecture, vol. 1, no. v.15.6.0 - Release 15, pp. 1–79 (2018)
6. Vintilă, C.-E., Patriciu, V.-V., Bica, I.: An analysis of secure interoperation of EPC and mobile equipments. In: 6th International Conference on Digital Telecommunications, pp. 1–6 (2011)
7. 3GPP TS 33.501: Technical Specification Group Services and System Aspects; Security architecture and procedures for 5G system, vol. 1, no. v.15.3.1 - Release 15, pp. 1–181 (2018)
8. 3GPP TS 38.300: Technical Specification Group Radio Access Network; NR; NR and NG-RAN Overall Description; Stage 2. vol. 0, no. v.15.4.0 - Release 15, pp. 1–97 (2018)
9. Shafer, G.: A Mathematical Theory of Evidence. Princeton University Press, Princeton (1976)
10. Bassil, R., Chehab, A., Elhajj, I., Kayssi, A.: Signaling oriented denial of service on LTE networks. In: Proceedings of the 10th ACM International Symposium on Mobility Management and Wireless Access, Ser. MobiWac 2012, pp. 153–158. ACM, New York (2012)
11. Jover, R.P., Lackey, J., Raghavan, A.: Enhancing the security of LTE networks against jamming attacks. EURASIP J. Inf. Secur. 1–14 (2014)
12. Li, X., Wang, Y.: Security enhanced authentication and key agreement protocol for LTE/SAE network. In: 2011 7th International Conference on Wireless Communications, Networking and Mobile Computing, September 2011, pp. 1–4 (2011)

13. Køien, G.M.: Mutual entity authentication for LTE. In: 2011 7th International Wireless Communications and Mobile Computing Conference, July 2011, pp. 689–694 (2011)

14. Zheng, Y., He, D., Yu, W., Tang, X.: Trusted computing-based security architecture for 4G mobile networks. In: Sixth International Conference on Parallel and Distributed Computing Applications and Technologies (PDCAT 2005), pp. 251–255 (2005)

15. Zheng, Y., He, D., Tang, X., Wang, H.: AKA and authorization scheme for 4G mobile networks based on trusted mobile platform. In: 2005 5th International Conference on Information Communications Signal Processing, December 2005, pp. 976–980 (2005)

16. 3GPP TS 22.278: Technical Specification Group Services and System Aspects; Service requirements for the Evolved Packet System (EPS), Network, vol. 0, no. v.16.1.0 - Release 16, pp. 1–50 (2018)

17. 3GPP TS 36.331: Technical Specification Group Radio Access Network; Evolved Universal Terrestrial Radio Access (E-UTRA); Radio Resource Control (RRC); Protocol specification, no. v.15.4.0 - Release 15 (2018)

18. Wetteroth, D.: OSI Reference Model for Telecommunications. McGraw-Hill Professional, New York (2001)

19. Cao, J., Ma, M., Li, H., Zhang, Y., Luo, Z.: A Survey on security aspects for LTE and LTE-A networks. IEEE Commun. Surv. Tutor. **16**(1), 283–302 (2014)

20. Lichtman, M., Reed, J.H., Clancy, T.C., Norton, M.: Vulnerability of LTE to hostile interference. In: 2013 IEEE Global Conference on Signal and Information Processing, December 2013, pp. 285–288 (2013)

21. Cho, J.-S., Kang, D., Kim, S., Oh, J., Im, C.: Secure UMTS/EPS authentication and key agreement. In: Park, J., Leung, V., Wang, C.L., Shon, T. (eds.) Future Information Technology. Application, and Service, pp. 75–82. Springer, Dordrecht (2012). https://doi.org/10.1007/978-94-007-5064-7_11

22. 3GPP TS 36.300: Technical Specification Group Radio Access Network; Evolved Universal Terrestrial Radio Access (E-UTRA) and Evolved Universal Terrestrial Radio Access Network (E-UTRAN); Overall description; Stage 2, no. v.14.0.2 - Release 14, 2017

23. Khan, M., Ahmed, A., Cheema, A.R.: Vulnerabilities of UMTS access domain security architecture. In: Ninth ACIS International Conference on Software Engineering, Artificial Intelligence, Networking, and Parallel/Distributed Computing, 2008, SNPD 2008, pp. 350–355 IEEE (2008)

24. Reineking, T.: Belief functions: theory and algorithms. Ph.D. dissertation, Mathematics and Informatics, University of Bremen (2014)

25. Escudero-Andreu, G.: Protection of Mobile and Wireless Networks Against Service Availability Attacks. Ph.D. dissertation, Loughborough University (2018)

26. Kyriakopoulos, K.G., Aparicio-Navarro, F.J., Parish, D.J.: Manual and automatic assigned thresholds in multi-layer data fusion intrusion detection system for 802.11 attacks. IET Inf. Secur. **8**(1), 42–50 (2014)

27. Olson, D.L., Delen, D.: Advanced Data Mining Techniques, 1st edn. Springer, Heidelberg (2008). https://doi.org/10.1007/978-3-540-76917-0

28. Elhamahmy, M.E., Elmahdy, H.N., Saroit, I.A.: A new approach for evaluating intrusion detection system. Artif. Intell. Syst. Mach. Learn. **2**(11), 290–298 (2010)

Anomaly Detection Using One-Class SVM for Logs of Juniper Router Devices

Tat-Bao-Thien Nguyen[1]($\boxtimes$), Teh-Lu Liao[2], and Tuan-Anh Vu[1]

[1] Posts and Telecommunications Institute of Technology,
Ho Chi Minh City, Vietnam
nguyentatbaothien@gmail.com
[2] Department of Engineering Science, National Cheng Kung University,
Tainan, Taiwan

Abstract. The article deals with anomaly detection of Juniper router logs. Abnormal Juniper router logs include logs that are usually different from the normal operation, and they often reflect the abnormal operation of router devices. To prevent router devices from being damaged and help administrator to grasp the situation of error quickly, detecting abnormal operation soon is very important. In this work, we present a new way to get important features from log data of Juniper router devices and use machine learning method (basing on One-Class SVM model) for anomaly detection. One-Class SVM model requires some knowledge and comprehension about logs of Juniper router devices so that it can analyze, interpret, and test the knowledge acquired. We collect log data from a lot of real Juniper router devices and classify them based on our knowledge. Before these logs are used for training and testing the One-Class SVM model, the feature extraction phase for these data was carried out. Finally, with the proposed method, the system errors of the routers were detected quickly and accurately. This may help our company to reduce the operation cost for the router systems.

Keywords: Anomaly detection · Juniper devices · One-Class SVM · Log feature extraction

1 Introduction

Nowadays, router devices are so important with Internet Service Provider (ISP) in core network. Juniper router that is used in SCTV core network enables a wide range of business and residential applications and services (e.g., high-speed transport, Virtual Private Network services, high-speed Internet) [1]. To detect and prevent abnormal operations of Juniper router devices, there are some solutions such as monitoring systems, checking syslog server [2].

Abnormal operations of devices may be cause of routing errors in the network, chassis errors, Distributed Denial-of-service (DDos) attack or some fail processes in router devices. These abnormal operations of router devices are quite reported to syslog server if they are configured. It is possible to detect abnormal operation by inspecting manually these logs. Anomaly detection has been a practical research topic and has

© ICST Institute for Computer Sciences, Social Informatics and Telecommunications Engineering 2019
Published by Springer Nature Switzerland AG 2019. All Rights Reserved
T. Q. Duong et al. (Eds.): INISCOM 2019, LNICST 293, pp. 302–312, 2019.
https://doi.org/10.1007/978-3-030-30149-1_24

great importance in many application domains in university and in industry (e.g., [3–5]). For conventional small systems, engineers manually define rules or check system logs to detect anomalies based on their domain knowledge. Additionally, they can use regular expression match or keyword searches (e.g., "error", "fail"). However, the extremely large sizes of log data generated (up to million logs) by hundreds of devices make manual analysis impossible.

As a result, automated log analysis methods for anomaly detection of Juniper router devices based on machine learning systems are highly in demand. However, we found no resources about this problem. This paper is the first work about anomaly detection using One-Class SVM for logs of Juniper router devices. We present a new way to get features from log data of Juniper router devices based on importance characteristics of log messages. The standard support vector machine (SVM) [6] is the machine learning method to classify two class or multiple class of data. But the log data of anomaly detection are very special, the abnormal logs are much less than the normal logs. Therefore, the standard SVM does not work well on our task, we use One-Class SVM described in [7] to handle the logs classification.

2 Collecting Log Data, Preprocessing and Feature Extraction

Collecting log data for training and testing is necessary because we use the real logs from real router Juniper devices. Logs must be made clean before they can be used for feature extraction.

2.1 Collecting Log Data and Text Preprocessing

Juniper router devices routinely generate logs to record device states and runtime information, each including a content indicating what has happened and a times-tamp. These devices are configured to send logs to syslog server that always receives data on corresponding port. The log messages have some common characteristics although the format of them is not fixed length. This valuable information could be utilized for anomaly detection and other purposes; thereby logs are collected first and saved as a file in syslog server for further usage. For example, Fig. 1 depicts some log messages of Juniper router devices.

As the example above, a timestamp is the beginning of each message, a name of Juniper router device with the same format and other characteristics: the log messages at syslog server are written in English and comprise digits, lower, upper case letters and many special characters. The raw logs are pre-processed by Python program. Our text normalization procedures are given below:

- Remove timestamps: Remove whole timestamp before each message (both date and time).
- Remove router device's name: Remove device's name.
- Remove digits, special characters: Remove any special character, including punctuations, all digits.

30-06-2018 07:00:07 AM HCM-Q12-MX5 last message repeated 19 times

30-06-2018 07:00:07 AM HCM-Q12-MX5 inetd[1380]: /usr/sbin/sshd[55491]: exited, status 255

30-06-2018 07:00:10 AM NBH-HED-MX5 tfeb0 MIC(1/0) link 0 SFP syslog throttling: enabling syslogs for receive power alarms and warnings. (0/0)

30-06-2018 07:00:12 AM HCM-Q12-MX5 mib2d[1474]: SNMP_TRAP_LINK_DOWN: ifIndex 533, ifAdminStatus down(1), ifOperStatus down(2), ifName ge-1/1/7

30-06-2018 07:00:15 AM HNI-DDA-MX5 sshd[76567]: error: Received disconnect from 172.16.123.224: 11: disconnected by user

30-06-2018 07:00:15 AM HNI-DDA-MX5 inetd[1368]: /usr/sbin/sshd[76567]: exited, status 255

30-06-2018 07:00:33 AM HNI-BDH-MX5 sshd[15489]: error: Received disconnect from 172.16.123.224: 11: disconnected by user

Fig. 1. Some log messages of Juniper router devices.

- Replace continuous spaces with a single space: The length of log message depends on how many spaces in log message, so that replacing continuous spaces with a single space is necessary.
- Lower cases: Replace all upper case letters by lower case.

Although timestamp information, which is the time the event happened, is very important according to the SCTV engineers, we still remove it from the log message. After anomaly detection, the abnormal original log messages (including timestamp information) will be sent to the engineers.

2.2   Feature Extraction

Some methods used to perform feature extraction for text classification [8–10] are not effective with logs of Juniper router devices. In our works, we use three characteristics of each log message to be three elements of the feature vector: the length of log message, the number of different words and the sum of TF-IDF in log message.

The Length of Log
The length of log message is a significant characteristic. In the process of manually defining abnormal logs, the length of logs shows that the logs with irregular length have a higher probability of being abnormal logs. We denote S_i as the length of log i. We use the spaces of each log message to calculate S_i.

The Number of Different Words in Log Message
The number of different words in log message: the number of words which are different from dictionary of each log message. Bag-of-Words (BoW) model and the length of log are used to calculate this characteristic. Some of new document representation methods [11–13] are developed based on BoW.

The dictionary is built based on normal logs that we classify from original logs. To get the number of words that are in dictionary, BoW model is the accordant and simple classical model for our purpose. Based on BoW, each vector represents a log message; each element denotes the normalized number of occurrence of a word in log. The words that do not appear in dictionary are not counted by BoW model.

Based on the dictionary and BoW model, each log is converted from text to a vector space. The log data become the matrix that is given by:

$$A = \begin{bmatrix} a_{11} & a_{12} & \cdots & a_{1m} \\ a_{21} & a_{22} & \cdots & a_{2m} \\ \vdots & \vdots & \vdots & \vdots \\ a_{n1} & a_{n2} & \cdots & a_{nm} \end{bmatrix} \tag{1}$$

where m is the length of dictionary, n is the size of log data. Each row of the matrix A represents a log message in log data. The number of words which are different from dictionary of each log message is given by the formula:

$$L_i = S_i - \sum_{j=1}^{m} a_{ij} \tag{2}$$

where S_i is length of the i-th log.

The Sum of TF-IDF

Term frequency-inverse document frequency (TF-IDF) is one of the most famous algorithms used in document mining research; it is used for calculating the weight of each word. The word frequency means the number of time a term is repeated in a log message, and Inverse Document Frequency is an algorithm used to calculate the inverse probability of finding a word in log data [14]. Some improvements of feature extraction based on TF-IDF are mentioned in [15, 16].

TF-IDF Formula:

$$g_{ij} = tf_{ij} \times \log\left(\frac{N}{df_j}\right) \tag{3}$$

where g_{ij} is the weight of the word j in the log i, N is the total number of log messages, tf_{ij} is the frequency of the word j in the log i, df_j is the number of logs containing the word j.

The sum of TF-IDF in log is the third element of feature vector. Equation 4 is the summary equation used to calculate the sum of TF-IDF:

$$G_i = \sum_{j=1}^{h} g_{ij} \tag{4}$$

where G_i is the sum of TF-IDF of log i, h is the number of words in log i.

The feature vector of log i:

$$x_i = (S_i, L_i, G_i) \tag{5}$$

The feature vectors are the input data of One-Class SVM model, which will be discussed below.

3   One-Class SVM

One famous machine learning method is the support vector machine (SVM), which was invented by Vladimir Vapnik and Alexey Ya. Chervonenkis in 1963, and is widely applied for pattern classification and data analyzing. However, Vladimir Vapnik and Corinna Cortes proposed the current standard incarnation in 1993 [6].

The labels associated with training data are based to group anomaly detection techniques for log data into two broad categories: one-class and multi-class anomaly detection technique. The labels in log messages of Juniper router devices are grouped into two types: abnormal log and normal log, so that one-class anomaly detection technique was chosen for our purpose.

For the case of one-class classification, Scholkopf et al. proposed a maximum margin based classifier that is an adaptation of the Support Vector Machine algorithm [7]. The data are separated from the origin by a separating hyperplane $\langle w, z \rangle - \rho = 0$ with maximum margin (where w and ρ are respectively the normal vector of the hyperplane and the distance from the hyperplane to the origin).

The maximum margin from the origin is found by solving the below quadratic optimization problem:

$$\min_{w,\rho,\xi} \quad \frac{1}{2}\|w\|^2 + \frac{1}{vl}\sum_i \xi_i - \rho \tag{6}$$

$$\text{subject to } (w \cdot \Phi(x_i)) \geq \rho - \xi_i, \ \xi_i \geq 0.$$

where ξ_i are so-called slack variables that are used to model the separation errors. The $v \in (0, 1]$ is a parameter that adjusts the balance between maximizing the distance from the origin and the region created by the hyperplane containing most of the data. $\Phi(\cdot)$ is a non-linear projection is evaluated through a kernel function that is used as a mapping from the original feature space to a possibly higher dimensional feature space: $k(x,y) = (\Phi(x) \cdot \Phi(y)$. In our works, we consider the kernel Radial Basis Function (RBF), linear kernel, polynomial kernel and sigmoid kernel. They are expressed respectively by these following Eqs. (7)–(10):

$$k_{(RBF)}(x,y) = e^{-\left(\frac{\|x-y\|^2}{2\sigma^2}\right)} \tag{7}$$

$$k_{(linear)}(x,y) = x^T y + C \tag{8}$$

$$k_{polynomial}(x,y) = \left(\gamma x^T y + C\right)^d \tag{9}$$

$$k_{sigmoid}(x,y) = \tanh\left(\gamma x^T y + C\right) \tag{10}$$

4   Experimental Results

In this section, we descript the log datasets, the evaluate method, perform an experimental evaluation and comparison of our anomaly detection method for log data of Juniper router devices based on One-Class SVM model. We have chosen Python programming language for our convenience purpose. Python is a powerful interpreted and popular language and supports some power libraries for data science, machine learning (numpy, matplotlib, scikit-learn, etc.) [17–19].

4.1   Log Datasets

Publicly available production logs are scarce data, especially log data of router devices because companies and ISPs rarely publish them to community due to confidential issues. So that we collected log data from real Juniper router devices in core network of SCTV (Saigontourist Cable Television Company). Logs from devices were sent through the internet network to syslog server and saved as txt files. These files can be read directly and easily by most popular programs. The log data contains 12907 log messages and 266 abnormal log messages, which are manually labeled by us and the SCTV experts. Using this log dataset, we evaluated the performance and compared the results of the models. More statistical information of the log dataset is provided in Table 1.

Table 1. Summary of log datasets.

Devices	Data size	Number of messages	Anomalies
Router Juniper	1,7 Mb	12 907	266

After the raw logs are collected using Python script, they are called from our main Python program. Feature extraction is applied to these log messages, each log is represented by a feature vector include: the length of log message, the number of words which are different from dictionary and the sum of TF-IDF value in log message. We choose 60% of log messages as the training data and the remainders as the testing data.

Figure 2 shows the training data and testing data on 3-dimensional feature space. Blue points are the normal data, red points are abnormal data. Due to the fact that the log data come from over a hundred router Juniper devices, we receive a lot of similar log messages. There are some differences between them such as the timestamp, device's name, etc. So a lot of log messages become the same after they are passed the preprocess step that we mentioned above. Because of that, the data points in the figure may look less than reality.

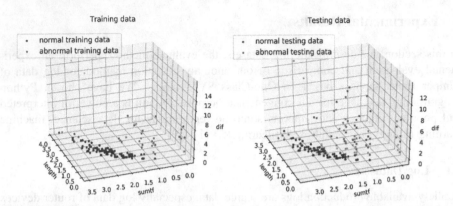

Fig. 2. Training data and testing data on 3-dimensional feature space (Color figure online).

Although the visualizations in Fig. 2 shows that the data are naturally well-separated after they are converted to feature vectors, we still propose the One-Class SVM method to separate the normal log data and abnormal log data automatically, and make sure the model still works efficiently in case there are more abnormal log data appeared when the hardware system is extended.

4.2   Method Evaluation

The Precision, Recall and F-measure, which are the most commonly used metrics, are used to evaluate the accuracy of One-Class SVM anomaly detection method using different kernels (Radial Basis Function (RBF), linear kernel, polynomial kernel and sigmoid kernel) as we have already the ground truth for the log data. As shown below, Precision shows the percentage of true anomalies among all anomalies detected, Recall measures the percentage of how many real anomalies are detected, and F-measure denotes the harmonic mean of precision and recall [20]. They are expressed respectively by the following Eqs. (11)–(13):

$$Precision = \frac{Anomalies\ detected}{Anomalies\ reported} \tag{11}$$

$$Recall = \frac{Anomalies\ detected}{All\ anomalies} \tag{12}$$

$$F - measure = \frac{2 \times Precision \times Recall}{Precision + Recall} \tag{13}$$

4.3   Result Evaluation

Based on evaluate methods that we mentioned, we show the results which evaluate of each model applied on training data and testing data. Figure 3 shows the accuracy of anomaly detection on training log data.

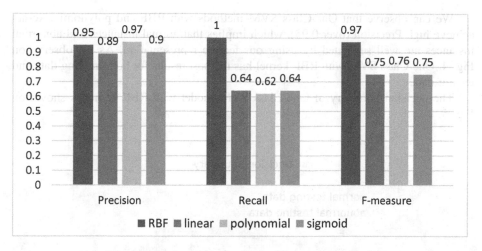

Fig. 3. Training accuracy of One-Class SVM method.

Three models (with linear, polynomial, sigmoid kernels) have no good performance on training data with the F-measure close to 0.75. We can observe that recall measures of these models are low (close to 0.65). We give priority to minimize the number of abnormal log messages which the model predicted wrongly, One-Class SVM method with RBG kernel is the best model in our case. Recall measure of this model is equal to 1, it shows that the model predicted rightly all abnormal logs. One-Class SVM RBF kernel model's performance on training data is better than others models. However, their accuracy on testing data varies with different kernels.

When applying these models for testing data, three models (with linear, polynomial, sigmoid kernels) become unacceptable. These kernels are not suitable for our purpose.

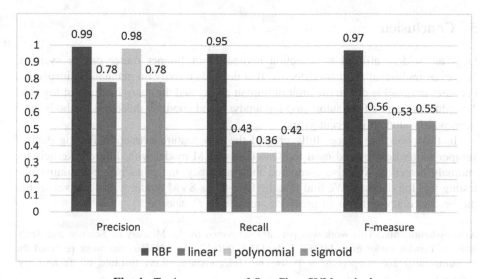

Fig. 4. Testing accuracy of One-Class SVM method.

We can observe that One-Class SVM methods with RBF and polynomial kernel achieve high Precision (over 0.95), which implies that normal instances and abnormal instances are well separated by using our feature representation. As we observe on Fig. 4, One-Class SVM with RBF kernel has the best results for both training data and testing data.

The decision boundary of One-Class SVM model with RBF kernel is shown on Fig. 5.

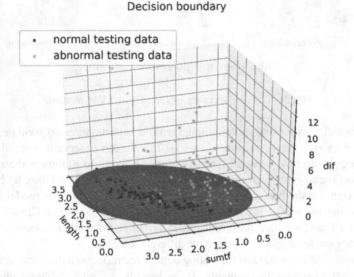

Fig. 5. The decision boundary of One-Class SVM model with RBF kernel.

5   Conclusion

Logs are widely utilized to detection anomalies in Juniper router device systems. However, traditional anomaly detection that depends heavily on manual log inspection becomes impossible due to the limit of human ability and the sharp increase of log size. To reduce manual effort, automated log analysis and anomaly detection methods have been widely studied in recent years.

In this paper, we successfully created the new feature extraction for log data of Juniper router devices and used the One-Class SVM model with different kernels for anomaly detection. We also compared their accuracy and efficiency on training and testing real log datasets. We find that the One-Class SVM model with RBF kernel has the best accuracy in terms of precision, recall and F-measure.

Acknowledgments. This work was partially supported by the Ministry of Science and Technology, Taiwan, under grant MOST 107-2221-E-006-222. In addition, this work received the encouragement and support from Posts and Telecommunications Institute of Technology, Vietnam.

References

1. MX960 3D Universal Edge Router Hardware Guide. https://www.juniper.net/documentati on/en_US/release-independent/junos/information-products/pathway-pages/mx-series/mx960/ index.pdf. Accessed 22 Jan 2018
2. Network Performance Monitor Getting Started Guide. https://support.solarwinds.com/Succ ess_Center/Network_Performance_Monitor_(NPM)/NPM_Documentation/Network_Perfor mance_Monitor_Getting_Started_Guide. Accessed 25 Jan 2018
3. Lin, Q., Lou, J.G., Zhang, H.: Log clustering based problem identification for online service systems. In: IEEE/ACM 38th International Conference on Software Engineering Companion (ICSE-C), Austin, TX, USA, pp. 102–111 (2016)
4. Macit, M., Delibaş, E., Karanlık, B.: Real time distributed analysis of MPLS network logs for anomaly detection. In: IEEE/IFIP Network Operations and Management Symposium (NOMS), Istanbul, Turkey, pp. 750–753 (2016)
5. Gao, Y., Ma, Y., Li, D.: Anomaly detection of malicious users' behaviors for web applications based on web logs. In: IEEE 17th International Conference on Communication Technology, Chengdu, China, pp. 1352–1355 (2017)
6. Cortes, C., Vapnik, V.: Support-vector networks. J. Mach. Learn. **20**(3), 273–297 (1995)
7. Scholkopf, B., Platt, J., Shawe-Taylor, J., Smola, A.J., Williamson, R.: Estimating the support of a high-dimensional distribution. Neural Comput. **13**(7), 1443–1471 (2001)
8. Chang, C.Y., Lee, S.J., Lai, C.C.: Weighted Word2Vec based on the distance of words. In: International Conference on Machine Learning and Cybernetics, Ningbo, China, pp. 563–568 (2017)
9. Tian, W., Li, J., Li, H.: A method of feature selection based on Word2Vec in text categorization. In: 37th Chinese Control Conference (CCC), Wuhan, China, pp. 9452–9455 (2018)
10. Van, T.P., Thanh, T.M.: Vietnamese news classification based on BoW with keywords extraction and neural network. In: Asia Pacific Symposium on Intelligent and Evolutionary Systems (IES), Hanoi, Vietnam, pp. 43–48 (2017)
11. Zhao, R., Mao, K.: Fuzzy bag-of-words model for document representation. IEEE Trans. Fuzzy Syst. **26**(2), 794–804 (2018)
12. Wu, L., Hoi, S.C.H., Yu, N.: Semantics-preserving bag-of-words models and applications. IEEE Trans. Image Process. **19**(7), 1908–1920 (2010)
13. Alahmadi, A., Joorabchi, A., Mahdi, A.E.: A new text representation scheme combining bag-of-words and bag-of-concepts approaches for automatic text classification. In: IEEE GCC Conference and Exhibition (GCC), Doha, Qatar, pp. 108–113 (2013)
14. Dadgar, S.M.H., Araghi, M.S., Farahani, M.M.: A novel text mining approach based on TF-IDF and Support Vector Machine for news classification. In: International Conference on Engineering and Technology (ICETECH), Coimbatore, India, pp. 112–116 (2016)
15. Huang, X., Wu, Q.: Micro-blog commercial word extraction based on improved TF-IDF algorithm. In: International Conference of IEEE Region 10 (TENCON 2013), Xi'an, China, pp. 1–5 (2013)
16. Guo, A., Yang, T.: Research and improvement of feature words weight based on TFIDF algorithm. In: IEEE Information Technology, Networking, Electronic and Automation Control Conference, Chongqing, China, pp. 415–419 (2016)
17. Dubosson, F., Bromuri, S., Schumacher, M.: A python framework for exhaustive machine learning algorithms and features evaluations. In: IEEE 30th International Conference on Advanced Information Networking and Applications (AINA), Crans-Montana, Switzerland, pp. 987–993 (2016)

18. Patterson, E., McBurney, R., Schmidt, H.: Dataflow representation of data analyses: toward a platform for collaborative data science. IBM J. Res. Dev. **61**(6), 9:1–9:13 (2017)
19. Hwang, C.P., Chen, M.S., Shih, C.M.: Apply Scikit-learn in python to analyze driver behavior based on OBD data. In: International Conference on Advanced Information Networking and Applications Workshops (WAINA), Krakow, Poland, pp. 636–639 (2018)
20. Sokolova, M., Lapalme, G.: A systematic analysis of performance measures for classification tasks. Inf. Process. Manage. **45**(4), 427–437 (2009)

Author Index

Printed in the United States
By Bookmasters